Logical design
of digital computers

DIGITAL DESIGN AND APPLICATIONS

Montgomery Phister, Jr., Consulting Editor

LOGICAL DESIGN OF DIGITAL COMPUTERS
by Montgomery Phister, Jr.

CIRCUIT DESIGN OF DIGITAL COMPUTERS (*in preparation*)
by J. K. Hawkins

New York · John Wiley & Sons, Inc.
London · Chapman & Hall, Limited

/ *Logical design of digital computers*

by **MONTGOMERY PHISTER, Jr.**

DIRECTOR OF ENGINEERING
THOMPSON-RAMO-WOOLDRIDGE PRODUCTS, INC.
LOS ANGELES, CALIFORNIA

LIBRARY OF CONGRESS CATALOG CARD NUMBER: 58-6082
PRINTED IN THE UNITED STATES OF AMERICA

To my mother and father

Preface

Of the many important developments in electrical engineering during the past decade, perhaps none has caught hold of the public and scientific imagination more than the electronic digital computer. The application of these machines to military and scientific computations, to the accounting problems of business, and more recently to the control of industrial tools and processes has shown how effectively engineers have applied their imaginations. It appears likely that the next decade will see digital techniques used more and more widely, not only in huge and powerful data-processing systems, but also in small, limited-purpose machines which handle jobs too difficult to be done any other way, or too dangerous, too expensive, or too exhausting to be done manually.

It is my intention that this book should form the basis of a one- or two-semester course in computer logic, introducing the student to the art and science of designing these digital systems. The book describes techniques, not computers, and should serve to present the practical application of these techniques to interested engineers, mathematicians, and physicists. No previous experience with computers is assumed, and the text is liberally sprinkled with worked examples and with exercises, so that a serious student should have no trouble in acquiring some dexterity in digital design as well as a knowledge of the problems and possibilities. A selected set of references for each chapter directs the reader to sources of further information.

In a text of this size and scope, it is inevitable that many important aspects of computer design are omitted, or are treated in less detail than might be desired. For example, parallel computers are treated only briefly; the operation of division receives little attention; "floating-point" number systems and arithmetic units are not discussed; the different computer memories are described in but little detail; and the many variations in general-purpose computer order codes are hardly mentioned. As was stated above, it is my intent to describe methods and techniques, and to illustrate their use with the simplest and (it is hoped) most pertinent examples. Once the techniques are learned, it is relatively easy for the designer to apply them to other computing configurations.

The design techniques described here require the extensive use of Boolean algebra, and the complete description of a computer designed in this way is given by a set of Boolean equations. Furthermore, the circuit components employed here are almost entirely synchronous—that is, the signal wires connecting circuit elements contain meaningful information only at discrete intervals of time, usually called clock-pulse times. A great many very successful computers have been constructed without the use of Boolean algebra, and many computers employ asynchronous circuits which require no clock pulses. Most of the current literature on computers refers to these other methods of design. However, several years of experience in computer work have convinced me that there are a great many advantages to be gained in carrying out design in the way described in this book. The computer systems now desired are so complicated and the logical properties of circuit components are so intricate that it is often difficult to derive computer circuits *without* using Boolean algebra; and although much work is being done on the synthesis and analysis of asynchronous computer circuits, such circuits have inherent timing problems which tend to make their interconnection difficult and which thus obscure their intended functions.

Because of the importance of Boolean algebra to the designer, a full chapter is devoted to a development of that subject. It is my belief that a basic and formal approach to Boolean algebra is useful if the designer is to make the best use of this tool, and so the subject is introduced as a deductive theory rather than as a list of theorems. This approach should be of special use to the reader who is interested in some of the extensive mathematical literature on this subject and who wants to apply facets of that work to computer design.

The contributions made here to the subject of logical design are principally those of interpretation and exploitation of work done by others. The Veitch diagram method of simplification of Boolean equations is systematized, extended to many dimensions, and applied to many different problems (Chapter 4). The "difference-equation" approach to memory elements is systematized and is developed in enough detail for the student to be able to adapt the mathematics to whatever logic the circuit designer finds most advantageous (Chapter 5). The Huffman-Moore model of digital systems is explained and is applied to a number of problems (Chapters 6 and 9). Finally, an attempt is made to provide an orderly approach to the design of a complete computer system, and two simple computers are designed in full (Chapter 11). All these techniques are useful, but the fact remains that the most efficient and interesting computer systems are the result of flashes of inventive genius, and no text can provide such flashes. It is hoped, however, that this book will supply a substantial foundation upon which originality and ingenuity can build structures of elegance and utility.

I owe a debt of gratitude to many individuals who supplied help and encouragement and who read the manuscript and contributed their ideas and suggestions. In particular, Lowell Amdahl, John Blankenbaker, William Frady, Joseph Hawkins, A. D. Scarbrough, and Willis Ware deserve special mention and thanks.

Something more than thanks are also due to my wife; without her help, this book could not have been written.

MONTGOMERY PHISTER, JR.

Los Angeles, Calif.
December 1957

Contents

1 / *Digital computers and their design*

DIGITAL COMPUTERS

It is the purpose of this chapter to explain in simple terms what a digital computer is, and to describe briefly the three basic steps involved in computer design—the steps of system design, logical design, and circuit design.

We can define a digital computer by stating that it is a device capable of automatically carrying out a sequence of operations on data expressed in discrete or digital form. Each operation is one of four kinds: an *input* of data from outside the computer; an *arithmetic operation* on data stored within the computer; an *output* of data from the computer to the outside world; or a *sequence-determining operation* which makes it possible for the computer to choose which of two (or more) alternative sequences of operations it will carry out, basing its choice on the results of previous operations.

A large-scale computer is able to carry out many different types of operation, each of one of these four kinds. It may "read" inputs from punched card readers, keyboards, or measuring devices. It may carry out the arithmetic operations of addition, subtraction, multiplication, and division. Its sequence of operations may be made to depend on the results of computations, upon switches set by an operator, or upon random numbers generated inside the machine. Finally, it may perform output operations which control printers, guided missile control surfaces, or machine tools.

The ability to perform extended sequences of these relatively simple operations very rapidly makes the digital computer an amazingly flexible and useful device and permits it to carry out, in a fairly short time, computations which would take months or years if done by hand.

In the chapters which follow, we shall study the characteristics of the fundamental building blocks from which a computer is constructed, and discuss various procedures and methods which can be used in design. In Chapter 6 we shall return again to the concept of a complete computer, and in Chapter 11 we shall design such a computer. In the intervening chapters, the description just given of a digital computing system should be kept in mind.

COMPUTER DESIGN

The designers of any kind of machinery, including electronic digital computers, are faced at the outset with some operation which they are to mechanize, and with a set of components which must be connected together to form the mechanism. In many situations, the components may be selected from the world at large, and the designer approaches his problem with an open mind, prepared to choose and apply whatever materials and techniques lead to efficient mechanization. Very often, however, a preliminary study will indicate that the operation to be performed can best be handled with a certain quite restricted set of components, and subsequent study is directed toward finding the most effective configuration of those components.

When the operation or operations to be carried out require the application of electronic digital computing and control techniques, the design effort is often divided among three groups of people. The first is the System Analysis or System Design Group. It is the function of this group to define carefully the operation which is to be performed and then to analyze it, attempting to enumerate and explain those features of the problem which will have the greatest influence on the design of the machine. The result of this study will be a set of specifications which tell in some detail what the machine should do and which specify in a general way how it should do it.

The second group of engineers, often called the Circuit Design or perhaps the Components Research Group, develops components which can be the basis of information-handling equipment—i.e., components which store information, and components which change information according to fixed rules. Where the system analyst starts with the definition of an operation and tries to systematize it, the circuit designer starts with a set of basic elements (resistors, transformers, vacuum tubes, diodes, wire, cams, gears, etc.) and connects them together in circuits and apparatus which reliably perform well-defined operations.

It is the function of the third organization, the Logical Design Group, to take the specifications set out by the System Design Group, and the components perfected by the Circuit Design Group, and produce a set of wiring diagrams which show how to fit the components together to realize the specifications.

With the function of the logical designer thus strictly defined, it is desirable to hedge a bit and point out that in practice the functions of the three groups overlap much more than has been indicated in the brief discussion above. Members of each of the three groups should know something of the problems and difficulties of the others. In particular, the logical designer should have a hand in the system study, for his knowledge of what is practical may modify the analyst's idea of what is best. He should also have a hand in the research and development of new components, for he knows what functional properties are valuable and can suggest the characteristics he would like components to have. The somewhat complex relationship between these three groups is perhaps best illustrated by Figure 1-1, where their functions and responsibilities are indicated.

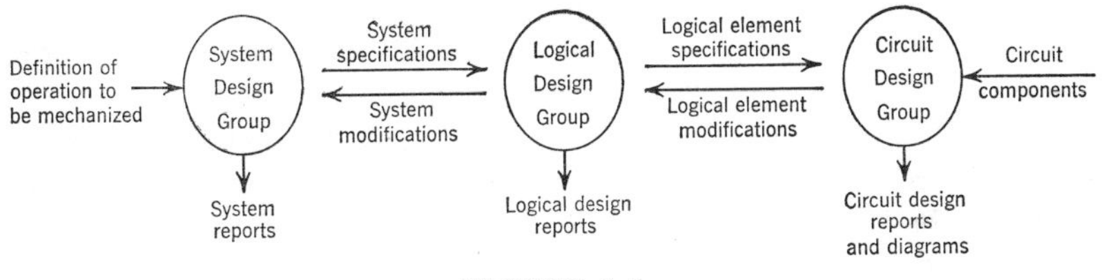

FIGURE 1-1

From what has been said it will be evident that the experienced logical designer is a Jack-of-all-trades. He should have some knowledge of the job the equipment is to perform. He should have some experience in using computers to perform similar jobs. He should be familiar with the way other designers have mastered similar problems of design. He should be familiar with the components—electronic or electromechanical or whatever—from which the equipment will be built. He should have some experience in the practical operation and maintenance of equipment of this kind. In view of all this, it should be clear that no one can become a logical designer merely by studying a book. There is, as always, no substitute for experience. Nevertheless, this book will supply some of the most directly applicable tools and techniques of logical design.

We shall repeatedly refer to the work done by the system analyst and the circuit designer, and the next two sections of this chapter are devoted to a description of their duties and methods.

SYSTEM DESIGN

The system analyst, in examining the operation which must be mechanized, first tries to redefine it in terms of information flow, for this will point out to him the outstanding characteristics of his problem. He therefore asks:

1. What is the source of the information, and how may it best be translated into the language of the machine?
2. What kind of information must be represented—numerical? alphabetic? other?
3. What is the rate of flow of information from the source?
4. What must be done to the information? Must it be altered, and if so, how? Must it be sorted or combined in some way with other information?
5. How much time is available for processing the information?
6. What must be the output rate of processed information?
7. What is the purpose of the output information? What form should it be in to accomplish this purpose most effectively?
8. What effect would a machine error have on the information flow, and how would it affect the operation being performed? May the operation be interrupted for emergencies or for regular periods of preventive maintenance?

Having defined the problem by answering the above questions, the system analyst brings his imagination and ingenuity to bear and draws up general plans for a system which will do what is required. The basic components of the system he works out must be circuits and devices which already exist or can reasonably be developed and perfected in the time available by the people associated with the program. If certain standard components are available, it may be worthwhile to plan the system around them, capitalizing on the man-hours which have been devoted to their development.

The system devised can in some sense be optimized, for it satisfies certain more or less conflicting requirements: it must do the required job and at the same time must cost as little as possible. The problem of adjusting system design to provide an optimum system is very difficult, as can well be imagined. Speed, precision, and flexibility are all, in general, proportional to cost, and it is important in laying out plans for the system to find a combination which is no faster, no more accurate, and no more flexible than is necessary. Furthermore, it is often difficult or even impossible to estimate the true value of various system features, in order to decide whether or not they are worth what they cost. How much extra would it be worth, for example, to build a given system so that it may be expanded easily to meet the demands of new situations which may arise? How much time should be devoted to refining and simplifying some part of the system, when such time is very expensive in salaries, overhead, and delays? The answers to such questions can only be guessed at in any particular situation, but they may have important effects on the success of the resulting system.

To show how the system analyst will proceed with a somewhat typical kind of problem, consider the following example. A commercial airline is seeking an automatic system to handle reservations on its regularly scheduled flights. The system analyst is asked to provide some means of listing seats on all flights, answering whether or not seats are available on particular flights, making reservations on planes if so requested, and canceling reservations previously made if so ordered. Examining the problem in the light of the preliminary questions listed above, the system designer would present the following answers.

1. There are two sources of information. The first is a question or a request from an agent in one of 100 ticket offices. The other source is the airline itself, which specifies the flights to be handled and indicates the number of seats available on each flight.

2. The information may be entirely numerical, consisting of flight numbers, number of seats available, and date of flight. Each flight number must refer to the passage of a plane between a take-off and a landing, so that a given plane which flies 1000 miles in ten flight "legs" of 100 miles each will have ten flight numbers. In making requests, a simple numerical code may be used to indicate whether a reservation is being made or canceled or whether an inquiry is being made.

3. The information from ticket agents may specify any one of 1000 flight "legs" on any one of ten days. Up to four seats may be requested at one time. There are 100 agents making requests, and the rate of requests will vary from one every two minutes to fifty per minute, depending on the time of day and the time of year. The

information from the airline company may enter the system once a day, and consists of all data on the 1000 flight "legs" scheduled ten days from the entering day. Each of these "legs" may have up to 100 seats available.

4. The information from the airline company is filed away where it may be referred to, as soon as it is received. The action taken on information from the ticket agents depends on whether the agent is making an inquiry, requesting a reservation, or canceling a reservation previously made. In each of the three cases, the number of seats still available on the particular flight "legs" involved must be looked up in the file. The action taken next depends on what the request was, as follows:

Kind of Request	*Action Taken*
Inquiry	Subtract number of seats requested from number available and examine result to determine whether it is nonnegative.
Reservation request	Subtract number of seats requested from number available and file away the result, if not negative, as the new "number of seats available" for that flight "leg."
Reservation cancellation	Add number of seats canceled to number available and file away the result as the new "number of seats available" for that flight "leg."

5. The processing speed is determined by the requirement that no customer should have to wait more than one minute, even at peak load, for an answer to his inquiry or a fulfillment of his request.

6. The outputs of the system consist only of replies to the customer, and each reply will consist of either "yes" or "no" (when an inquiry has been made) and "request fulfilled" or "request not fulfilled" (when a seat reservation or cancellation has been requested). Therefore, one decimal digit will more than suffice as an answer to the request. Furthermore, replies must be made as fast as requests are made, so the output rate will vary between one decimal digit every two minutes and fifty digits per minute.

7. The output information is simply the answer to a request and need only be in the form of a light on a panel, which may be interpreted by the ticket salesman to the customer.

8. A machine error may have no effect more serious than giving a single incorrect reply to an inquiry. On the other hand, a serious

failure could destroy the entire list of seats available for all flights. The first error either sells a seat which is not available, which is extremely inconvenient for the customer, or refuses to sell a seat which *is* available, which does not please the airline. The second error would cause the breakdown of the entire reservation system until the equipment was repaired and the information replaced, and the resulting loss of revenue and customer good will would be intolerable. To help prevent such troubles, the airline is willing and able to allow a regular period of preventive maintenance during the small hours of each morning, when only a few reservations are made.

The final result of this study might be the following system description: The list of available seats will be stored in an automatic memory of some kind, which will be connected to the various ticket agents over private lines. The memory must be capable of storing the number of seats available (a two-digit decimal number, since up to 100 seats may be available on a single flight "leg") for each of 1000 flight "legs" on each of ten days. The memory must therefore be able to store at least 20,000 decimal digits, or their equivalent. This much information may be stored conveniently and cheaply on a magnetic drum, and that form of storage is therefore chosen.

At peak load, fifty requests are made every minute. If no customer is to wait longer than one minute, the computer must service a request in at most 1.2 seconds. Computation this slow indicates that the central computer may be built of cheap, reliable, and flexible relays rather than of some faster acting components.

Information will be stored and computations carried out in the binary number system (see Chapter 2). Binary operation is cheaper than any other, and its principal disadvantage—conversion of input information from and output information into the decimal number system—will not be much of a handicap because the input and output rates are relatively low.

A rotary stepping switch will enable the computer to "scan" each of the 100 input stations in turn, stopping on a contact when the corresponding agent has a request, and continuing when the request has been fulfilled. This switch will, of course, always scan the input lines in the same sequence, so it will not be possible for any agent to be serviced twice while some other agent is waiting.

Each ticket agent will be provided with a small control panel on which he can set up his requests by means of buttons and switches, and from which he may read the replies sent by the computer. The most complicated part of this input/output panel will be the device which permits the ticket agent to choose from the 10,000 flight "legs" on file. Note that in many instances a passenger will want a seat on two or more adjacent "legs," as when he is flying from Minneapolis to New York on a plane which stops at Chicago and Cleveland. The agent should in this example be able to request seats on all three "legs" with one setup, if the system is to be fast and convenient, and this means that his *effective* choice of flights is from a number considerably larger than 10,000.

The desired reliability will be obtained by duplicating the entire computer, including the magnetic drum memory. All operations will be carried out in duplicate, and before an output is dispatched, or a change is made in the contents of the memory, the two answers are compared. If they disagree, all operations immediately stop, an "out-of-order" light goes on on each agent's panel, and maintenance technicians are notified. In order to make possible a quick and orderly resumption of operation after a catastrophic failure of some kind, the entire contents of the magnetic drum memory are read out and recorded three times a day. If the seats-available list recorded on the drums is accidentally destroyed, it may be replaced by inserting the most recently copied list and then requiring the individual ticket agents to re-enter all those operations they had made since the last time the memory was recorded. It is hoped that preventive maintenance techniques will anticipate and prevent such catastrophes, but a procedure must nevertheless be outlined for use if they do occur.

The foregoing description would of course form only part of the system specifications for the airline computer. The complete report might contain many other details, including layout of information on the drum; determination of gross drum characteristics, such as circumference and speed; description of system interlocks, including those which detect mistakes made by the ticket agent in setting up a request; details of the method used for entering information about next week's flights; suggested methods for making the system expandable, so that the airline may add more ticket agents or more flight "legs" or both as time goes on; description of additional error-checking equipment or procedures, if any; and specifications of the method to be used by the agent in setting up a customer's request.

It should be obvious that a system study of the kind described above is a necessary beginning for a coherent program of design and development, simply because its end result is a preliminary design. The report on the system study should, in addition, be much more than that: it is the first and in many ways the most important of a series of documents describing the system being designed. It will be referred to, as time goes on, for at least three purposes.

First, it will be referred to by the logical designers and circuit designers when they propose changes in the system. Such changes are very often made after detailed design is begun when it is discovered that the system can be improved by modifying slightly the specifications proposed by the system analyst. Changes of this kind can only be made, of course, if the system analysis contains a complete description of the job being mechanized, for otherwise it will not be possible to evaluate the proposed changes.

Second, when detailed design is complete and the equipment is under construction, the report on the system study will be consulted by the individual whose job it is to describe and justify the final system.

And third, the report on the system study will be referred to by later designers who are planning new systems or improving the old one. In this regard the study may save an enormous amount of time and effort, by pointing out what avenues of design have been explored, by recalling that certain decisions had been made because of technical difficulties which have now been surmounted, or by suggesting new and profitable lines of research.

Taken as a whole, these are powerful reasons for insisting that a very complete study and report be made in the early stages of design. The arguments also emphasize the importance of interim reports by the logical designer and circuit designer, as they study their problems and solve them with decisions to proceed in one way or another.

CIRCUIT DESIGN

Like the system analyst, the circuit designer has conflicting requirements which he must satisfy. The components he invents and perfects must be as cheap as possible to produce and operate consistent with the functions they perform. They must also be as reliable and as fast-operating as possible, and these two characteristics are often expensive.

The functions of the circuit designer impinge on those of the logical designer at three points. First, the circuit and logical designers between them determine the logical and functional characteristics of the components which are to be used in the system. This is a give-and-take sort of operation, with the logical designer describing what he thinks are the ideal components and the circuit designer compromising the ideal into something practical and specifying what the operating parameters should be. For example, the logical designer may state that he would like a memory element which has certain logical properties, operates instantaneously, and is capable of supplying an indefinitely large amount of power to other such elements. The circuit designer may be able to provide the logical properties unchanged, but the circuit element he designs may be capable of operating in nothing less than two millionths of a second, and will only supply power to twenty other such elements. The logic must then be planned with these facts in mind.

The second important interchange between the circuit designer and the logical designer has to do with the establishment of comparison ratios, which help the logical designer choose between different logical mechanisms which perform the same function. This subject is discussed in Chapter 6, pages 148–149.

Finally, the circuit and logical designers must get together when the machine nears completion in order to set up test and maintenance procedures. This subject is discussed in Chapter 12.

The most important single responsibility of the circuit designer has little or nothing to do with logical design. It is his responsibility to provide *reliable* circuits. His raw materials are a vast number of more or less unreliable and fragile components: resistors, capacitors, inductors, transformers, rectifiers, tubes, transistors, plugs and sockets, wire, magnetic materials, etc., each available from a number of different manufacturers. First the designer must choose from among these, and this is often difficult because the manufacturers themselves are unable to answer all his questions about their products. The next step is detailed circuit and structural design. Not only must the circuits carry out certain functions, but also they must be capable of carrying them out even when all of the component parameters have drifted in a random way, as they will when in service over a period of time. As they are designed, prototype circuits are built and tested under conditions as close to actual operating conditions as possible. The third step in the circuit designer's procedure is the supervision of construction. Here, great care must be taken that the wiring diagrams are carefully followed, that the proper components are used, and that mechanical and especially electrical connections are well made. Finally, as has been mentioned before, the circuit designer helps set up test and maintenance procedures.

From the complexity and extent of the work outlined above, it should be evident that the importance of a good circuit designer can hardly be overemphasized. No matter how careful the system analysis, no matter how clever the logical design, a project stands or falls—and many have fallen—on the ability of those who design and construct the circuits and related equipment.

LOGICAL DESIGN

As has been mentioned before, it is the duty of the logical designer to fill the void lying between the system analysis, which describes the problem, and the circuit components which must be connected together to solve it. The remainder of this book will be devoted to a discussion of these duties and of the techniques which may be usefully applied in carrying them out. However, as is often true when a general technique is used to solve a family of problems, the really elegant solutions are found by the designer who is able to bring new ideas and new approaches to bear, not by the man who is most skilled in adapting the technique. The same situation exists in most fields of industrial design. It is certainly not very difficult to manufacture a paper clip or a safety pin using modern wire-forming techniques, but those techniques were not of much use to the inventor who wanted to fasten two papers together and had the *idea* of the paper clip, or to the one who first conceived of the loop, bends, head, and point which are a safety pin. The inventors may have been influenced by the knowledge that wire having certain properties was available and could be bent in such and such a way, but that knowledge did not automatically lead to their very excellent gadgets. In the same way, the techniques presented here need to be augmented by the inspiration of a creative imagination if they are to be most useful.

In concluding this introduction, it might be appropriate to say a word about the adjective "logical." One might well ask why this adjective is reserved for the logical designer, and whether it implies that the system analyst and circuit designer are entitled to proceed in a haphazard or random or generally unreasonable way. The answer to the second part of the question is, of course, easy: circuit and system designers are encouraged to work in as logical a fashion as they know how. The answer to the first part is more difficult, for the origin of the term "logical design" is somewhat obscure. It was probably derived from the fact that some of the mathematical aspects of logic bear on the design problem. However, the title seems appropriate and has a fortunate history. It was applied by Oliver Wendell Holmes to the design of a famous and remarkable horse-drawn carriage which was so reliable that it operated for a century without a single failure of any kind. The carriage was, of course, the "wonderful one-horse shay," which was described in a poem beginning:

> Have you heard of the wonderful one-horse shay,
> It was built in such a logical way,
> It lasted a hundred years to the day . . .

BIBLIOGRAPHY

Books on Digital Computer Design

E. C. Berkeley, *Giant Brains*, John Wiley & Sons, New York, 1949.

D. R. Hartree, *Calculating Instruments and Machines*, University of Illinois Press, Urbana, 1949.

Staff of Engineering Research Associates, Inc., *High-Speed Computing Devices*, McGraw-Hill Book Company, New York, 1950.

Staff of Harvard Computation Laboratory, *Synthesis of Electronic Computing and Control Circuits*, Harvard University Press, Cambridge, 1951.

W. Keister, A. E. Ritchie, and S. H. Washburn, *The Design of Switching Circuits*, D. Van Nostrand Company, New York, 1951.

B. V. Bowden, *Faster Than Thought*, Sir Isaac Pitman and Sons, London, 1953.

A. D. Booth and K. H. V. Booth, *Automatic Digital Calculators*, Butterworth Scientific Publications, London, 1953.

Proc. I.R.E., **41**, no. 10 (Oct. 1953) (the computer issue).

R. K. Richards, *Arithmetic Operations in Digital Computers*, D. Van Nostrand Company, New York, 1955.

M. V. Wilkes, *Automatic Digital Computers*, John Wiley & Sons, New York, 1956.

E. M. Grabbe, *Automation in Business and Industry*, John Wiley & Sons, New York, 1957.

Airline Reservation Computer

C. Andrews and H. R. Quick, "Magnetic Memory Inventory," *Electrical Manufacturing*, **53**, no. 4, 124–129 (Oct. 1953).

E. L. Schmidt and M. L. Haselton, "Automatic Inventory System for Air Travel Reservations," *Electrical Engineering*, **73**, 641–646 (July 1954).

2 / *Circuit components and binary numbers*

The first chapter served to introduce the reader to the organization often set up to carry out digital computer design, and to the functions of members of that organization. This chapter, which is also of a preliminary nature, will discuss in some detail the circuit components needed, the restrictions placed on these components by the circuit designer in his quest for reliability, and the binary number system whose use stems from these restrictions. The intention is partly to provide advance justification for the mathematical methods developed in Chapter 3, and partly to supply some specific examples of the kinds of circuits which may be used to implement the logic found in this book. However, this is not a book on circuit design, and it is not necessary to understand circuit operation in order to learn and use the techniques which will be developed in subsequent chapters.

COMPONENT CHARACTERISTICS

To obtain some idea of what component characteristics are necessary for a typical computing system, it will be useful to consider a particular example and to try to make some deductions about it.

Suppose we examine that portion of a decimal computer which carries out the arithmetic operation of addition, and try to figure out the important characteristics of this computing element simply by observing what it does. We note that the "black box" which carries out additions has two inputs, one for each of the numbers being added, and an output upon which the sum appears. If the adder is of the type known as a *serial* adder, successive digits of the augend and addend are supplied to the inputs at successive time intervals. As each pair of digits reaches the inputs, the corresponding sum digit may be read at the output. Suppose that the two numbers 4743 and 2825 are to be added together; the state of the adder at each of the four necessary digit times is given in Figure 2-1. The addition is carried out starting with the least significant digit of each number, just as we add a series of numbers on paper by starting in the right-hand column.

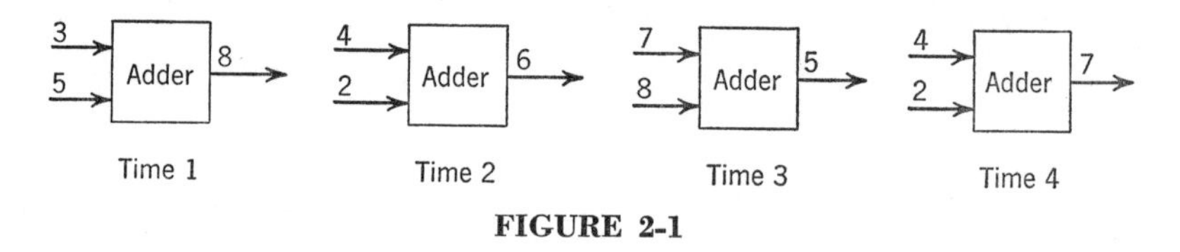

FIGURE 2-1

Two interesting deductions can be made by examining the operation of this black box. In the first place, note that the output at a given time is not only a function of the input at that time, but also of the input digits which have come before. At time 2, for example, a 4 and a 2 on the input establish a 6 at the output; but at time 4, a 4 and a 2 establish a 7. Since the inputs are exactly the same at these two times, but the outputs are different, there must exist within the box at time 4 some condition which did not exist at time 2. Furthermore, that condition must have been set up at time 3, when it was discovered that the sum of the input digits was greater than 9; thus there would have to be a "carry" into the next column. Another way of describing this capability of the black box is to say that it has a memory, or that it stores information. In this instance, the memory need not be very large, for only the presence or absence of a "carry" from the previous column need be stored; but the interesting and significant thing is that the circuit designer must supply some kind of *memory element*.

The other deduction which can be made by observing the black box is a little less subtle. The box must contain some kind of equipment for making decisions based on the input digits. In other words, there must exist some kind of circuit components which act on the input data in a way which is defined by the rules for addition. It is evident that these rule-following or *decision elements* must somehow read decimal digits in the same *form* as they occur at the input to the box, but without taking the box apart we cannot discover anything else about their nature. For example, we cannot deduce whether these elements are somehow inherently decimal in operation, or whether they can be used in the design of an adder which must form the sum of, say, two Roman numerals.

This cursory discussion has unearthed two important and fundamental digital components, the memory element and the decision element. Together, they are called computer *logical elements*. No matter what the design details of the circuits in the black box of Figure 2-1 may be, these memory and decision functions must somehow be carried out.

CIRCUIT CHARACTERISTICS

Having considered the functional aspects of a black box, let us now examine it with the eye of a circuit designer. It must be evident that the operation of the circuits will depend entirely on the way the decimal digits are represented electrically. Suppose, to begin with, that a decimal digit is to be represented by a voltage between a pair of wires, as follows:

Voltage greater than or equal to	but less than	represents the digit
$-\infty$	10 volts	0
10	20	1
20	30	2
30	40	3
40	50	4
50	60	5
60	70	6
70	80	7
80	90	8
90	$+\infty$	9

A *range* of voltages, rather than a single voltage, is specified for each digit because it is impossible to maintain exact voltage levels with electronic circuits. The decision circuit elements must then be able to distinguish between these ten possible voltage levels, and provide accurate output levels within the same range. Similarly, each of the memory elements must be capable of "remembering" which of ten states it was last required to store, and of providing corresponding output levels just as the logical elements do.

All of these requirements could be met by a competent and ingenious circuit designer. There is, however, another requirement that is very important and very difficult to satisfy: both decision and memory elements must operate reliably over long periods of time with little or no attention. It is easy to understand why such reliability is hard to come by. A 6% change in a circuit which has been designed to provide an 85-volt level to represent the digit eight may increase or decrease the voltage to 90.1 or 79.9, with the result that a nine or a seven is represented instead of an eight. Inasmuch as the characteristics of a vacuum tube, for example, very commonly change by 50% over a period of time, it is reasonable to expect that there would be some difficulty in preventing parameter changes of the order of 5–6%. In practice, the difficulties are so great that the designer turns to another way of representing numbers electronically.

Suppose now that a pair of wires, instead of representing one of ten possible digits, is made to represent only one of two. The digit "zero" then might be represented by any voltage less than 50 volts, and the digit "one" by any voltage greater than 50. The nominal representation for "zero" might be 25 volts, and the nominal value for "one" could be 75 volts, and a change of 33⅓% could take place in the larger figure before the level would change enough to be misinterpreted as a "zero." Alternatively (and more practically), the power level of the signals may be lowered, and a reliable system devised wherein a "zero" is nominally represented by 0 volts and a "one" by 20 volts, with 10 volts being the dividing line between the two digits.

Most of the digital computers which have been built represent numbers in the two-valued or *binary* manner just described. The next section will explain how binary circuits can be made to represent decimal numbers.

THE BINARY NUMBER SYSTEM

The decimal number system, with which everyone is familiar, represents numbers as a sum of powers of ten, where each power of ten is weighted by a digit between zero and nine inclusive. When we write a decimal number, we note down only the weights to be attached to the various powers, and a decimal point which tells which powers of ten are to be weighted. For example,

$$433. = 4 \times 10^2 + 3 \times 10^1 + 3 \times 10^0$$

$$9751.68 = 9 \times 10^3 + 7 \times 10^2 + 5 \times 10^1 + 1 \times 10^0 + 6 \times 10^{-1} + 8 \times 10^{-2}$$

and in general

$$\cdots a_2 a_1 a_0 \cdot a_{-1} a_{-2} a_{-3} \cdots = \cdots + a_2 r^2 + a_1 r^1 + a_0 r^0 + a_{-1} r^{-1} + a_{-2} r^{-2} + a_{-3} r^{-3} \cdots$$

where a_2, a_1, a_0, a_{-1}, a_{-2}, a_{-3}, etc., are each digits between zero and $(r - 1)$, r is the *base* of the number system, and $r = 10$ for a decimal number.

The number ten was probably chosen as the base for the numbers we use because there are that many fingers and thumbs on the hands. However, there is no reason why some other base might not be chosen. Using a subscript to indicate the base being used, we see, for example, that:

$$
\begin{aligned}
(433.)_{10} &= \mathbf{6} \times 8^2 + \mathbf{6} \times 8^1 + \mathbf{1} \times 8^0 = (661.)_8 \\
&= \mathbf{3} \times 5^3 + \mathbf{2} \times 5^2 + \mathbf{1} \times 5^1 + \mathbf{3} \times 5^0 = (3213.)_5 \\
&= \mathbf{1} \times 3^5 + \mathbf{2} \times 3^4 + \mathbf{1} \times 3^3 + \mathbf{0} \times 3^2 + \mathbf{0} \times 3^1 + \mathbf{1} \times 3^0 \\
&\qquad = (121001.)_3 \\
&= \mathbf{1} \times 2^8 + \mathbf{1} \times 2^7 + \mathbf{0} \times 2^6 + \mathbf{1} \times 2^5 + \mathbf{1} \times 2^4 + \mathbf{0} \times 2^3 \\
&\qquad + \mathbf{0} \times 2^2 + \mathbf{0} \times 2^1 + \mathbf{1} \times 2^0 = (110110001.)_2
\end{aligned}
$$

The last number system used, having the base two, is called the binary number system, and it catches our eye immediately because the weights consist entirely of ones and zeros. It is apparent that, using the base two, any number may be expressed by a sequence of high and low voltages, or by a group of memory devices, each of which is capable of storing either a one or else a zero. Figure 2-2 indicates how the

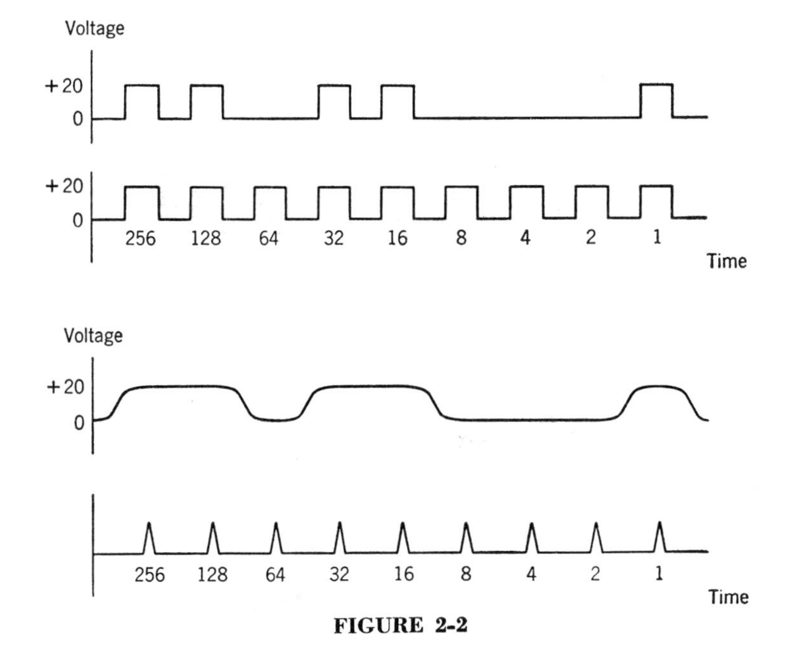

FIGURE 2-2

number $(433.)_{10}$ may be represented by a train of pulses, or by a voltage level which is sampled periodically. In each instance, the first waveform is the signal waveform, and the second a sampling pulse, often called a clock pulse, which appears somewhere in the system and is used as a reference to determine when the signal waveform contains significant information.

Binary numbers, and especially numbers less than sixteen, occur so often that they soon become as familiar to the logical designer as decimal numbers are. Table 2-1 and Examples 2-1, 2-2, and 2-3 are provided to help familiarize the beginner with these fundamentals.

Table 2-1

Decimal	Binary
.0625	.0001
.125	.001
.25	.01
.5	.1
1.	1.
2.	10.
3.	11.
4.	100.
5.	101.
6.	110.
7.	111.
8.	1000.
9.	1001.
10.	1010.
11.	1011.
12.	1100.
13.	1101.
14.	1110.
15.	1111.
16.	10000.
32.	100000.
64.	1000000.
100.	1100100.
128.	10000000.
256.	100000000.

Example 2-1. What is the decimal value of the binary number 101101110?
Ans. 101101110.

$$= 1 \times 2^8 + 0 \times 2^7 + 1 \times 2^6 + 1 \times 2^5 + 0 \times 2^4 + 1 \times 2^3 + 1 \times 2^2 + 1 \times 2^1 + 0 \times 2^0$$

$$= 256 + 64 + 32 + 8 + 4 + 2 = 366$$

Example 2-2. Convert the decimal number 9318 into a binary number.

Ans. The highest power of two which is less than 9318 is $2^{13} = 8192$. Therefore

$$9318 = 2^{13} + x_1 \qquad x_1 = 1126$$

and we must express x_1 as a sum of powers of two. We thus have

$$\begin{array}{lll}
2^{10} = 1024 & x_1 = 1126 = 2^{10} + x_2 & x_2 = 102 \\
2^{6} = 64 & x_2 = 102 = 2^{6} + x_3 & x_3 = 38 \\
2^{5} = 32 & x_3 = 38 = 2^{5} + x_4 & x_4 = 6 \\
2^{2} = 4 & x_4 = 6 = 2^{2} + x_5 & x_5 = 2 \\
2^{1} = 2 & x_5 = 2 = 2^{1} + x_6 & x_6 = 0
\end{array}$$

Therefore $9318 = 2^{13} + x_1 = 2^{13} + 2^{10} + x_2 = 2^{13} + 2^{10} + 2^6 + x_3$

$$= 2^{13} + 2^{10} + 2^6 + 2^5 + x_4 = 2^{13} + 2^{10} + 2^6 + 2^5 + 2^2 + x_5$$

$$= 2^{13} + 2^{10} + 2^6 + 2^5 + 2^2 + 2^1$$

$$= 10010001100110.$$

Example 2-3. Convert the decimal number 80.286 into a binary number.

Ans.

$$\begin{array}{lll}
2^6 = 64 & 80.286 = 2^6 + x_1 & x_1 = 16.286 \\
2^4 = 16 & x_1 = 16.286 = 2^4 + x_2 & x_2 = 0.286 \\
2^{-2} = .25 & x_2 = .286 = 2^{-2} + x_3 & x_3 = 0.036 \\
2^{-5} = .03125 & x_3 = .036 = 2^{-5} + x_4 & x_4 = .00475
\end{array}$$

Therefore $80.286 = 2^6 + 2^4 + 2^{-2} + 2^{-5} + x_4 = 1010000.01001 \cdots$

Since $x_4 \neq 0$, the binary number is not exactly equal to the decimal one. There is, in fact, no exact binary equivalent to 80.286, just as there is no exact decimal equivalent for $4\frac{1}{3}$.

The binary number system provides one method of representing numbers with electronic circuits which only recognize two voltage levels and can, therefore, be made reliable. However, in order to be interpreted and understood properly by human beings, numbers must be expressed in decimal form, and if they are represented in a computer in the binary number system, a decimal-binary conversion must be made whenever information is inserted into the machine and a binary-decimal conversion whenever the operator wants a result. This conversion is, of course, not particularly difficult, as we have seen, and it is quite possible to arrange that the computer itself perform the arithmetic. It may nevertheless be inconvenient to use the binary number system (because computer users are unfamiliar with binary numbers), and the designer may decide to turn to some kind of coded decimal scheme instead.

In a binary-coded decimal system, each decimal digit is replaced by a group of binary digits (or *bits*, as they are usually called). Since each group of bits must represent one of ten digits, there must be at least four bits in a group, for three bits can only represent eight different states and four bits, sixteen. The code which is perhaps most often used is that given in Table 2-2. These binary numbers are, of course,

Table 2-2

Decimal Digit	Binary Code
0	0000
1	0001
2	0010
3	0011
4	0100
5	0101
6	0110
7	0111
8	1000
9	1001

the true binary translation of the decimal digits, and this is often called the 8-4-2-1 code because those are the weights given to the four bit positions. The number 433, which was expressed in the binary number system by 110110001, is written 010000110011 in the 8-4-2-1 code, each set of four bits representing one decimal digit. All coded decimal systems are somewhat wasteful in that some combinations of bits can never occur, e.g., the combinations 1010, 1011, 1100, 1101, 1110, 1111 in the 8-4-2-1 code. A four-bit decimal code requires $4n$ bits to represent 10^n different numbers. To find the number of bits required to represent 10^n numbers with a straight binary code, we set $2^x = 10^n$, and find $x = \log_2 10^n = 3.32n$. Thus $3.32n$ bits could do the work of $4n$, and the waste is $[(4n - 3.32n)/4n] \times 100 = 17\%$, expressed as a percentage of the number of decimal digits necessary.

Other codes besides the 8-4-2-1 have been used, some of them requiring more than four bits per digit. Some of these other codes will be discussed in Chapter 9. All such codes have as their chief advantage the fact that the translation from decimal to binary code and back again can be done very easily, one decimal digit at a time. No arithmetic is necessary. This advantage is paid for in wasted space, as indicated above, and in increased complexity of computing logic.

BINARY CIRCUITS

We have already decided that computer circuits must be capable of performing decision-like operations and of storing information. Two other features of digital computer circuits are important enough to deserve special mention. First is the notion of using a few different kinds of circuits as "building blocks." It would be quite possible to build a computer having a great many storage and decision elements, where the circuit design for each element is different from that of all other elements. The logical designer might say that it is convenient to use this kind of element here and that kind there, and lay out different specifications for each circuit. Although this would appear to make things a little easier for the logical designer, it does so by making a great deal of extra work for the circuit designer. A much better scheme, the one which is used almost exclusively in computer design, is to decide on a small number of different kinds of circuits—e.g., a memory element and two or three decision elements—and make the logical designer adapt his design to them. This means that the circuit designer can devote all his time to perfecting the few components which are to be used, instead of having to spread his talent and efforts among many different circuits. It also has the important advantage that only a few spare parts need be kept on hand by the operator of a system built of "building blocks" in this way.

The other important component characteristic follows more or less from the first: the output of each component must be at the proper voltage and impedance level so that it may be applied to the input of any other component. This requirement is obviously necessary if the "building block" system is to work efficiently. The circuit designer must make interconnections standard, as well as the circuits themselves.

We shall now describe briefly some circuits which are widely used to mechanize logical elements. However, an understanding of these circuits is not essential to an understanding of the fundamentals of logical design, and the reader who is not interested in circuits may go on to Chapter 3.

Decision Circuit Elements

A decision element, it will be remembered, is a circuit which allows the designer to mechanize certain rules. The circuits presented in this section operate on two-valued voltage levels, and mechanize rules which seem so simple that one may well ask how they can be useful in carrying out a complicated operation, e.g., decimal division. Subsequent chapters will answer this quite reasonable question and justify the existence of the decision elements described here.

Perhaps the most widely used elements are constructed of diodes and resistors, as shown in Figure 2-3 and 2-4. Figure 2-3*a* shows an

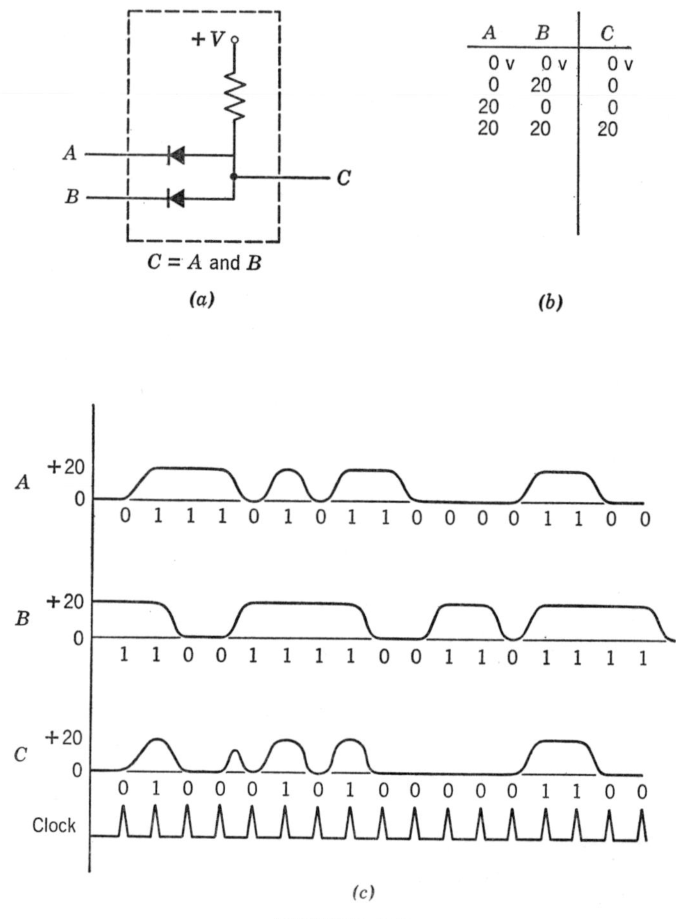

A	B	C
0 v	0 v	0 v
0	20	0
20	0	0
20	20	20

FIGURE 2-3

"and" circuit having two inputs, A and B, and one output, C. If the nominal voltage levels of A and B are 0 volts (representing "zero") and +20 volts (representing "one"), then the output line is at +20 volts if and only if both A *and* B are at +20 volts—hence the name, "and" circuit. The operation of the circuit may be explained by pointing out that if line A, say, is at 0 volts while line B is at 20 volts, current flows from the $+V$ supply and the resistor, through the low forward resistance of the A diode to point A. Since point C is at ground potential and point B at 20 volts, there is a 20-volt drop across the high back resistance of the B diode, and the B voltage has little or no effect on point C. When A is at 20 volts and B at 0, the A diode is back-biased and the B diode holds point C down to 0 volts. It is only when both A and B are at 20 volts that the output line can rise.

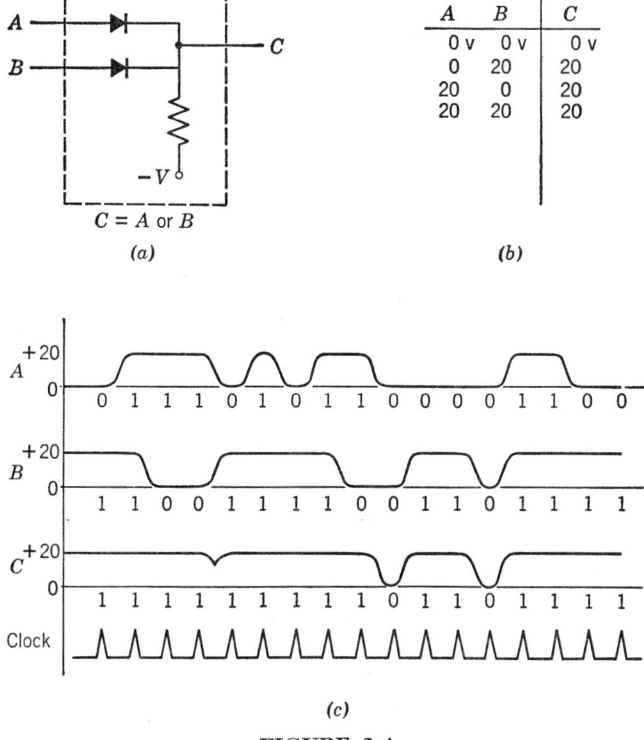

FIGURE 2-4

Figure 2-3*b* is a table showing the state of the output for each of the four possible input combinations, and Figure 2-3*c* shows the output waveform for the circuit as a function of two arbitrary input waveforms.

Figure 2-4*a* shows another widely used combination of diodes and resistor called an "or" circuit. An analysis similar to that given in the last paragraph should convince the reader that in this circuit point C is always at the potential of the higher of A or B. Therefore C represents a "one" whenever A *or* B *or* both represent a "one," and the circuit is called an "or" circuit.

Two other important features of these circuits are illustrated in Figure 2-5. Figure 2-5*a* shows how they can be cascaded, with the outputs of two "and" circuits becoming the inputs to an "or" circuit. The output signal E will be 20 volts (representing a "one") only when both A and B are 20 volts, or when C and D are both 20 volts. In a similar way, the outputs of two "or" circuits could be made the input to an "and" circuit. Figure 2-5*b* indicates that a single "and" or "or"

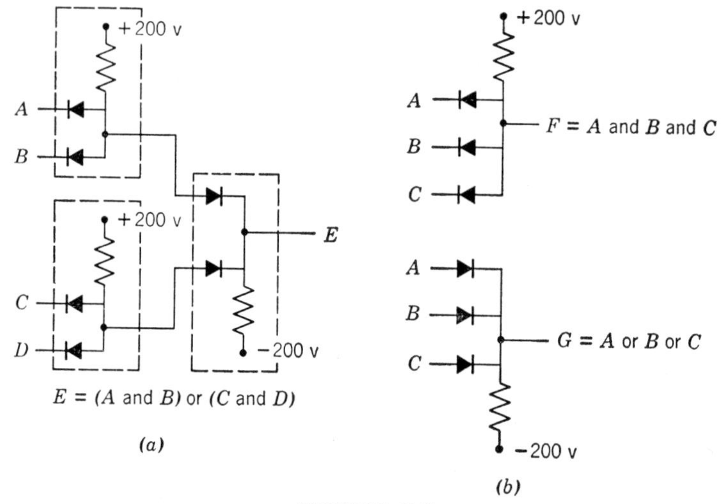

FIGURE 2-5

circuit may have three or more inputs. F in Figure 2-5*b* is high only if A, B, and C are all high; G is high whenever any of A, B, or C are high.

The "and" and "or" circuits may be thought of as specific mechanizations for functions of two variables, since for each combination of values for the inputs there is defined a value for the output. Although the two functions "and" and "or" are very widely used, there are other circuits, containing tubes, relays, transistors, etc., which mechanize these and other functions. For example, in Figure 2-6 a circuit is shown which provides two new functions which may be described as follows: C is "one" whenever A and B represent different binary digits, and D is "one" when A and B represent the same digit. For example, when B is 110 volts and A 190, current flows through the left half of the lower tube, through the right half of the upper left tube, and then through the resistor connected to point D. This reduces the voltage at D, so that a "zero" is represented; but it does not affect the voltage at C, and a high voltage there represents a "one." The disadvantages of this circuit are that it uses a great many tubes and that input lines have two different voltage levels. The circuit is intended to show what can be done rather than what is done in practice.

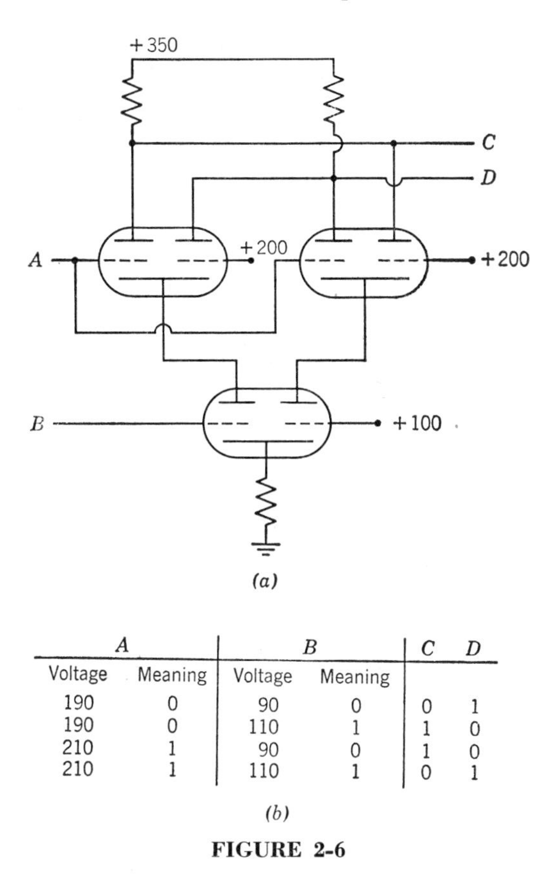

A		B		C	D
Voltage	Meaning	Voltage	Meaning		
190	0	90	0	0	1
190	0	110	1	1	0
210	1	90	0	1	0
210	1	110	1	0	1

(b)

FIGURE 2-6

There are still other functions of two variables. They will all be discussed in the next chapter, and though no more circuits will be presented here, there are many published descriptions of mechanizations of decision elements.

Memory Circuit Elements

The circuit element most widely used today for storing a single bit of information is the flip-flop, illustrated in Figure 2-7. This circuit is said to store one bit because it has two stable states: when V_1 is conducting, its plate voltage is low and that low voltage is coupled through a resistor chain to the grid of V_2, which is thereby cut off; since V_2 is cut off, its plate voltage is high and that high voltage is coupled to the grid of V_1, so that V_1 conducts as originally assumed. When the flip-flop is in its other stable state, V_2 conducts and V_1 is cut off.

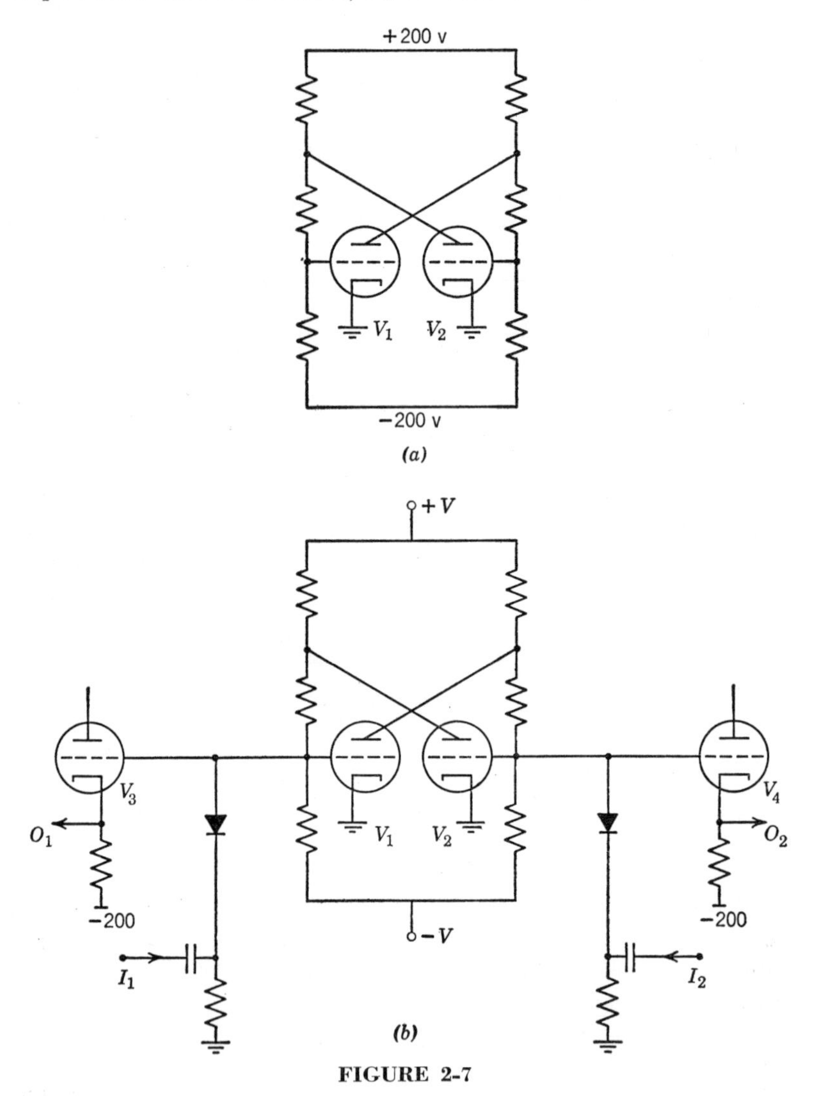

FIGURE 2-7

Typical input and output circuits for the flip-flop are shown in Figure 2-7*b*. If tube V_1 is conducting, a negative pulse of correct magnitude and duration applied at point I_1 will cut off V_1, causing its plate voltage to rise and turning on V_2. V_2 will then continue to conduct until a negative pulse is applied to I_2, and then the flip-flop will return to its original state. Cathode followers V_3 and V_4 isolate the flip-flop from the circuits it supplies information to, so that those circuits do not affect flip-flop stability. They also act as power amplifiers, so that a single flip-flop may control a great many other circuits.

There are of course many other similar bistable circuits (including transistor versions of Figure 2-7), and many other methods for attaching inputs and obtaining outputs. There are also a number of other possible methods for storing and reading information electronically. Special vacuum tubes have been developed which can be made to have two or more stable states. Magnetic materials, which may be magnetized in one of two directions, can be incorporated into circuits which allow one or more bits of information to be written, stored, and read. Chapter 7 will describe several methods which have been used for the storage of large numbers of bits—hundreds or thousands or even millions. Each of these various storage schemes has its advantages and disadvantages, and a complete discussion and evaluation of them all would be out of place here. However, it may be said that the widespread use of the flip-flop stems primarily from four basic facts about that circuit: it can be built from readily available components; it can be made to change states in as little as a few millimicroseconds; it is a power-amplifying device; and its input logic can be combined with a signal from its output to provide a circuit element whose logical properties are very subtle and useful. This last feature is particularly important, and will be discussed in some detail in Chapter 5.

BIBLIOGRAPHY

Mechanization of Decision Elements

L. W. Hussey, "Semiconductor Diode Gates," *Bell System Technical J.*, **32,** 1137–1154 (1953).

J. D. Goodell, "Testing Magnetic Decision Elements," *Electronics*, 200–203 (Jan. 1954).

B. Moffat, "Saturable Transformers as Gates," *Electronics*, 174–176 (Sept. 1954).

M. Karnaugh, "Pulse-Switching Circuits Using Magnetic Cores," *Proc. I.R.E.*, **43,** 570–584 (1955).

B. J. Yokelson and W. Ulrich, "Engineering Multistage Diode Logic Circuits," *A.I.E.E. Trans.*, **74,** Part 1, 466–475 (1955).

R. B. Trousdale, "The Symmetrical Transistor as a Bilateral Switching Element," *Communications and Electronics*, 400–403 (Sept. 1950).

J. L. Moll, M. Tanenbaum, J. M. Goldey, and N. Holonyak, "*P-N-P-N* Transistor Switches," *Proc. I.R.E.*, **44,** 1174–1182 (1956).

R. A. Cola, "Low Voltage Beam Switching Tube," *Electronic Design*, 22–25 (Sept. 1956).

G. W. Booth and T. P. Bothwell, "Basic Logic Circuits for Computer Applications," *Electronics*, 190–193 (March 1957).

Mechanization of Memory Elements

R. D. Kodis, S. Ruhman, and W. D. Woo, "Magnetic Shift Register Using One Core per Bit," *Convention Record of the I.R.E.*, Part 7, 38–42 (1953).

R. H. Baker, I. L. Lebow, et al., "The Phase-Bistable Transistor Circuit," *Proc. I.R.E.*, **41,** 1119–1124 (1953).

R. S. Mackay, "Switching in Bistable Circuits," *J. Applied Physics*, **25,** no. 4, 424–429 (1954).

R. L. Brock, "Transistor Flip-Flop Using Two Frequencies," *Electronics*, 175–179 (June 1954).

R. C. M. Barnes, "Relay Scale-of-two Circuits," *Electronic Engineering*, 493–497 (Nov. 1954).

R. B. Koehler and R. K. Richards, "Decade Counter Tube for Accounting Machines," *Electronics*, 151–153 (Nov. 1954).

S. S. Guterman and W. M. Carey, "A Transistor-Magnetic Core Circuit; a New Device Applied to Digital Computer Techniques," *Convention Record of the I.R.E.*, Part 4, 84–94 (1955).

C. L. Wanlass, "Transistor Circuitry for Digital Computers," *I.R.E. Trans. on Electronic Computers*, **EC-4,** no. 1, 11–15 (1955).

W. Renwick and M. Phister, Jr., "A Design Method for Direct-Coupled Flip-Flops," *Electronic Engineering*, 246–250 (June 1955).

J. A. Rajchman and A. W. Lo, "The Transfluxor—A Magnetic Gate with Storage Variable Setting," *RCA Review*, 302–311 (June 1955).

H. Aharoni et al., "A New Active Circuit Element Using the Magnetostrictive Effect," *J. Applied Physics*, 1411–1415 (1955).

A. S. Fitzgerald, "Asymmetrical Magnetic Amplifier Sequential Circuits," *Communications and Electronics*, 685–690 (Jan. 1956).

N. R. Scott, "Temporary Storage Elements and Special-Purpose Tubes," *Control Engineering*, **3,** no. 3, 93–98 (1956).

C. E. Gremer, "An Analysis and Design of the Ferroelectric Resonant Trigger Pair," *Communications and Electronics*, 404–407 (Sept. 1956).

EXERCISES

1. Convert the following numbers into equivalent decimals:

a. $(8643.)_9$ *d.* $(3500.)_7$
b. $(1.22)_3$ *e.* $(110.010)_2$
c. $(40.6)_8$ *f.* $(110.010)_5$

2. Convert the following decimal numbers into numbers having the base indicated:

a. $(999.)_{10}$ into $(x)_9$ *c.* $(82{,}321.)_{10}$ into $(x)_8$
b. $(63.)_{10}$ into $(x)_5$ *d.* $(503.)_{10}$ into $(x)_2$

3. Convert the following decimal numbers into binary numbers. Carry the result out to five binary digits past the binary point.

a. 1.347
b. 0.07783
c. 17.692
d. 8.431
e. 12.53125
f. $3\frac{1}{3}$

4. Convert the number $(62437)_{10}$ into an octal (i.e., base eight) number. Now convert it into a binary number. What is the relationship between binary and octal numbers? State the rule for converting binary numbers into octal form, and vice versa.

5. Convert $(0.1956)_{10}$ into a binary number. Does the decimal have an exact binary equivalent? Under what conditions does a decimal number have an exact binary equivalent?

6. Convert $(0.1101011)_2$ into a decimal number. Is there an exact decimal equivalent? Does every binary number have an exact decimal equivalent?

7. Draw a circuit diagram, similar to that of Figure 2-5*a*, for each of the following functions:

a. (A or C) and (B or E)
b. A or (B and C) or (C and D and E)

3 / *Boolean algebra*

The heuristic arguments of the last chapter pointed out the necessity for decision and memory circuit elements, and indicated how demands for reliability lead the circuit designer to insist on two-valued elements. In the latter part of the chapter it was shown that any number may be represented by a series of binary symbols; and the operation of several decision and memory devices was discussed. Beginning with Chapter 5, we shall learn how to combine these circuit elements to carry out complicated operations, but before we can use them effectively we must become familiar with a kind of algebra which enables us to combine and manipulate binary signals. Chapters 3 and 4 introduce this new branch of mathematics. It is called Boolean algebra in honor of its founder, George Boole (1815–1864), who first published an account of it in 1854. This introduction is intended to bring out the important characteristics of the algebra quickly and effectively, and for this reason many of the fine points essential to mathematical rigor will be ignored here.

HUNTINGTON'S POSTULATES

Deductive Theories

A deductive theory may be established by the following procedure: first, choose a set of fundamental concepts; second, decide on a group of postulates relating the concepts in some way, and imposing certain conditions on them. The postulates should have the following three properties:

1. They must be consistent. Putting it another way, they must not be self-contradictory.
2. They should be simple statements, not decomposable into two or more parts. (This is an elusive property, being very difficult to define properly.)
3. For elegance, they should be independent. That is to say, it should not be possible to derive any of the postulates from any of the other postulates.

Having a set of concepts and another of postulates, it should be possible to deduce a number of theorems which show relationships among the concepts in much the same way as the postulates do.

One deductive theory we are all familiar with is Euclidean geometry. Another is ordinary algebra, though it is usually not introduced to us in schools in this formal way. Boolean algebra is still another deductive theory, and the development given here is one suggested by E. V. Huntington.

We begin by defining a class, K, and two rules of combination denoted by "+" and "·". (We also take for granted certain principles of logical procedure, which enable us to understand, for example, that the "=" sign means that the symbols which it separates represent identical quantities, and that one may anywhere be substituted for the other.) The postulates may then be stated as follows:

P1a. If an element A is a member of K, and an element B is also a member of K, then $(A + B)$ is also a member of K.

P1b. If an element A is a member of K, and an element B is also a member of K, then $(A \cdot B)$ is also a member of K.

P2a. There is an element $\mathbf{0}$ such that $A + \mathbf{0} = A$ for every element A in K.

P2b. There is an element $\mathbf{1}$ such that $A \cdot \mathbf{1} = A$ for every element A in K.

P3a. Whenever elements A and B are in the class K, $(A + B) = (B + A)$.

P3b. Whenever elements A and B are in the class K, $(A \cdot B) = (B \cdot A)$.

P4a. Whenever elements A and B are in the class K, $A + (B \cdot C) = (A + B) \cdot (A + C)$.

P4b. Whenever elements A and B are in the class K, $A \cdot (B + C) = (A \cdot B) + (A \cdot C)$.

P5. If the elements **0** and **1** of P2 are unique, then for every element A in the class K, there is an element $\bar{A}$ such that $A \cdot \bar{A} = \mathbf{0}$, and $A + \bar{A} = \mathbf{1}$.

P6. There are at least two elements, X and Y, in class K such that $X \neq Y$.

The first eight postulates are grouped together in four sets of two each, and this grouping illustrates the perfect symmetry, or duality, of the algebra with respect to the operations $\cdot$ and $+$: if in any of these eight postulates, **0** is replaced by **1**, **1** by **0**, each $+$ is replaced by a $\cdot$, and each $\cdot$ by a $+$, the result will be the dual of the original postulate. This duality will appear next when we begin to derive theorems, for if we are given some theorem which can be proved from the postulates, we can immediately construct a proof of a dual theorem justifying each step by the dual postulates.

Consistency

To show that the postulates written above are consistent, we need only exhibit one example of a system in which K, $+$, and $\cdot$ are so interpreted that all postulates are satisfied. This is quite a standard

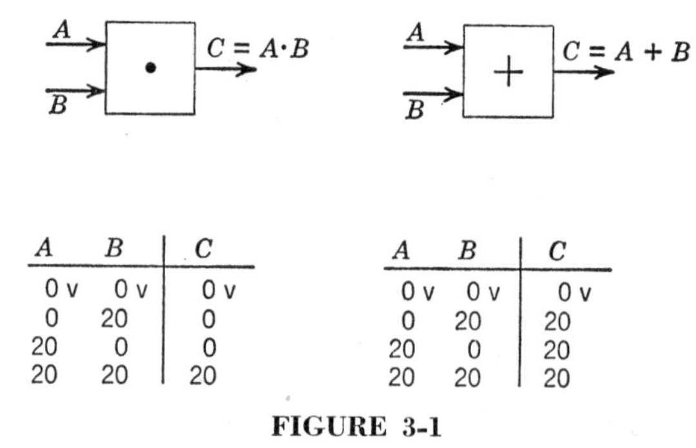

A	B	C
0 v	0 v	0 v
0	20	0
20	0	0
20	20	20

A	B	C
0 v	0 v	0 v
0	20	20
20	0	20
20	20	20

FIGURE 3-1

method for proving consistency, and may be justified by the following simple argument: if the postulates are contradictory, any application of them to some real situation will also lead to a contradiction. In other words, it will not be possible to find a physical interpretation of the class and of the rules of combination in which all the postulates are satisfied. Conversely, if such an interpretation is found and leads to no contradiction, the postulates must be consistent.

One interpretation of the universe K could be a set of wires A, B, C, etc., each of which is at either 0 or +20 volts. The symbol $\cdot$ could then be represented by a diode circuit like the one shown in Figure

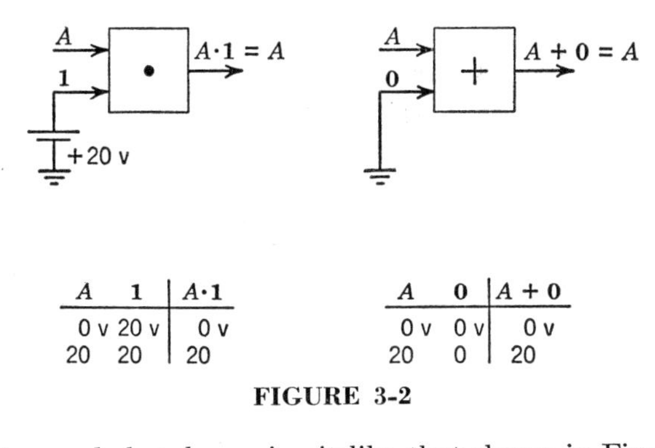

A	$\mathbf{1}$	$A \cdot \mathbf{1}$
0 v	20 v	0 v
20	20	20

A	$\mathbf{0}$	$A + \mathbf{0}$
0 v	0 v	0 v
20	0	20

FIGURE 3-2

2-3, and the symbol + by a circuit like that shown in Figure 2-4. If we abbreviate these circuit diagrams by introducing the symbols of Figure 3-1, we see at once that P1a, P1b, P3a, P3b, and P6 are satisfied. Furthermore, P2a will be satisfied if we interpret $\mathbf{0}$ as meaning 0 volts, and P2b if we interpret $\mathbf{1}$ as meaning +20 volts, as shown in Figure 3-2. Next, Figure 3-3 shows that P4a is true under this inter-

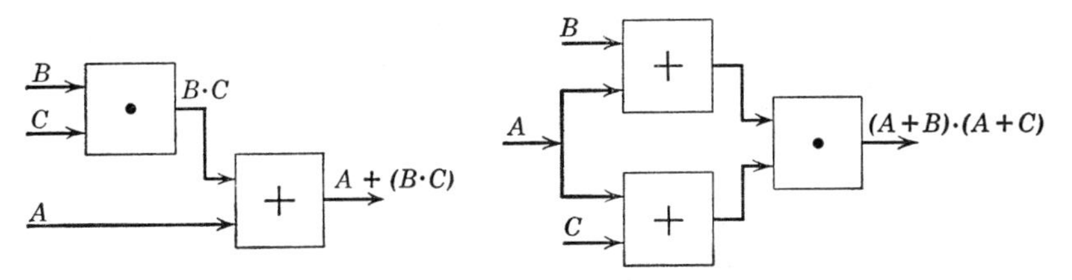

A	B	C	$B \cdot C$	$A+(B \cdot C)$
0 v	0 v	0 v	0 v	0 v
0	0	20	0	0
0	20	0	0	0
0	20	20	20	20
20	0	0	0	20
20	0	20	0	20
20	20	0	0	20
20	20	20	20	20

A	B	C	$A+B$	$A+C$	$(A+B) \cdot (A+C)$
0 v	0 v	0 v	0 v	0 v	0 v
0	0	20	0	20	0
0	20	0	20	0	0
0	20	20	20	20	20
20	0	0	20	20	20
20	0	20	20	20	20
20	20	0	20	20	20
20	20	20	20	20	20

FIGURE 3-3

pretation. The two sides of the equation correspond to the two circuit diagrams shown, and for each of the $2^3 = 8$ possible input combinations, the output voltage is seen to be the same for the two circuits.

A similar pair of diagrams can be drawn and interpreted to verify the consistency of P4b. Finally, Figure 3-4 shows how P5 holds, when $\bar{A}$ is interpreted as meaning "the opposite of the voltage A." That is, when A is at 0 volts, $\bar{A}$ must be at +20, and when A is at +20, $\bar{A}$ is at 0 volts.

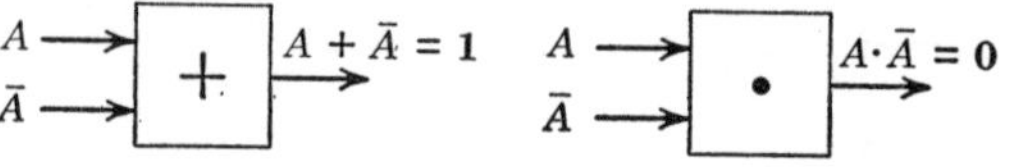

A	$\bar{A}$	$A + \bar{A}$
0 v	20 v	20 v
20	0	20

A	$\bar{A}$	$A \cdot \bar{A}$
0 v	20 v	0 v
20	0	0

FIGURE 3-4

The interpretation of the postulates in terms of binary voltages, with the rules of combination becoming diode circuits, proves the postulates are consistent. Many other interpretations are possible, and one other is interesting enough and useful enough to warrant presentation here. Suppose now the class K is defined as all possible regions within a square. Any element A within the class is a certain set of points within the square, e.g., all the points within a given closed curve. The element $(A + B)$ may then be defined as the smallest region containing both A and B, and the element $(A \cdot B)$ as the largest region contained in both A and B. This interpretation of the rules of combination is indicated in Figure 3-5, and diagrams of this kind are called Venn diagrams, after another 19th century British mathematician.

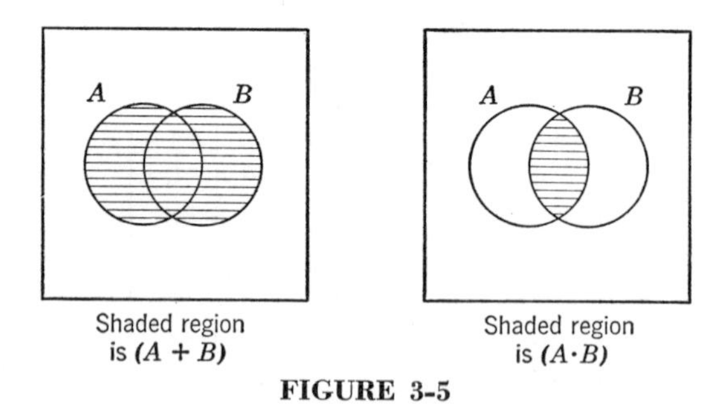

Shaded region is $(A + B)$

Shaded region is $(A \cdot B)$

FIGURE 3-5

Examining the postulates in the light of this particular interpretation, we see at once that P1a, P1b, P3a, P3b, and P6 are all satisfied. Furthermore, if **0** is interpreted as meaning the class of no points at all, and **1** the class of all points within the square, then obviously P2a and P2b are consistent. Next, consider Figure 3-6, which shows the consistency of P4a. In Figure 3-6*a*, the region A is shaded with horizontal lines, and the region $(B \cdot C)$ with vertical lines, so that $A + (B \cdot C)$

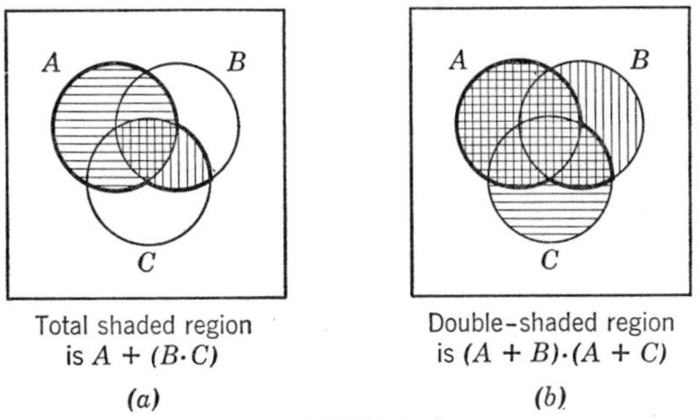

Total shaded region is $A + (B \cdot C)$ (*a*)

Double-shaded region is $(A + B) \cdot (A + C)$ (*b*)

FIGURE 3-6

is represented by the smallest area which contains both shaded areas, i.e., by the total shaded area. In Figure 3-6*b* the region $(A + B)$ is shaded with vertical lines, and the region $(A + C)$ with horizontal lines. The term $(A + B) \cdot (A + C)$ is then represented by the largest area included in both $(A + B)$ and $(A + C)$, i.e., by the area which is shaded both vertically and horizontally. In an exactly analogous way, P4b can be shown to be consistent. Finally, P5 is obviously noncontradictory if $\bar{A}$ is interpreted to be all those points in the square which are not members of A. Figure 3-7 shows A and $\bar{A}$; it is evident that $A + \bar{A} = \mathbf{1}$.

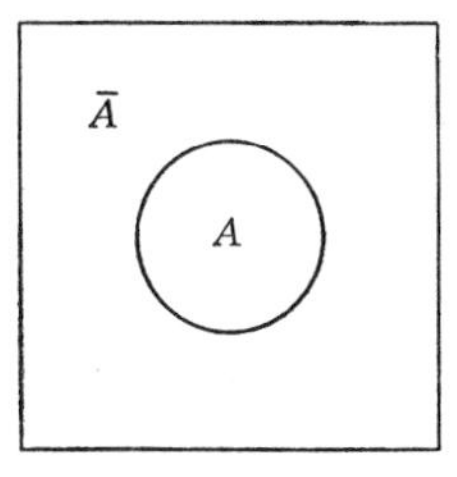

FIGURE 3-7

Independence

The method and philosophy of the usual proof of independence is as follows: if a given postulate—say, postulate n—is *not* independent, then it is possible to derive it from the other postulates. If it can be derived from the other postulates, then in any interpretation under which all the other postulates are true, postulate n must also be true. It therefore follows that if one can find an interpretation of some kind in which postulate n is false while all the other postulates are true, one has proved the independence of n. The independence of all the postulates can be proved by finding a contradictory interpretation like this for each one.

As a typical example of this technique, consider Huntington's proof for the independence of P1a. Suppose that the class K consists of only two elements, **1** and **0**. And suppose these two elements satisfy the following rules: $\mathbf{0} + \mathbf{0} = \mathbf{0}$, $\mathbf{0} + \mathbf{1} = \mathbf{1}$, $\mathbf{1} + \mathbf{0} = \mathbf{1}$, $\mathbf{1} + \mathbf{1} = X$ (X is not in the class K), $\mathbf{0} \cdot \mathbf{0} = \mathbf{0}$, $\mathbf{0} \cdot \mathbf{1} = \mathbf{0}$, $\mathbf{1} \cdot \mathbf{0} = \mathbf{0}$, and $\mathbf{1} \cdot \mathbf{1} = \mathbf{1}$. Careful application of the postulates to this simple class shows that all postulates except P1a hold true, and therefore P1a must be independent. Similar proofs can be devised for the other postulates.

Formal Theorems

A series of theorems will now be deduced from the fundamental postulates. These theorems express other Boolean relationships which, like the postulates, must become as familiar to the logical designer as the rules of ordinary algebra are already.

The theorems, like the postulates, are presented in pairs where each theorem is the dual of the one paired with it. At the right-hand side of the page, opposite each step in the proof, the number of the postulate or theorem which justifies it is given in brackets. Finally, the notation is simplified by omitting the $\cdot$ wherever this does not lead to confusion. For example, $A + (B \cdot C)$ is written $A + BC$ as in ordinary algebra.

T1a. The element $\mathbf{0}$ in P2a is unique.

Proof: Suppose there are two $\mathbf{0}$ elements, called $\mathbf{0}_1$ and $\mathbf{0}_2$. Then for every element A,

$$A + \mathbf{0}_1 = A \qquad \text{and} \qquad A + \mathbf{0}_2 = A \qquad \text{[P2a]}$$

Now substitute $A = \mathbf{0}_2$ in the first of these equations, and $A = \mathbf{0}_1$ in the second. Then

$$\mathbf{0}_2 + \mathbf{0}_1 = \mathbf{0}_2 \qquad \text{and} \qquad \mathbf{0}_1 + \mathbf{0}_2 = \mathbf{0}_1$$

But $$\mathbf{0}_2 + \mathbf{0}_1 = \mathbf{0}_1 + \mathbf{0}_2 \qquad \text{[P3a]}$$

Therefore $$\mathbf{0}_2 = \mathbf{0}_1 \qquad \text{Q.E.D.}$$

T1b. The element $\mathbf{1}$ in P2b is unique.

Proof: Suppose there are two $\mathbf{1}$ elements, called $\mathbf{1}_1$ and $\mathbf{1}_2$. Then for every element A,

$$A \cdot \mathbf{1}_1 = A \qquad \text{and} \qquad A \cdot \mathbf{1}_2 = A \qquad \text{[P2b]}$$

Now substitute $A = \mathbf{1}_2$ in the first of these equations, and $A = \mathbf{1}_1$ in the second. Then

$$\mathbf{1}_2 \cdot \mathbf{1}_1 = \mathbf{1}_2 \qquad \text{and} \qquad \mathbf{1}_1 \cdot \mathbf{1}_2 = \mathbf{1}_1$$

But $$\mathbf{1}_1 \cdot \mathbf{1}_2 = \mathbf{1}_2 \cdot \mathbf{1}_1 \qquad \text{[P3b]}$$

Therefore $$\mathbf{1} = \mathbf{1}_2$$

T2a. $A + A = A.$

Proof:

$$
\begin{aligned}
A + A &= (A + A)\cdot\mathbf{1} && \text{[P2b]}\\
&= (A + A)(A + \bar{A}) && \text{[P5]}\\
&= A + A\bar{A} && \text{[P4a]}\\
&= A + \mathbf{0} && \text{[P5]}\\
&= A && \text{[P2a]}
\end{aligned}
$$

T2b. $A \cdot A = A.$

Proof:

$$
\begin{aligned}
A \cdot A &= AA + \mathbf{0} && \text{[P2a]}\\
&= AA + A\bar{A} && \text{[P5]}\\
&= A(A + \bar{A}) && \text{[P4b]}\\
&= A \cdot \mathbf{1} && \text{[P5]}\\
&= A && \text{[P2b]}
\end{aligned}
$$

Note that the dual theorems T1b and T2b may be derived mechanically from T1a and T2a by starting with dual expressions and then justifying the steps of the proof by the application of dual postulates. (It will be remembered that two relations are the dual of one another if one can be formed from the other by interchanging $+$ and $\cdot$ signs, and also interchanging **0** and **1**, wherever any of these appear.) The dual theorems are thus so easily derived that their derivations will be omitted from now on.

The results $A + A = A$ and $AA = A$ can of course be interpreted either in terms of diode circuits or as areas on Venn diagrams. The Venn diagram interpretation is perfectly obvious: The smallest region containing both A and A is the same as the largest region included in both A and A, and is equal to A. The circuit interpretation is a little less obvious, and is illustrated by the diagrams and tables of Figure 3-8.

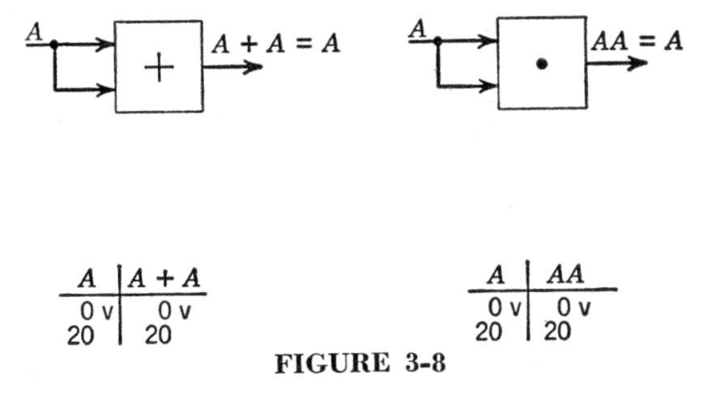

A	A + A
0 v	0 v
20	20

A	AA
0 v	0 v
20	20

FIGURE 3-8

There it is shown that a single signal applied to both inputs of either an "and" or an "or" circuit causes an output which is the same as the original signal.

T3a. $A + \mathbf{1} = \mathbf{1}$.
Proof:

$$
\begin{aligned}
A + \mathbf{1} &= (A + \mathbf{1})\cdot\mathbf{1} && \text{[P2b]} \\
&= (A + \mathbf{1})(A + \bar{A}) && \text{[P5]} \\
&= A + (\bar{A}\cdot\mathbf{1}) && \text{[P4a]} \\
&= A + \bar{A} && \text{[P2b]} \\
&= \mathbf{1}
\end{aligned}
$$

T3b. $A\cdot\mathbf{0} = \mathbf{0}$.

Again the Venn diagram meaning of these theorems is clear. When it is remembered that **0** represents a wire at 0 volts, and **1** a wire at

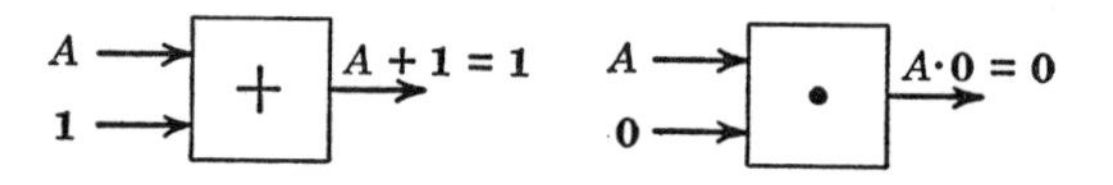

A	$\mathbf{1}$	$A + \mathbf{1}$
0 v	20 v	20 v
20	20	20

A	$\mathbf{0}$	$A\cdot\mathbf{0}$
0 v	0 v	0 v
20	0	0

FIGURE 3-9

+20 volts, Figure 3-9 shows that the diode circuit interpretation of these theorems is also reasonable and consistent.

T4a. $A + AB = A$.
Proof:

$$
\begin{aligned}
A + AB &= (A\cdot\mathbf{1}) + (A\cdot B) && \text{[P2b]} \\
&= A(\mathbf{1} + B) && \text{[P4b]} \\
&= A\cdot\mathbf{1} && \text{[P3a, T3a]} \\
&= A && \text{[P2b]}
\end{aligned}
$$

T4b. $A(A + B) = A$.

Figure 3-10 indicates the Venn diagram representation of these theorems, and Figure 3-11 indicates that they are also consistent as circuit connections.

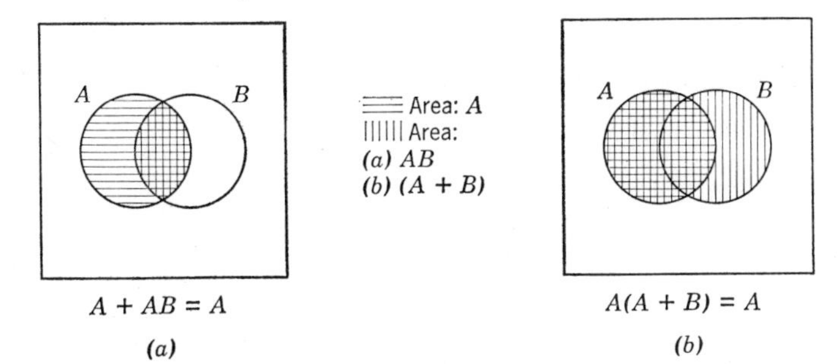

FIGURE 3-10

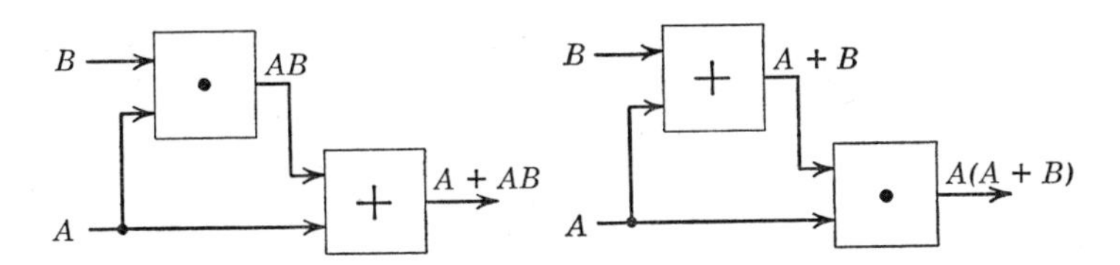

A	B	AB	A + AB
0 v	0 v	0 v	0 v
0	20	0	0
20	0	0	20
20	20	20	20

A	B	A + B	A(A + B)
0 v	0 v	0 v	0 v
0	20	20	0
20	0	20	20
20	20	20	20

FIGURE 3-11

T5. $\bar{A}$ is uniquely determined.

Proof: Suppose A has $\bar{A}_1$ and $\bar{A}_2$ such that $A + \bar{A}_1 = A + \bar{A}_2 = \mathbf{1}$, and $A\bar{A}_1 = A\bar{A}_2 = \mathbf{0}$. Then

$$\begin{aligned} \bar{A}_2 &= \mathbf{1}\cdot\bar{A}_2 && \text{[P2b]} \\ &= (A + \bar{A}_1)\bar{A}_2 && \text{[P5]} \\ &= A\bar{A}_2 + \bar{A}_1\bar{A}_2 && \text{[P4b]} \\ &= \mathbf{0} + \bar{A}_1\bar{A}_2 && \text{[P5]} \\ &= \bar{A}_1 A + \bar{A}_1\bar{A}_2 && \text{[P5]} \\ &= \bar{A}_1(A + \bar{A}_2) && \text{[P4b]} \\ &= \bar{A}_1\cdot\mathbf{1} && \text{[P5]} \\ &= \bar{A}_1 && \text{[P2b]} \end{aligned}$$

Therefore there is only one $\bar{A}$ and this quantity is defined as the *complement* of A.

T6. $\overline{(\bar{A})} = A.$

Proof: We seek the complement of $\bar{A}$. But since

$$\bar{A} + A = \mathbf{1} \quad \text{and} \quad \bar{A}A = \mathbf{0} \qquad [P5]$$

one complement of $\bar{A}$ is A. Hence

$$A = \overline{(\bar{A})} \qquad [T5]$$

T7a. $\overline{(A + B)} = \bar{A}\bar{B}.$

Proof: The plan for the proof is as follows: We will prove that $(A + B) + \bar{A}\bar{B} = \mathbf{1}$, and that $(A + B)\cdot\bar{A}\bar{B} = \mathbf{0}$. We will then be able to deduce that $(A + B)$ and $\bar{A}\bar{B}$ are the complement of one another by applying P5 and T5. In order to carry out this plan, we need to prove two lemmas.

L1a. $A + (\bar{A} + C) = \mathbf{1}\cdot[A + (\bar{A} + C)]$ [P2b]

$= (A + \bar{A})[A + (\bar{A} + C)]$ [P5]

$= A + \bar{A}(\bar{A} + C)$ [P4a]

$= A + \bar{A}$ [T4b]

$= \mathbf{1}$ [P5]

L1b. $A\cdot(\bar{A}C) = \mathbf{0}.$

(The proof is the dual of the proof of L1a.)

We are now ready to proceed with the plan:

$(A + B) + \bar{A}\bar{B} = [(A + B) + \bar{A}][(A + B) + \bar{B}]$ [P4a]

$= \mathbf{1}\cdot\mathbf{1}$ [L1a]

$= \mathbf{1}$ [P2b]

$(A + B)\cdot\bar{A}\bar{B} = A(\bar{A}\bar{B}) + B(\bar{A}\bar{B})$ [P4b]

$= \mathbf{0} + \mathbf{0}$ [L1b]

$= \mathbf{0}$ [P2b]

This completes the proof.

T7b. $\overline{AB} = \bar{A} + \bar{B}.$

Theorems T7a and T7b are called *De Morgan's theorems*, and are especially useful and important. Interpretations in terms of our two models are given in Figures 3-12 and 3-13; Figure 3-12 is the Venn diagram for T7a, and Figure 3-13 illustrates the diode circuits for T7b. The following example is provided to show how the theorems may be applied in practice.

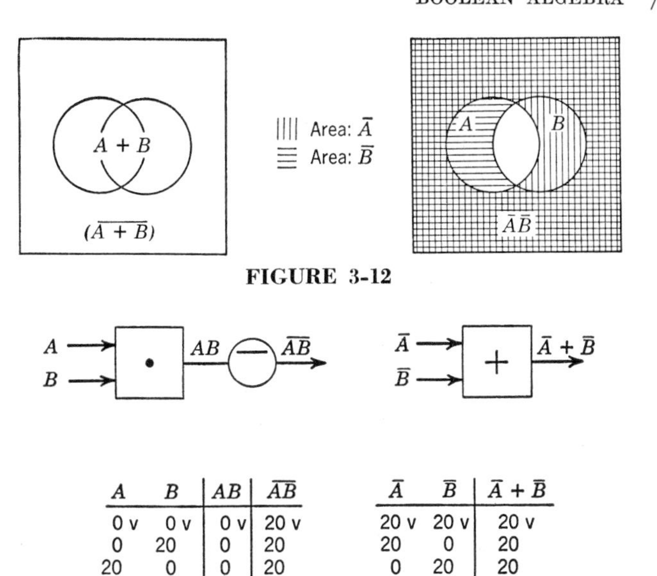

FIGURE 3-12

A	B	AB	$\overline{AB}$
0 v	0 v	0 v	20 v
0	20	0	20
20	0	0	20
20	20	20	0

$\bar{A}$	$\bar{B}$	$\bar{A} + \bar{B}$
20 v	20 v	20 v
20	0	20
0	20	20
0	0	0

FIGURE 3-13

Example 3-1. Find the complement of $A[\bar{B} + (C\bar{D} + \bar{E}F)]$.

$$\overline{A[\bar{B} + (C\bar{D} + \bar{E}F)]} = \bar{A} + \overline{[\bar{B} + (C\bar{D} + \bar{E}F)]} \qquad \text{[T7b]}$$

$$= \bar{A} + B \cdot \overline{(C\bar{D} + \bar{E}F)} \qquad \text{[T7a]}$$

$$= \bar{A} + B \cdot [\overline{(C\bar{D})} \cdot \overline{(\bar{E}F)}] \qquad \text{[T7a]}$$

$$= \bar{A} + B \cdot [(\bar{C} + D)(E + \bar{F})] \qquad \text{[T7b]}$$

The following simple rule embodies the applications of T7a and T7b while eliminating a great many formal steps: *To find the complement of a Boolean expression, change all + signs to ·, all · signs to +, and replace each letter by its complement.*

T8a. $(A + B) + C = A + (B + C)$.

Proof: The proof for this important theorem is somewhat difficult, and Huntington proposed that it be proved by altering its form as follows: Let $(A + B) + C = X$ and $A + (B + C) = Y$. Then show that $Y + \bar{X} = \mathbf{1}$ and $Y\bar{X} = \mathbf{0}$. If these relations are true, $\bar{X}$ and Y are complements of one another, and $\bar{\bar{X}} = X = Y$.

Now $$\bar{X} = (\bar{A}\bar{B}) \cdot \bar{C} \qquad \text{[T7a]}$$

So $$Y + \bar{X} = Y + [(\bar{A}\bar{B}) \cdot \bar{C}]$$

$$= [(Y + \bar{A})(Y + \bar{B})] \cdot (Y + \bar{C}) \qquad \text{[P4a]}$$

We now evaluate $(Y + \bar{A})$, $(Y + \bar{B})$, and $(Y + \bar{C})$ as follows:

$$\begin{aligned} Y + \bar{A} &= \bar{A} + [A + (B + C)] = \mathbf{1} && \text{[L1a]} \\ Y + \bar{B} &= (\bar{B} + B)(\bar{B} + Y) && \text{[P2b, P5]} \\ &= \bar{B} + BY && \text{[P4a]} \\ &= \bar{B} + B[A + (B + C)] && \\ &= \bar{B} + [AB + B(B + C)] && \text{[P4b]} \\ &= \bar{B} + [AB + B] && \text{[T4b]} \\ &= \bar{B} + B && \text{[T4a]} \\ &= \mathbf{1} && \text{[P5]} \end{aligned}$$

Using exactly the same argument, $(Y + \bar{C}) = \mathbf{1}$. Therefore

$$\begin{aligned} (Y + \bar{X}) &= [(Y + \bar{A})(Y + \bar{B})] \cdot (Y + \bar{C}) \\ &= [\mathbf{1} \cdot \mathbf{1}] \cdot \mathbf{1} \\ &= [\mathbf{1}] \cdot \mathbf{1} = \mathbf{1} && \text{[P2b]} \end{aligned}$$

Next it is necessary to show that $Y\bar{X} = \mathbf{0}$.

$$\begin{aligned} Y\bar{X} &= \bar{X}[A + (B + C)] \\ &= \bar{X}A + [\bar{X}B + \bar{X}C] \end{aligned}$$

$\bar{X}A$, $\bar{X}B$, and $\bar{X}C$ are evaluated in the following way:

$$\begin{aligned} \bar{X}C &= [(\bar{A}\bar{B})\bar{C}] \cdot C = \mathbf{0} && \text{[L1b]} \\ \bar{X}A &= A\bar{A} + A\bar{X} && \text{[P2a, P5]} \\ &= A(\bar{A} + \bar{X}) && \text{[P4b]} \\ &= A[\bar{A} + (\bar{A}\bar{B})\bar{C}] && \\ &= A[(\bar{A} + \bar{A}\bar{B})(\bar{A} + \bar{C})] && \text{[P4a]} \\ &= A[\bar{A}(\bar{A} + \bar{C})] && \text{[T4a]} \\ &= A[\bar{A}] && \text{[T4b]} \\ &= \mathbf{0} && \end{aligned}$$

Using precisely the same argument, $(\bar{X}B) = \mathbf{0}$. Therefore

$$\begin{aligned} Y\bar{X} &= \bar{X}A + [\bar{X}B + \bar{X}C] \\ &= \mathbf{0} + [\mathbf{0} + \mathbf{0}] = \mathbf{0} && \text{[P2a]} \end{aligned}$$

Since we have proved that $(Y + \bar{X}) = \mathbf{1}$ and $(Y\bar{X}) = \mathbf{0}$, then $X = Y$ and therefore

$$(A + B) + C = A + (B + C) \qquad \text{Q.E.D.}$$

T8b. $(AB)C = A(BC)$.
Proof:

$$(\bar{A} + \bar{B}) + \bar{C} = \bar{A} + (\bar{B} + \bar{C}) \qquad \text{[T8a]}$$

Taking complements of both sides of this equation, we see that

$$\overline{(\bar{A} + \bar{B})}C = A\overline{(\bar{B} + \bar{C})} \qquad \text{[T7a]}$$

Therefore $\qquad (AB)C = A(BC) \qquad$ Q.E.D. $\qquad$ [T7a]

As a consequence of these two theorems, we can write Boolean sums and products without parentheses just as we write arithmetic sums and products. The solution to Example 3-1 may thus be written:

$$\overline{A(\bar{B} + C\bar{D} + \bar{E}F)} = \bar{A} + B(\bar{C} + D)(E + \bar{F})$$

Figure 3-14 shows the voltages which exist on three-input "and" and "or" circuits (see Figure 2-5), and indicates how they are equivalent to pairs of two-input circuits.

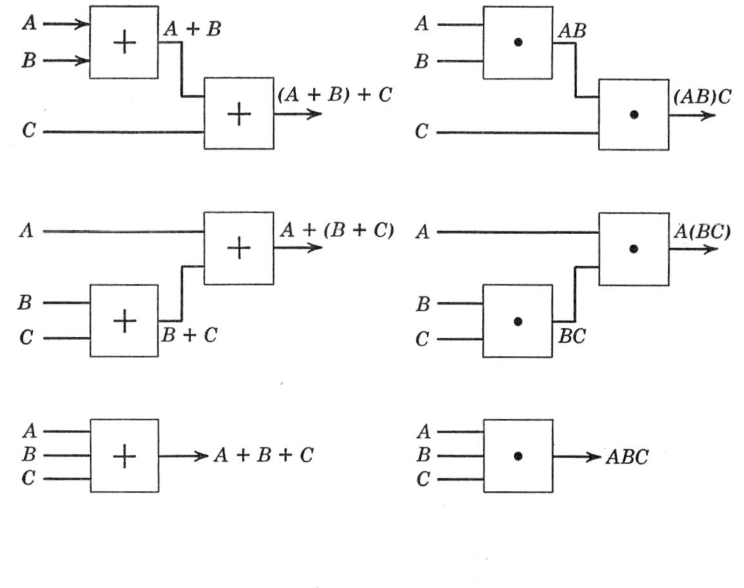

A	B	C	$A + B + C$	ABC
0 v	0 v	0 v	0 v	0 v
0	0	20	20	0
0	20	0	20	0
0	20	20	20	0
20	0	0	20	0
20	0	20	20	0
20	20	0	20	0
20	20	20	20	20

FIGURE 3-14

The theorems which have so far been presented provide the backbone of the everyday use of this algebra. The next five theorems also find frequent application in this work.

T9a. $A + \bar{A}B = A + B.$
Proof:

$$A + \bar{A}B = (A + \bar{A})(A + B) \qquad [P4a]$$
$$= A + B \qquad [P5, P2b]$$

T9b. $A(\bar{A} + B) = AB.$

T10. $(A + B)(\bar{A} + C) = AC + \bar{A}B.$
Proof:

$$(A + B)(\bar{A} + C) = A\bar{A} + AC + \bar{A}B + BC \qquad [P4b]$$
$$= AC + \bar{A}B + BC(A + \bar{A}) \qquad [P5]$$
$$= AC(\mathbf{1} + B) + \bar{A}B(\mathbf{1} + C) \qquad [P2b, P4b]$$
$$= AC + \bar{A}B \qquad [T3a]$$

This interesting theorem, which is its own dual, says it is permissible to omit two cross products when "multiplying out" (or two cross sums when "multiplying in") an expression where A appears in one term and $\bar{A}$ in the other.

T11a. $\overline{(AC + B\bar{C})} = \bar{A}C + \bar{B}\bar{C}.$
Proof:

$$\overline{(AC + B\bar{C})} = (\bar{A} + \bar{C})(\bar{B} + C) \qquad [T7a, T8b]$$
$$= \bar{A}C + \bar{B}\bar{C} \qquad [T10]$$

T11b. $\overline{(A + C)(B + \bar{C})} = (\bar{A} + C)(\bar{B} + \bar{C}).$

These theorems are very often useful when complements must be found. Note that the complemented expression is formed in each instance simply by complementing two letters in the original expression.

The foregoing development of Boolean algebra has been carried out in a precise and formal way. We have seen how certain interpretations (in Venn diagrams and diode circuits) can be given to the elements of the universe K and to the operations $+$ and $\cdot$, but these interpretations are not essential and the theorems we have derived are independent of them. The theorems which follow can be proved just as rigorously, but we will here drop the formal development of the algebra and turn to a particular application of it: the two-valued circuit interpretation introduced in Figure 3-1. It is hoped that the formal development has provided a substantial foundation for the work which follows, and for further study of the literature of Boolean and other algebras. In the remainder of the chapter, theorems will be developed intuitively rather than formally, and it will be possible to proceed at a faster pace than a more precise development would permit. Where the previous development was essentially Huntington's, the following one is largely Serrell's.

FURTHER DEVELOPMENT OF BOOLEAN ALGEBRA

In this section we will continue to apply Boolean algebra to digital computer circuits. It will be remembered that the Boolean variables may be interpreted as wires at one of two voltages, 0 or +20 volts. The operation + is to be thought of as the result of connecting two (or more) such wires to a circuit of the kind shown in Figure 2-4*a* (or 2-5*b*), and the operation · as the result of connecting wires to a circuit of the kind shown in Figure 2-3*a* (or 2-5*b*). Furthermore, the symbols **0** and **1** correspond to wires of constant voltage 0 and +20 respectively; and the symbol $\bar{A}$ may be interpreted as a wire whose voltage is the opposite of wire A.

There is of course nothing sacred about the voltages 0 and +20. Circuits can equally well be designed to operate on signal voltages of −20 and +20, or +50 and +100. It is quite reasonable, and very convenient, to represent one voltage by the numeral 0 and the other by the numeral 1, and this will be done hereafter. Furthermore, since the following formulas are correct when they are constructed of diode circuits

$$A + 0 = A, \qquad A \cdot 1 = A, \qquad A + \bar{A} = 1, \qquad A \cdot \bar{A} = 0$$

we can, and will from now on, replace the symbol **1** by 1, and the symbol **0** by 0.

Boolean Functions

The Boolean expressions with which the logical designer is concerned are always functions of a finite number of Boolean variables, and we will now study some characteristics of these functions. The very simplest expression is a function of only one variable (e.g., $f = \bar{A}$), but let us consider a slightly more complicated function defined by the following equation.

$$f = A + \bar{B} \tag{3-1}$$

First of all, suppose we construct a circuit—a simple "or" circuit is all that is needed—whose output is f. Then for each possible combination of voltages for wires A and $\bar{B}$, f will of course assume some voltage. Putting it another way, for each "value" of A and $\bar{B}$, f has a value. The only possible values the input and output wires can have are 1 and 0, and Table 3-1 sets out their relationship. Since $\bar{B}$ is defined when

Table 3-1

A	$\bar{B}$	$f = A + \bar{B}$
0	0	0
0	1	1
1	0	1
1	1	1

B is known, we could equally well have written Table 3-2. In fact, we

Table 3-2

A	B	$f = A + \bar{B}$
0	0	1
0	1	0
1	0	1
1	1	1

could have used either of these tables—which are known as *truth tables*—to define the function, instead of defining it by means of an equation.

The next point of interest is that f can be expressed in several other ways. For example

$$
\begin{aligned}
f &= A + \bar{B} \\
&= A + \bar{A}\bar{B} \\
&= \bar{B} + BA \\
&= AB + A\bar{B} + \bar{B} \\
&= AB + A\bar{B} + \bar{B}(A + \bar{A}) \\
&= AB + A\bar{B} + \bar{A}\bar{B} \qquad (3\text{-}2)
\end{aligned}
$$

Each of these expressions is realized by a different combination of diode circuits, but each of these circuits would have exactly the same output, f, for each of the four combinations of values of A and B.

Minterms and Maxterms

Equations 3-1 and 3-2 are very special ways of writing the function f, and it will be worthwhile to examine these two equations with some care and to make some generalizations. Taking equation 3-2 first, we see that it is composed of the Boolean sum of three terms. Each term is the product of the two variables A and B, and the three differ from one another only in the presence or absence of complements. Each of these three terms is called a *minterm*, where a minterm is defined in general by the following statement: a minterm of n variables is a Boolean product of these n variables, with each variable present in either its true or its complemented form. There are four minterms of two variables, and they are $\bar{A}\bar{B}$, $\bar{A}B$, $A\bar{B}$, and AB.

Now examining equation 3-1, we note that it is composed of a single Boolean sum, the sum of the two variables A and B with B appearing in complemented form. This sum is called a *maxterm*, where a maxterm is defined in general by the following statement: a maxterm of n variables is a Boolean sum of these n variables, where each variable is present in either its true or its complemented form. There are four maxterms of two variables, and they are $(\bar{A} + \bar{B})$, $(\bar{A} + B)$, $(A + \bar{B})$, and $(A + B)$.

We see, then, that equation 3-2 represents the function f as a sum of minterms, and equation 3-1 as a product of maxterms (in this case a single maxterm). We shall see in a moment that it is always possible to represent any function in these two ways, but first let us become familiar with minterms and maxterms, and introduce some notation.

The notation to be used represents a minterm by a lower case m and a maxterm by a capital M, with a subscript attached which identifies the particular minterm or maxterm. The subscript is chosen as follows: First write down the term with the variables appearing in some standard sequence. (A very convenient standard sequence, and one which will always be used in this book, is alphabetical order.) Next, write a zero below each complemented variable and a one below each uncomplemented variable. If this sequence of zeros and ones is interpreted as an ordinary binary number, the decimal equivalent of that number is then the subscript for the minterm or maxterm. For example, $AB\bar{C} = m_6$, and $(\bar{A} + \bar{B} + \bar{C}) = M_0$.* Table 3-3 lists all the minterms and maxterms of three variables.

* It has been suggested that the superscript n be attached to all minterms and maxterms to indicate how many variables are involved. Thus $B\bar{C} = m_2{}^2$ but $\bar{A}B\bar{C} = m_2{}^3$. It is sometimes useful to do this, but usually we are concerned with only one value of n at a time, so there is no ambiguity.

Table 3-3

	Minterms	Maxterms
0 0 0	$m_0 = \bar{A}\bar{B}\bar{C}$	$M_0 = \bar{A} + \bar{B} + \bar{C}$
0 0 1	$m_1 = \bar{A}\bar{B}C$	$M_1 = \bar{A} + \bar{B} + C$
0 1 0	$m_2 = \bar{A}B\bar{C}$	$M_2 = \bar{A} + B + \bar{C}$
0 1 1	$m_3 = \bar{A}BC$	$M_3 = \bar{A} + B + C$
1 0 0	$m_4 = A\bar{B}\bar{C}$	$M_4 = A + \bar{B} + \bar{C}$
1 0 1	$m_5 = A\bar{B}C$	$M_5 = A + \bar{B} + C$
1 1 0	$m_6 = AB\bar{C}$	$M_6 = A + B + \bar{C}$
1 1 1	$m_7 = ABC$	$M_7 = A + B + C$

There are eight different minterms and eight different maxterms of three variables, as is shown in Table 3-3. In fact, there are 2^n different minterms and 2^n different maxterms of n variables. We can show this simply by noting that there are 2^n different binary numbers having n binary places, and that one maxterm and one minterm correspond to each number, as described in the preceding paragraph.

The Venn diagram of Figure 3-15a provides a useful graphical interpretation of minterms. If the area enclosed by the left-hand circle is

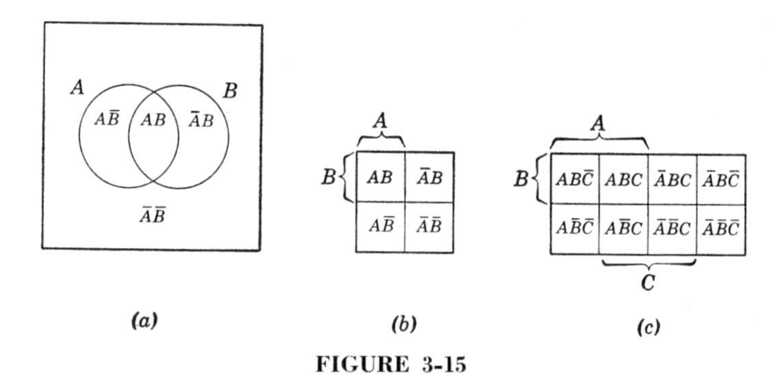

FIGURE 3-15

A and that by the right-hand one B, then each of the four areas defined by the boundaries of the circles and the square corresponds to one minterm as shown. Figure 3-15b shows a special kind of Venn diagram called a Veitch diagram, in which the four areas are delineated by straight lines. Note that the left-hand half of the square of Figure 3-15b is the area defined by the variable A, and the top half is the area defined by B. A Veitch diagram for three variables is shown in Figure 3-15c. Diagrams of this kind can be extended for use with any number of variables, and will be employed extensively in the chapters which follow in interpreting Boolean equations.

A *min*term is thus seen to be represented by a *min*imum distinguishable area on a Venn or Veitch diagram. Similarly, a *max*term is represented by a *max*imum distinguishable area, and this is illustrated in the next figure. The shaded area of Figure 3-16*a* represents the function $f = A + \bar{B}$, which is the maxterm M_2. The shaded area of Figure 3-16*b* represents a maxterm of three variables, namely $(A + \bar{B} + C)$. We see at once the meaning of "maximum distinguishable area"—in each example, all squares but one are shaded.

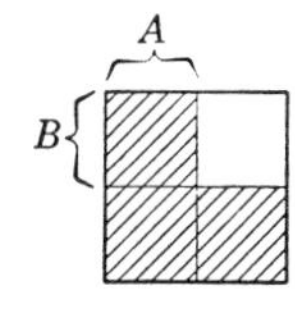

Shaded area $= A + \bar{B} = M_2$
Unshaded area $= \bar{A}B = m_1$
$\bar{m}_1 = M_{2^2-1-1} = M_2$

(*a*)

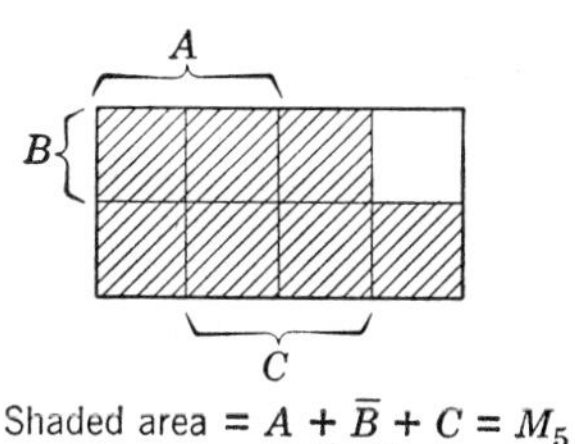

Shaded area $= A + \bar{B} + C = M_5$
Unshaded area $= \bar{A}B\bar{C} = m_2$
$\bar{M}_5 = m_{2^3-1-5} = m_2$

(*b*)

FIGURE 3-16

This leads us naturally to an elementary and obvious relationship between minterms and maxterms: the complement of any minterm is a maxterm, and vice versa. It can easily be verified that the proper relationship between subscripts is given by the following formulas:

$$\bar{m}_i = M_{2^n-1-i} \qquad \bar{M}_i = m_{2^n-1-i} \tag{3-3}$$

Two examples of the truth of these equations are given by Figure 3-16.

It is also easy to show that the Boolean sum of all minterms of any n variables is equal to "one", i.e.,

$$\sum_{i=0}^{2^n-1} m_i = 1 \tag{3-4}$$

This is in the first place evident from a Venn or Veitch diagram. It is the equivalent to saying that the whole is equal to the sum of its parts. It may also be deduced as follows. First,

$$\bar{A} + A = 1$$

This corresponds to equation 3-4 with $n = 1$. Now multiply this equation by $(\bar{B} + B)$

$$(\bar{A} + A)(\bar{B} + B) = \bar{A}\bar{B} + \bar{A}B + A\bar{B} + AB = 1$$

This is the theorem for $n = 2$. Now multiplying by $(\bar{C} + C)$,

$$(\bar{A}\bar{B} + \bar{A}B + A\bar{B} + AB)(\bar{C} + C)$$
$$= \bar{A}\bar{B}\bar{C} + \bar{A}\bar{B}C + \bar{A}B\bar{C} + \bar{A}BC$$
$$+ A\bar{B}\bar{C} + A\bar{B}C + AB\bar{C} + ABC = 1$$

Evidently for each additional variable we double the number of terms and continue to get all possible combinations of variables.

The dual of equation 3-4 may be found by taking the complement of both sides of that equation and applying De Morgan's theorem, i.e.,

$$\sum_{i=0}^{2^n-1} m_i = 1$$

$$\prod_{i=0}^{2^n-1} M_i = 0 \tag{3-5}$$

For example, with $n = 2$,

$$M_0M_1M_2M_3 = (\bar{A} + \bar{B})(\bar{A} + B)(A + \bar{B})(A + B) = 0$$

Still another pair of useful relations involving minterms and maxterms are given by equations 3-6 and 3-7.

$$m_im_j = 0 \qquad i \neq j \tag{3-6}$$

$$M_i + M_j = 1 \qquad i \neq j \tag{3-7}$$

The truth of these equations should be evident, for if two minterms (maxterms) are different, then at least one variable in one is present in its true state, whereas in the other it appears in its complemented form. For example, with $n = 4$,

$$m_{14}m_3 = ABC\bar{D} \cdot \bar{A}\bar{B}CD = 0$$

$$M_6 + M_0 = (\bar{A} + B + C + \bar{D}) + (\bar{A} + \bar{B} + \bar{C} + \bar{D}) = 1$$

The Basic Theorem

We will now develop what might be called the basic theorem of Boolean algebra. This theorem (and its dual) states that any function of n variables can be expressed as a Boolean sum of a set of minterms (or as a Boolean product of a set of maxterms). We have already mentioned this theorem while discussing equations 3-1 and 3-2, and have seen that these two equations are expressions of a particular function f in *maxterm* and *minterm form* respectively. To derive the minterm form for *any* function, we proceed in the following way.

First make out a truth table for the function. That is, write down the n variables at the top of n columns in the same order as they appear in the minterms, and below write the 2^n different combinations of values the variables may have. Opposite each of these 2^n values, write the value of f which is desired for that particular combination of the variables. Each of these values of f is given a name f_i, where the subscript is the decimal number corresponding to the binary number opposite it.

For example, consider the function of three variables defined by the truth table shown as Table 3-4. Once the function has been defined

Table 3-4

A	B	C	f
0	0	0	$0 = f_0$
0	0	1	$1 = f_1$
0	1	0	$0 = f_2$
0	1	1	$0 = f_3$
1	0	0	$0 = f_4$
1	0	1	$1 = f_5$
1	1	0	$0 = f_6$
1	1	1	$1 = f_7$

by the truth table, the eight symbols $f_0, f_1, f_2, \cdots, f_7$ are determined as shown. Now suppose we want to write a Boolean expression for this new function f. We see that f must be 1 only when A, B, and C have the values 001, 101, or 111. But when A, B, and C have the values 001, for example, then $\bar{A}\bar{B}C = 1$. Similarly, when they have the values 101, then $A\bar{B}C = 1$; and when all three are 1, then $ABC = 1$. Since these are the only circumstances under which f should be 1, it is evident that we may express f as the Boolean sum of these three minterms; i.e.,

$$f = \bar{A}\bar{B}C + A\bar{B}C + ABC = m_1 + m_5 + m_7$$

This expression is correct because it has the value 1 for the correct combinations of A, B, and C, *and for no other combinations*. If this is not clear, it may be verified by substituting into the equations each of the eight possible combinations of values for A, B, and C, and thus determining which combinations make $f = 1$.

Let us now review the procedure we have followed. We simply searched the truth table for those combinations for which $f_i = 1$, and included in the expression for f the corresponding minterms m_i. This procedure may be translated into Boolean language very nicely by noting that the Boolean sum of Boolean products of f_i and m_i will eliminate the undesired minterms and retain the desired ones; i.e.,

$$\begin{aligned} f &= 0\cdot m_0 + 1\cdot m_1 + 0\cdot m_2 + 0\cdot m_3 + 0\cdot m_4 + 1\cdot m_5 + 0\cdot m_6 + 1\cdot m_7 \\ &= f_0 m_0 + f_1 m_1 + f_2 m_2 + f_3 m_3 + f_4 m_4 + f_5 m_5 + f_6 m_6 + f_7 m_7 \end{aligned}$$

We can put this into a more general form by writing

$$f = \sum_{i=0}^{2^n-1} f_i m_i \tag{3-8}$$

and this is the mathematical expression of the basic theorem.

To derive the dual of this theorem, which says that any function may be expressed as a product of maxterms, we begin by expressing the *complement* of f in terms of minterms, using equation 3-8. This is very easy to do, for wherever f was "one," $\bar{f}$ will be "zero" and vice versa. Therefore

$$\bar{f} = \sum_{i=0}^{2^n-1} \bar{f}_i m_i \tag{3-9}$$

Applying this to our previous example, we see that

$$\begin{aligned} \bar{f} &= 1\cdot m_0 + 0\cdot m_1 + 1\cdot m_2 + 1\cdot m_3 + 1\cdot m_4 + 0\cdot m_5 + 1\cdot m_6 + 0\cdot m_7 \\ &= m_0 + m_2 + m_3 + m_4 + m_6 \end{aligned} \tag{3-10}$$

Taking the complement of equation 3-9 and using equation 3-3, we get the desired result, namely:

$$f = \prod_{i=0}^{2^n-1} (f_i + \bar{m}_i) = \prod_{i=0}^{2^n-1} (f_i + M_{2^n-1-i}) \tag{3-11}$$

For example, taking the complement of equation 3-10 we find

$$f = \bar{m}_0\bar{m}_2\bar{m}_3\bar{m}_4\bar{m}_6 = M_7M_5M_4M_3M_1$$

Of course, applying equation 3-11 directly to the truth table, we must get the same result; i.e.,

$$\begin{aligned} f &= (f_0 + M_7)(f_1 + M_6)(f_2 + M_5)(f_3 + M_4)(f_4 + M_3)(f_5 + M_2) \\ &\quad (f_6 + M_1)(f_7 + M_0) \\ &= (0 + M_7)(1 + M_6)(0 + M_5)(0 + M_4)(0 + M_3)(1 + M_2) \\ &\quad (0 + M_1)(1 + M_0) \\ &= M_7M_5M_4M_3M_1 \\ &= (A + B + C)(A + \bar{B} + C)(A + \bar{B} + \bar{C})(\bar{A} + B + C) \\ &\quad (\bar{A} + \bar{B} + C) \end{aligned}$$

Finally, returning to the original example of this section, we see now that equation 3-1 may be written $f = M_2$, and equation 3-2, $f = m_3 + m_2 + m_0$.

The basic theorem, as stated by equations 3-8 and 3-11, is important because it expresses the fact that any Boolean function may be written in both minterm and maxterm form. These two ways of expressing functions are not very useful themselves. The entire next chapter, in fact, will be devoted to finding alternative, simpler ways of writing Boolean functions. Nevertheless, minterm and maxterm forms are the starting points for much of the work that follows, and are essential to a thorough understanding of the workings of Boolean algebra.

OTHER DEVELOPMENTS OF BOOLEAN ALGEBRA

The postulates of page 31 are only one set which can be used to define Boolean algebra. In the same paper in which Huntington proposed these postulates, he showed that two other sets were exactly equivalent; and in a very interesting paper in 1913 Sheffer developed the algebra from only five independent postulates.

Other developments, however, do not restrict themselves to the operations "and," "or," and "complement." In fact, as Table 3-5 shows,

Table 3-5

A	B	F_0	F_1	F_2	F_3	F_4	F_5	F_6	F_7	F_8	F_9	F_{10}	F_{11}	F_{12}	F_{13}	F_{14}	F_{15}
0	0	0	0	0	0	0	0	0	0	1	1	1	1	1	1	1	1
0	1	0	0	0	0	1	1	1	1	0	0	0	0	1	1	1	1
1	0	0	0	1	1	0	0	1	1	0	0	1	1	0	0	1	1
1	1	0	1	0	1	0	1	0	1	0	1	0	1	0	1	0	1

there are sixteen possible functions of two variables A and B. Using the basic theorem, we can express each of these functions in terms of $+$, $\cdot$, and $\bar{\ }$ as follows:

$F_0 = 0$.

$F_1 = AB$. This is the familiar function "A and B."

$F_2 = A\bar{B}$.

$F_3 = A$.

$F_4 = \bar{A}B$.

$F_5 = B$.

$F_6 = A\bar{B} + \bar{A}B$. This expression is sometimes written $A \oplus B$, and is known as the "exclusive or" function. It may be read "A or B but not both."

$F_7 = A + B$. Again, this is the familiar "or" function. It is sometimes called the "inclusive or" to distinguish it from F_6.

$F_8 = \bar{A}\bar{B}$. This expression is sometimes written $A \downarrow B$, and is read "A Peirce B," after C. S. Peirce, the American logician.

$F_9 = \bar{A}\bar{B} + AB$. If this expression is set equal to "one," it is equivalent to the statement "A equals B."

$F_{10} = \bar{B}$.

$F_{11} = A + \bar{B}$. This function has two interesting and useful associations, neither of them having much to do with computer design. The first is the logical notion of implication, and will be discussed in the next section. The second is the notion of inclusion, and may be interpreted by referring to a Venn diagram. Figure 3-17, for example, shows two elements of a universe, having the peculiar relationship that one of them (B) is entirely contained in the other. This may be expressed by a familiar-looking relation, $B < A$; or by the statement, "B is included in A"; or by the Boolean equation $F_{11} = A + \bar{B} = 1$.

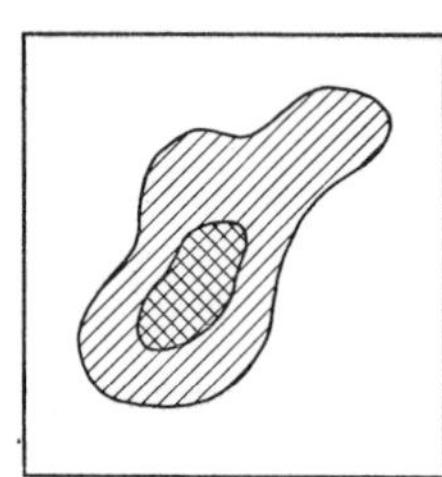

//// Area: A
\\\\ Area: B

FIGURE 3-17

$F_{12} = \bar{A}$.

$F_{13} = \bar{A} + B$. This is simply F_{11}, with A and B transposed.

$F_{14} = \bar{A} + \bar{B}$. This function is sometimes written $A|B$, and is called the Sheffer stroke function.

$F_{15} = 1$.

Now suppose that, instead of the three operations $+$, $\cdot$, and $\bar{\ }$, we are given some other decision elements. For example, suppose the "or" circuit has not been invented. Is it possible to write *any* Boolean function using only the "and" circuit and a complementing circuit? We know, from the basic theorem, that we can express any function, given $+$, $\cdot$, and $\bar{\ }$. Therefore, if we can find a way of constructing an "or" circuit from "and" and complementing circuits, we will be able to construct a circuit for any Boolean function from those two circuits. De Morgan's theorem provides a way of solving the problem.

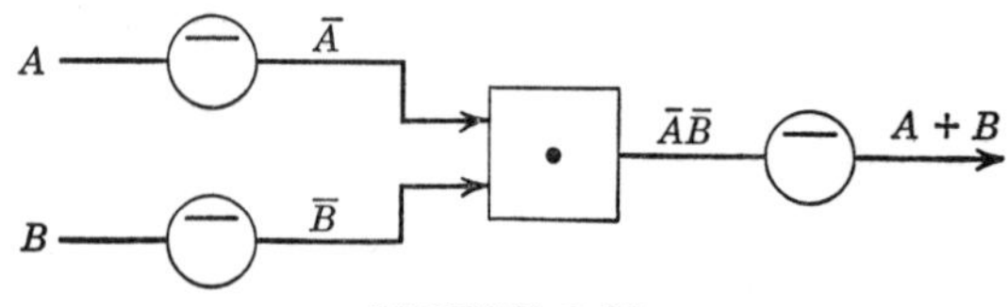

FIGURE 3-18

Since

$$A + B = \overline{\bar{A}\bar{B}}$$

we can get the function $(A + B)$ by using three complementing circuits and an "and" circuit, as shown in Figure 3-18.

Table 3-6 shows five possible combinations of decision elements, any one of which is complete in itself in the sense that it is capable of being

Table 3-6

	Decision Elements Allowed	Method Used for Forming: "and"	"or"	"complement"
1	$\cdot, \bar{\ }$		$A + B = \overline{\bar{A}\bar{B}}$	
2	$+, \bar{\ }$	$AB = \overline{\bar{A} + \bar{B}}$		
3	$\downarrow$	See 2	$A + B = (A \downarrow B) \downarrow (A \downarrow B)$	$\bar{A} = A \downarrow A$
4	$\mid$	$AB = (A \mid B) \mid (A \mid B)$	See 1	$\bar{A} = A \mid A$
5	$\oplus$, $\cdot$, and 1	See 2	$A + B = 1 \oplus [(1 \oplus A) \cdot (1 \oplus B)]$	$\bar{A} = 1 \oplus A$

used to represent any Boolean function. For example, line 3 in the table indicates that $\bar{A}$ may be formed by using a single Peirce circuit (if one is available), because $A \downarrow A = \bar{A} \cdot \bar{A} = \bar{A}$. Furthermore, since

$$A + B = \overline{\bar{A}\bar{B}} = (\bar{A}\bar{B}) \downarrow (\bar{A}\bar{B}) = (A \downarrow B) \downarrow (A \downarrow B)$$

an "or" circuit may be constructed from three Peirce circuits. Now by line 2 of the table, we know that we can form $(A \cdot B)$ if we can construct $\bar{A}$ and $(A + B)$, and we can therefore represent any Boolean function using Peirce circuits alone.

The five combinations given in Table 3-6 are by no means all the possible complete combinations. Furthermore, any complete group of operations could have been used to form the basic "rules of combination" for a set of postulates which would define Boolean algebra. We began by defining the operations $+$, $\cdot$, and $\bar{\ }$, deduced a number of properties of these operations, and now have defined other operations in terms of them. Although there is nothing wrong with this procedure, there also is nothing unique about it. We could just as well have started out with the operation $\mid$, and later defined $+$, $\cdot$, and $\bar{\ }$ in terms of it.

There are nevertheless good reasons for introducing Boolean algebra to the prospective computer designer in the way it has been introduced here. First, the concepts "and," "or," and "not" (complement) are familiar ones, used by all of us in expressing everyday ideas. We tell the waitress we would like some apple pie *and* cheese, *or* else some strawberries *and* whipped cream. (The "or" here may be either exclusive or inclusive, depending on how hungry we are.) Second, the operations $+$, $\cdot$, and $^{-}$ are easily realized by reliable circuits, and have been used in the design of most present-day digital computers. Third, the postulates used here reflect the beautiful and useful dual nature of the algebra, emphasizing from the first the symmetry of $+$ and $\cdot$ with respect to one another and to the complement. Fourth and finally, Huntington's development is a more or less classical one, very often referred to in literature on the subject of Boolean algebra.

It might be well to end this section with a warning about notation. Unfortunately, no notation has as yet been universally adopted by computer people. When the student ventures to read other mathematical or engineering literature which involves Boolean algebra, he will find that there are almost as many ways of expressing Boolean operations as there are authors. It is always necessary to proceed very cautiously when studying a new source until the meaning of the symbols is absolutely clear. The natural difficulty is compounded by the dual nature of the algebra. If the operation "*" is introduced, and it is stated that $A * B = B * A$, one can deduce that the author is not referring to F_2, F_3, F_4, F_5, F_{10}, F_{11}, F_{12}, or F_{13}, but he may mean any of the others; and only further reading will determine which he means.

Table 3-7

A and B	A or B	Not A
$A \cdot B$	$A + B$	$\bar{A}$, A', $\sim A$, A
AB		
$A \cap B$	$A \cup B$	
$A \wedge B$	$A \vee B$	
$A + B$	$A \cdot B$, AB	

Table 3-7 shows several alternative ways the functions "and," "or," and "not" have appeared in published works.

OTHER APPLICATIONS OF BOOLEAN ALGEBRA

Boole's original work was entitled *An Investigation of the Laws of Thought*, and it is the application of his algebra to the realm of logic that has been, historically, of most significance. Because of the importance of this aspect of Boolean algebra, and because much of the literature on the subject is directed toward these applications, it seems worthwhile to mention them here, at least.

The *calculus of propositions* is an interpretation of Boolean algebra in which the universe K is a "universe of discourse," the elements in the universe are particular members of the universe, and the relations between elements become statements or propositions about the members which are true or false according to whether the corresponding Boolean functions have the value "one" or "zero."

An example out of Lewis Carroll will illustrate this application and show its efficacy. The following three statements are given as premises, and some conclusion is to be found:

1. Babies are illogical.
2. Nobody is despised who can manage a crocodile.
3. Illogical persons are despised.

If the universe of discourse is taken to be all persons, and

A = persons able to handle a crocodile
B = babies
C = despised persons
D = logical persons

then the statements may be rewritten:

1. Every member of B is a member of $\bar{D}$.
2. Every A is a $\bar{C}$.
3. Every $\bar{D}$ is a C.

The premises may also be stated in the form of implications:

1. The fact that a person *is* a B implies that he is also a $\bar{D}$, *or* B implies $\bar{D}$.
2. A implies $\bar{C}$.
3. $\bar{D}$ implies C.

Now the statement "every x is a y," or "x implies y," may be represented, as was mentioned in the last section, by the Boolean equation $\bar{x} + y = 1$. It may also be restated "x is included in y," as we can see by referring again to Figure 3-17.

In order to use Boolean algebra to solve Carroll's problem, we need two more theorems.

T12. If x implies y, then $\bar{y}$ implies $\bar{x}$.
Proof:

Given	$\bar{x} + y = 1$	
	$y + \bar{x} = 1$	[P3a]
Therefore	$\bar{y}$ implies $\bar{x}$	

T13. If x implies y and y implies z, then x implies z.
Proof:

Given $\quad \bar{x} + y = 1, \quad \bar{y} + z = 1$

$$\bar{x} + y + z\bar{z} = 1, \quad x\bar{x} + \bar{y} + z = 1 \qquad \text{[P2a, P5]}$$

$$(\bar{x} + y + z)(\bar{x} + y + \bar{z}) = 1, \quad (\bar{x} + \bar{y} + z)(x + \bar{y} + z) = 1 \quad \text{[P4a]}$$

$$(\bar{x} + y + \bar{z})(\bar{x} + y + z)(\bar{x} + \bar{y} + z)(x + \bar{y} + z) = 1$$

Applying P4a and P5 to the two middle terms, we get

$$(\bar{x} + y + \bar{z})(\bar{x} + z)(x + \bar{y} + z) = 1$$

Now if any of the three terms in parentheses were "zero", the equation would not hold. Therefore, each parenthetical expression must equal "one", and so

$$\bar{x} + z = 1$$

Therefore x implies z.

We can now return to the problem, and solve it handily. First applying T12 to the second implication, we see that "A implies $\bar{C}$" is the same as "C implies $\bar{A}$." Next we may state the three premises in a slightly different sequence: "B implies $\bar{D}$, $\bar{D}$ implies C, C implies $\bar{A}$"; and applying T13 we see at once that "B implies $\bar{A}$." Finally translating this formula back into a statement, we have the proposition

"All babies are incapable of managing crocodiles."

Or, if we prefer to apply T12 and revise the result slightly,

"A person able to manage a crocodile is no baby."

The calculus of propositions forms a part of a more general subject known as *symbolic logic.* Where propositional calculus is concerned with the truth or falsity of statements, symbolic logic is concerned more with the arguments by means of which we come to conclusions about the statements. Scientists in general are concerned with the question of whether or not the arguments they use in coming to conclusions are valid arguments, and it is the function of symbolic logic to investigate the question and to help reassure them—or not, as the case might be. Paradoxes, for example, have always worried logicians, for they seem to indicate that at times we make deductive errors, and one cannot help but wonder whether we err at other times when we do not realize it. Symbolic logic is an attempt to put logic on a firm mathematical foundation so that some of the perplexing questions of logic may be answered; and Boolean algebra is a tool which may be employed in arriving at those answers.

BIBLIOGRAPHY

George Boole, *An Investigation of the Laws of Thought,* Dover Publications, New York, 1954.

E. V. Huntington, "The Algebra of Logic," *Trans. American Mathematical Society,* **5,** 288–309 (1904).

H. M. Sheffer, "A Set of Five Independent Postulates for Boolean Algebras, with Applications to Logical Constants," *Trans. American Mathematical Society,* **14,** 481–488 (1913).

C. E. Shannon, "Symbolic Analysis of Relay and Switching Circuits," *A.I.E.E. Trans.,* **57,** 713–723 (1938).

Lewis Carroll, *The Complete Works,* The Nonesuch Press, London, 1939.

H. Reichenback, *Elements of Symbolic Logic,* The Macmillan Company, New York, 1947.

P. C. Rosenbloom, *The Elements of Mathematical Logic,* Dover Publications, New York, 1950.

N. M. Martin, "On Completeness of Boolean Element Sets," *J. Computing Systems,* **1,** no. 3, 150–154 (1953).

R. Serrell, "Elements of Boolean Algebra for the Study of Information-Handling Systems," *Proc. I.R.E.,* **41,** 1366–1379 (1953). [Corrections, *Proc. I.R.E.,* **42,** 475 (1954.)]

A. W. Burks, D. W. Warren, and J. B. Wright, "An Analysis of a Logical Machine Using Parenthesis-Free Notation," *Mathematical Tables and Other Aids to Computation* (Apr. 1954).

F. E. Hohn and L. R. Schissher, "Boolean Matrices and the Design of Combinational Relay Switching Circuits," *Bell System Technical J.,* **34,** 177–202 (1955).

EXERCISES

1. Using the tabular method of Figure 3-3, prove the consistency of postulate P4b.

2. Show that postulate P4b is consistent using Venn diagrams.

3. Plot $\overline{AB}$, $\bar{A}\bar{B}$, and $(\bar{A} + \bar{B})$ on three Venn diagrams, showing that the first and third are equivalent.

4. Demonstrate the truth of theorems T9, T10, and T11 using Venn diagrams.

5. Given: $AB = AC$; $A + B = A + C$. Prove: $B = C$.

Note that neither division nor subtraction has meaning in Boolean algebra, and it is therefore impossible to prove the conclusion from either one of the two premises alone.

Draw a Venn diagram containing three elements A, B, and C, having B and C different but with $AB = AC$.

6. Write a truth table for each of the following functions.

a. $f_a = AB + \bar{B}C$
b. $f_b = A + \bar{B} + C$
c. $f_c = (\bar{A} + B)(\bar{B} + C)$
d. $f_d = \bar{A}B + C\bar{D} + A\bar{C}D$

7. List the minterms and maxterms for functions of four variables.

8. Express each of the following functions as a sum of minterms; as a product of maxterms.

a. $f_a = AB + \bar{A}\bar{B}$
b. $f_b = \bar{B}\bar{C} + AB + BC + \bar{A}\bar{C}$
c. $f_c = \bar{C}\bar{D}\bar{E} + CDE + \bar{A}\bar{B}D + ABCD + BCE + A\bar{B}\bar{C}E$
d. $f_d = (A + \bar{B})(\bar{A} + B)$
e. $f_e = (\bar{A} + B)(AB + \bar{B}C)$
f. $f_f = (A\bar{B} + \bar{C}D)(B + D)(A + \bar{C} + \bar{D})$

9. Express each of the four functions of Exercise 6 as a sum of minterms; as a product of maxterms.

10. Prove the following theorems:

a. $(A \downarrow C) \downarrow (A \downarrow B) = A + (\bar{B} \downarrow \bar{C})$
b. $A \oplus B = B \oplus A$
c. $A|(B|\bar{B}) = \bar{A}$
d. $(A < B) \cdot (A < C) = A < BC$
e. $(A < C) \cdot (B < C) = (A + B) < C$
f. $(A \oplus B) \oplus C = A \oplus (B \oplus C)$

11. Express the function $f_1 = A\bar{B} + \bar{A}BC$, using

a. The operations "+" and "complement" only.
b. The operations "·" and "complement" only.
c. The operation "$\downarrow$" only.
d. The operation "|" only.

12. Suppose the class K consists of all positive divisors of 30: 1, 2, 3, 5, 6, 10, 15, 30. Suppose further that: $(A \cdot B)$ is the least common multiple of A and B; $(A + B)$ is the greatest common divisor of A and B; $\bar{A}$ is the quotient of 30 and A. Show that the postulates of Boolean algebra are consistent when given this interpretation.

4 / *The simplification of Boolean functions*

THE MEANING AND IMPORTANCE OF SIMPLIFICATION

A Boolean function obtained by a logical designer is usually derived from some form of truth table by means of the basic theorem. In this form, it may often be simplified. That is to say, some other expression may be found which represents the same function but may be constructed with less equipment. For example, the designer may arrive at a function $f = \bar{A}BC\bar{D} + \bar{A}B\bar{C}\bar{D} + AB\bar{C}\bar{D} + ABC\bar{D}$, but when the function is built into his computer, he will prefer the simplified form $f = B\bar{D}$. It is therefore obviously of great importance that the designer have available methods for simplifying functions, and this chapter is devoted to an explanation of such methods.

The fact that there are alternative ways of representing functions opens up the possibility that some form may be the simplest of all, and that there may be some method for arriving at this minimal expression. However, any such method must hinge on the exact definition of the minimum form; and since the designer is really interested in minimizing the amount of equipment he must use, the definition may be different for different circuits. The diode "and" and "or" circuits which we have referred to in the last two chapters (see, e.g., Figure 2-5*b*) will be chosen as the basis for minimization for three principal reasons: because they are widely used today in computer design; because they can be made to correspond exactly to a particularly convenient definition of "minimum"; and because they may easily be compared for cost with other similar circuits.

How may we calculate the cost of a diode circuit? We first make the standardizing assumption that circuit costs depend only on diode costs—that the costs of other materials, and of assembling the circuits, are comparatively negligible. We can then quickly become familiar with the simple rules which must be followed in order to count the number of diodes needed to implement a given equation. For example, examine the functions f_1 through f_6 of Figure 4-1. The circuits for $f_1 = AB$ and for $f_2 = A + B$ require two diodes each, one for each letter in the expression. The function $f_3 = AB + C + D$ requires five diodes: two in an "and" circuit to form AB, and three more in an "or" circuit whose three inputs are C, D, and the previously formed AB. Similarly, $f_4 = (A + B)CD$ also requires five diodes: two for the "or" circuit $A + B$, and three more in the "and" circuit used to combine $(A + B)$ with C and D. Next, f_5 (and f_6) employs seven diodes: the five necessary to form $(AB + C + D)$, and two more to combine this in an "and" circuit with E. It is thus easy to see that the number of diodes necessary to form a given expression may be calculated by adding up the number of inputs necessary in successive "and" and "or" circuits.

Now examine Figure 4-1 again with the object of noting the hierarchy of "and" and "or" circuits. The functions f_1 and f_2 each correspond to a single circuit, and are therefore called *first-order* expressions. f_3 and f_4 each contain a diode circuit whose output becomes the input of another. They are called *second-order* expressions. f_5 and f_6, each of which contains *three* circuits in tandem, are called *third-order* expressions. We could go on constructing equations of higher and higher order. For example,

$f_7 = (AB + C + D)E + F$ is a fourth-order expression, while

$f_8 = [(A + B)CD + E]F + G$ is of fifth order. Note particularly that the order of an expression has nothing to do with the number of inputs there are in a given "and" or "or" circuit. For example, $A + B + C + D + E + F$ is a first-order expression, whereas $A + BC$ is second-order. Note also that it is always possible to represent a given function by means of several expressions, each of a different order. The function $f = AB$, which is of the first order, may be written $f = ABC + AB\overline{C}$ or $f = (AB + C)(AB + \overline{C})$, which are second- and third-order expressions respectively.

We will concentrate most of our attention, in this chapter and throughout the book, on first- and second-order expressions. Two facts may be stated in justification of this somewhat severe restriction. The first and most important is that it is possible to describe a straightforward and practical procedure for arriving at that second-order ex-

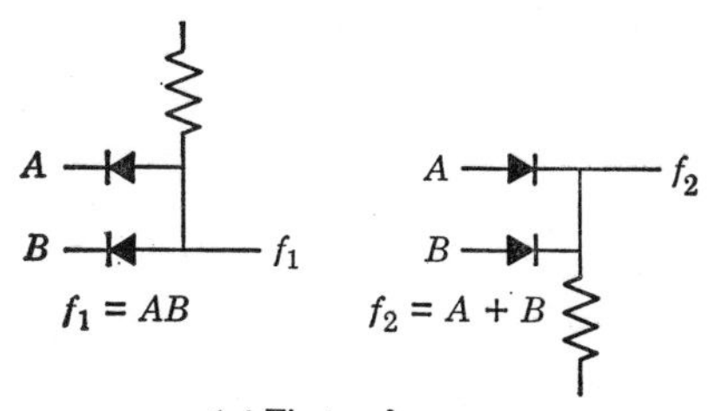

(a) First order

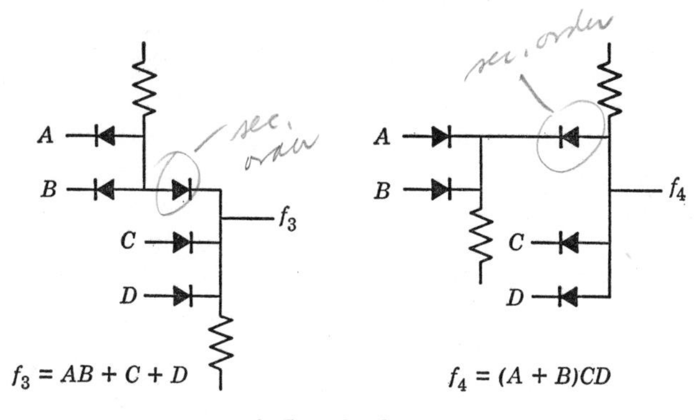

(b) Second order

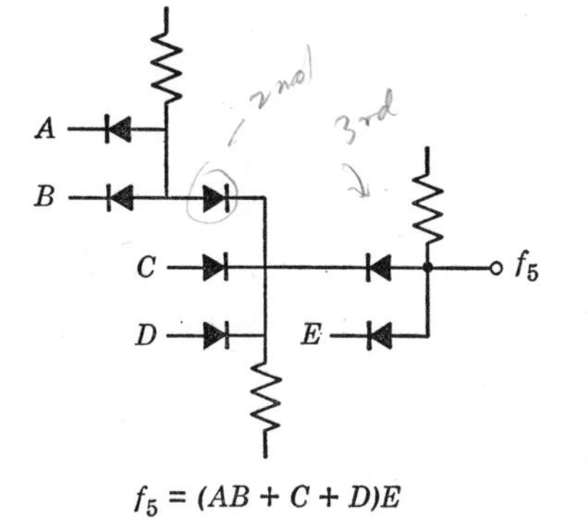

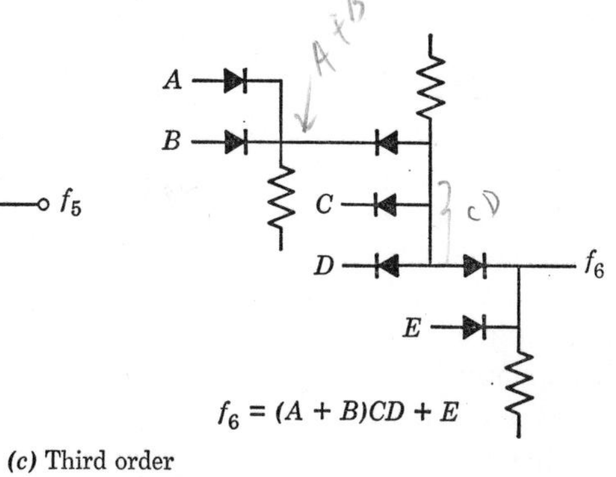

(c) Third order

FIGURE 4-1

pression of a given function which is simpler than (or as simple as) any other second-order expression for that function in terms of the number of diodes used. This is the "particularly useful definition of minimum" referred to above. No one has yet been able to devise a practical procedure which will always result in the simplest expression of *any* order. The other justification for our interest in first- and second-order functions is that there are engineering reasons for preferring first- and second-order expressions to higher order ones. These engineering restrictions are partly peculiar to the diode circuits we have been using; but they are inherent in any system of circuits in which the output of one decision element is the input to another. Each element has a delay associated with it, and every time the "order" of an expression is increased, one circuit is added in series with all the rest, and another unit of delay is added between input and output. If this delay gets too big, the output may not change in time to have the desired effect—and the delay may always be reduced if the circuit is third order or higher by rewriting the function as a lower order expression. However, this decreased delay is in general paid for by an increase in diodes. For example, the fifth-order expression $f_8 = [(A + B)CD + E]F + G$ requires eleven diodes. Its order may be reduced by "multiplying out" the innermost terms. The result is $f_8 = (ACD + BCD + E)F + G$, which is a fourth-order expression requiring thirteen diodes.

Before going on to describe some of the methods which may be used to simplify functions, let us review briefly the argument which has been followed, and the restrictions which have been imposed. We restricted ourselves to diode circuits because they are widely used and permit a very definite and easily calculated measure of simplicity: the number of diodes necessary. We further added the restriction that all simplifications would be in the form of second-order expressions, primarily because there exist techniques for performing these simplifications, but also because engineering considerations often require that no circuits of greater than second order be used. Despite these restrictions, the simplification techniques presented here have both direct and indirect applications to other circuits (e.g., to relay circuits) and to higher order expressions (see the next-to-last section of this chapter).

The basic theorem states that any function may be represented by at least two different second-order expressions: as a Boolean sum of minterms, and as a Boolean product of maxterms. Similarly, we will see that the *simplified* expression for a Boolean function may take on either of two second-order forms: the Boolean sum of Boolean products, or the Boolean product of Boolean sums. For definiteness, all the methods of this section will be aimed at obtaining expressions in the form of sums of products, and such expressions will be called *minterm-type expressions* (e.g., $f = A\bar{B} + B\bar{C}$). We will see later how it is possible to express the same function in the form of a product of sums, and will call such expressions *maxterm-type* [e.g., $f = (A + B)(\bar{B} + \bar{C})$].

Each product in a minterm-type expression will be called a *term;* and since all minterm-type expressions are sums of terms, it is natural that we should take a moment to examine the terms of n variables a little more carefully. In Table 4-1 all possible terms are listed for

Table 4-1

n	All Possible Terms	Number of Terms
1	$A, \bar{A}.$	2
2	$\bar{A}\bar{B}, \bar{A}B, A\bar{B}, AB,$	
	$A, \bar{A}, B, \bar{B}.$	8
3	$\bar{A}\bar{B}\bar{C}, \bar{A}\bar{B}C, \bar{A}B\bar{C}, \bar{A}BC, A\bar{B}\bar{C}, A\bar{B}C, AB\bar{C}, ABC,$	
	$\bar{A}\bar{B}, \bar{A}B, A\bar{B}, AB,$	
	$\bar{A}\bar{C}, \bar{A}C, A\bar{C}, AC,$	
	$\bar{B}\bar{C}, \bar{B}C, B\bar{C}, BC,$	
	$A, \bar{A}, B, \bar{B}, C, \bar{C}.$	26

$n = 1$, 2, and 3. In each instance we begin by listing all terms containing n letters. These are of course the minterms, and we already know there are 2^n of them. Then we list all possible combinations of $(n - 1)$ letters, $(n - 2)$ letters, $(n - 3)$ letters, etc., until we finally write down the terms containing only one letter—and there must be $2n$ of them. There are evidently a finite number of possible terms for each n, and any minterm-type expression can only contain terms from the list corresponding to the particular n involved. There are several ways of deriving the formula for the total number of terms possible for n variables, but perhaps this is the simplest: each term may be associated with a n-digit number, written to the base three as follows:

Letter appears in complemented form, write 0
Letter appears uncomplemented, write 1
Letter does not appear, write 2

For example, with $n = 3$,

$A\bar{B}C$ is associated with the number 101
$\bar{A}\bar{C}$ is associated with the number 020
$\bar{B}C$ is associated with the number 201
A is associated with the number 122

Now, since there are 3^n different numbers (base three) having n digits, and since all of these except the number $3^n - 1 = 222 \cdots 2$ correspond to a term, we see there must be $(3^n - 1)$ different terms of n variables. Since $3^1 - 1 = 2$, $3^2 - 1 = 8$, and $3^3 - 1 = 26$, we see that this formula checks with the particular examples shown in Table 4-1.

The simplification methods which will now be presented all have the objective of selecting from the $(3^n - 1)$ possible terms one group which defines the function and employs as few diodes as possible.

METHODS OF SIMPLIFICATION

Cut-and-Try Methods

The first method to be described requires that the designer employ judgment, experience, and ingenuity to simplify an expression by applying appropriate postulates and theorems of Boolean algebra. For example, the left-hand side of each of the following equations requires more diodes than the right-hand side; therefore, if the left-hand side appears in some equation, a saving may be effected by replacing it with the right-hand side.

P4a. $(A + B)(A + C) = A + BC$ **P4b.** $AB + AC = A(B + C)$

P5. $A + \bar{A} = 1$ **P5.** $A\bar{A} = 0$

T2a. $A + A = A$ **T2b.** $AA = A$

T4a. $A + AB = A$ **T4b.** $A(A + B) = A$

T9a. $A + \bar{A}B = A + B$ **T9b.** $A(\bar{A} + B) = AB$

T10. $AC + \bar{A}B + BC = (A + B)(\bar{A} + C) = AC + \bar{A}B$

The application of any of these simplifying theorems to a given function may not be obvious, as will be shown in a moment by three examples. It is usually necessary to rearrange and even to modify the original function before any of the rules can be used. For example, it may be necessary to complicate the original function by adding $X\bar{X}$ to it, or by multiplying it by $(X + \bar{X})$, if an X can be chosen which can subsequently be eliminated. These general principles will now be illustrated by some examples.

Example 4-1. Simplify $f = A\bar{B} + C + \bar{A}\bar{C}D + B\bar{C}D$.
Applying T9a to the last three terms, we find

$$f = A\bar{B} + C + \bar{A}D + BD$$

Rearranging and applying P4b

$$f = A\bar{B} + C + D(\bar{A} + B)$$

$$= A\bar{B} + C + D\overline{A\bar{B}}$$

And applying T9a again,

$$f = A\bar{B} + C + D$$

Example 4-2. Simplify $f = ABC + AB\bar{D} + A\bar{C} + \bar{A}\bar{B}\bar{C}\bar{D} + \bar{A}C.$ Factoring and applying T9a,

$$f = A(BC + \bar{C}) + \bar{A}(C + \bar{B}\bar{C}\bar{D}) + AB\bar{D}$$

$$= A(B + \bar{C}) + \bar{A}(C + \bar{B}\bar{D}) + AB\bar{D}$$

$$= AB + A\bar{C} + \bar{A}C + \bar{A}\bar{B}\bar{D} + AB\bar{D}$$

The last term may now be eliminated by combining it with the first and using T4a, giving

$$f = AB + A\bar{C} + \bar{A}C + \bar{A}\bar{B}\bar{D}$$

Example 4-3. Simplify $f = A\bar{B} + B\bar{C} + \bar{B}C + \bar{A}B.$

No obvious simplification exists. However, if the last two terms are multiplied by $(A + \bar{A})$ and $(C + \bar{C})$ respectively, the resulting terms may be rearranged and combined as follows:

$$f = A\bar{B} + B\bar{C} + \bar{B}C(A + \bar{A}) + \bar{A}B(C + \bar{C})$$

$$= A\bar{B} + A\bar{B}C + B\bar{C} + \bar{A}B\bar{C} + \bar{A}\bar{B}C + \bar{A}BC$$

$$= A\bar{B}(1 + C) + B\bar{C}(1 + \bar{A}) + \bar{A}C(\bar{B} + B)$$

$$= A\bar{B} + B\bar{C} + \bar{A}C$$

Note first that in all three examples the final expression *is* simpler than the original. In Example 4-1 the number of diodes necessary were reduced from twelve to five; in 4-2 from nineteen to thirteen; and in 4-3 from twelve to nine. However, though these reductions are satisfying, the methods used probably will not appeal to the beginning student. The steps taken to accomplish the simplifications are fairly obvious to the expert in Boolean algebra. They are not so obvious to the beginner, and even more important it is not obvious to the beginner that the final expression in each of the three examples is actually the simplest. Is it not possible, perhaps, to effect an even greater simplification in one or all of these examples by continuing in the same way? The answer is that no minterm-type expression can be found for any of the three functions which describes the function with fewer diodes. We will see how other approaches to the simplification problem always lead to the simplest possible expression. Nevertheless, the cut-and-try method is often useful in the preliminary stages of simplification, or in the simplification of elementary functions.

The Quine Simplification Method

W. V. Quine's method for simplifying Boolean functions is a little cumbersome to use, but can be considered to be the basis for all other methods to be mentioned here. The Quine procedure will be described by specifying a set of rules, as follows:

a. Express the function as a sum of minterms, i.e., express it in the form of equation 3-8. Of course, the function may already be in this form, but if it contains terms which are not minterms they must be expanded.

b. The next step is to derive a set of *prime implicants* from the minterms. First compare each minterm with every other minterm. Whenever two minterms differ by only one variable (X in one, $\bar{X}$ in the other), write down the term formed by omitting that variable from the minterm, and put a check mark opposite each minterm. This is the equivalent of employing the theorem $XY + \bar{X}Y = Y$. If the minterms each comprise n variables, this first group of terms will each contain $(n - 1)$ variables. Now the same procedure is followed with them; i.e., each term is compared with all the other terms of $(n - 1)$ variables, and whenever two terms differ by only one variable, the term formed by omitting that variable is written down and each of the parent terms is marked with a check. There are now three groups of terms: minterms, terms containing $(n - 1)$ variables, and terms containing $(n - 2)$ variables. Groups containing $(n - 3)$, $(n - 4)$, $(n - 5)$, $\cdots$ variables are now formed in the same way from the groups containing $(n - 2)$, $(n - 3)$, $(n - 4)$, $\cdots$ variables respectively, until finally a group of terms is formed, none of which can be combined with any other. Step b is now complete, and all the terms having no checks beside them are identified as prime implicants. (Note that a minterm may itself be a prime implicant if it does not combine with any other minterm.)

Before going on to step c, let us examine the meaning of the prime implicants. In the first place, note that any of the terms we have written down, checked and unchecked, could be included in the final expression for the desired function f. Furthermore, if we write down the Boolean sum of all the unchecked terms—all the prime implicants—we must have an expression for f. This follows because every time we derived a term in step b we could have used that term in the original expression in place of the two terms checked off, and so the final unchecked terms *must* include all the minterms we started out with. However, if we write down the sum of *all* the prime implicants we shall, in general, employ more terms than are necessary to specify the function. The remaining steps, c, d, e, and f, will show how to choose from the list derived in step b a set of prime implicants whose Boolean sum will be the simplest expression of f.

c. Next, prepare a table (Table A) having as many rows as there are prime implicants, and as many columns as there are minterms in step a. Identify each row with a prime implicant, and each column with a minterm. Place a check mark in a square on the table wherever the prime implicant to the left of the square is derived, in part, from the minterm above the square. This will be true wherever the letters in the prime implicant all appear in the same form as they do in the minterm.

d. Now examine the columns on Table A. If any column in the table contains only one check mark, the corresponding prime implicant is called an *essential term* and must be included in the final expression for f. Encircle the essential terms, together with all check marks on the same row with essential terms. A new table, Table B, may now be drawn up having the same form as Table A, but containing only the prime implicants (rows) that have not been encircled and only the minterms (columns) which contain no encircled checks. The minterms omitted from Table B need not be considered further because they will be included in the final expression for f by virtue of the fact that the essential terms appear in that expression.

e. Wherever in Table B there are columns m_i and m_j such that m_i has check marks on every row that m_j has check marks, eliminate column m_i.

f. Form Table C from Table B by omitting the columns eliminated by step e and any rows which remain but contain no check marks.

g. Examine Table C, and choose from it the simplest set (or sets) of prime implicants which, taken together, include at least one check mark in each column. The Boolean sum of these prime implicants, together with the essential terms of step d, forms the simplest expression for f. Note that this last step may be a very difficult or a very trivial one, depending on the nature of and number of entries in Table C.

The best way to make these rules clear is to work out a few examples.

Example 4-4. Simplify $f = A\bar{B} + B\bar{C} + \bar{B}C + \bar{A}B$.
a. Expanding f to obtain a sum of minterms:

$$\begin{aligned} f &= A\bar{B}(C + \bar{C}) + B\bar{C}(A + \bar{A}) + \bar{B}C(A + \bar{A}) + \bar{A}B(C + \bar{C}) \\ &= A\bar{B}C + A\bar{B}\bar{C} + AB\bar{C} + \bar{A}B\bar{C} + A\bar{B}C + \bar{A}\bar{B}C + \bar{A}BC + \bar{A}B\bar{C} \\ &= A\bar{B}C + A\bar{B}\bar{C} + AB\bar{C} + \bar{A}B\bar{C} + \bar{A}\bar{B}C + \bar{A}BC \end{aligned}$$

b. The set of prime implicants is found as follows:

$A\bar{B}C$ ✓
$A\bar{B}\bar{C}$ ✓
$AB\bar{C}$ ✓
$\bar{A}B\bar{C}$ ✓
$\bar{A}\bar{B}C$ ✓
$\bar{A}BC$ ✓

$A\bar{B}$ (combining $A\bar{B}C$ with $A\bar{B}\bar{C}$)
$\bar{B}C$ (combining $A\bar{B}C$ with $\bar{A}\bar{B}C$)
$A\bar{C}$ (combining $AB\bar{C}$ with $A\bar{B}\bar{C}$)
$B\bar{C}$ (combining ABC with $\bar{A}B\bar{C}$)
$\bar{A}B$ (combining $\bar{A}B\bar{C}$ with ABC)
$\bar{A}C$ (combining $\bar{A}\bar{B}C$ with $\bar{A}BC$)

None of the two-letter terms combine with one another, so there are no more prime implicants. The function f *may* be written as the Boolean sum of these prime implicants, but we continue with the rules, hoping to find a set of prime implicants which will form a simpler expression.

c. Table A may be now prepared, as shown in Table 4-2.

Table 4-2

	$A\bar{B}C$	$A\bar{B}\bar{C}$	$AB\bar{C}$	$\bar{A}B\bar{C}$	$\bar{A}\bar{B}C$	$\bar{A}BC$
$A\bar{B}$	✓	✓				
$\bar{B}C$	✓				✓	
$A\bar{C}$		✓	✓			
$B\bar{C}$			✓	✓		
$\bar{A}B$				✓		✓
$\bar{A}C$					✓	✓

d, e, and f. No column contains only one check mark. There are therefore no essential terms, and Tables B and C are the same as Table A.

g. Examining the table, we see there are two ways of choosing prime implicants so as to include a check mark in every column. These two ways are indicated in Table 4-3, where one set of terms is identified by triangles and

Table 4-3

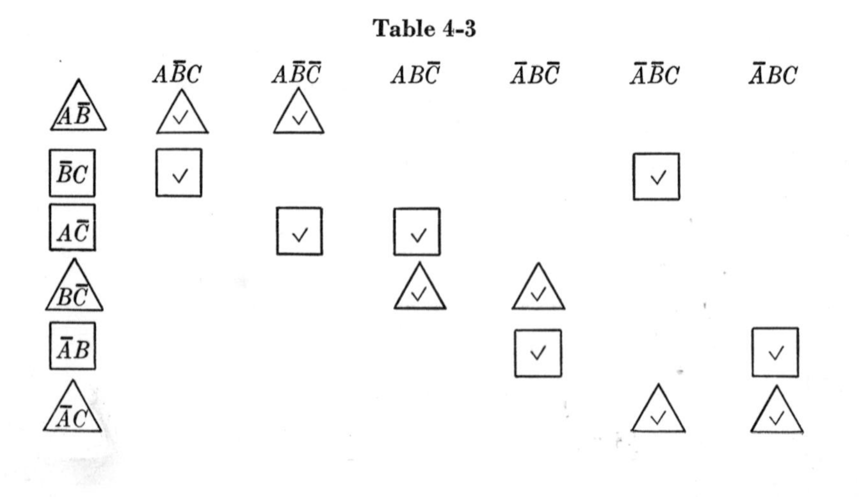

	$A\bar{B}C$	$A\bar{B}\bar{C}$	$AB\bar{C}$	$\bar{A}B\bar{C}$	$\bar{A}\bar{B}C$	$\bar{A}BC$
△ $A\bar{B}$	△✓	△✓				
□ $\bar{B}C$	□✓				□✓	
□ $A\bar{C}$		□✓	□✓			
△ $B\bar{C}$			△✓	△✓		
□ $\bar{A}B$				□✓		□✓
△ $\bar{A}C$					△✓	△✓

the other by squares. Since each column contains a triangle,

$$f = A\bar{B} + B\bar{C} + \bar{A}C$$

and since each column also contains a square,

$$f = \bar{B}C + A\bar{C} + \bar{A}B$$

The first of these expressions for f is the same as the result obtained in Example 4-3, using the cut-and-try method. The second, which employs the same number of diodes, could also be derived that way. However, not every function has alternative simplest forms, and the Quine method provides an easy way for finding any and all of them where the cut-and-try scheme gives no indication as to whether or not there may be more than one.

The fact that there can be more than one simplest expression is sometimes very useful. For example, if the designer must have the function $g = \bar{B}C + \bar{A}B$ as well as the function f above, he can save diodes by using the second of the above two representations for f; i.e., he can take

$$g = \bar{B}C + \bar{A}B \qquad \text{(six diodes)}$$

$$f = g + A\bar{C} \qquad \text{(four diodes)}$$

whereas if he had not known about the second expression he would have had to use nine diodes for f instead of only four.

Example 4-5. Simplify the following four-variable function:

$$f = m_3 + m_4 + m_5 + m_7 + m_9 + m_{11} + m_{12} + m_{13}$$

b. The set of prime implicants is now found:

$\bar{A}\bar{B}CD$	✓	$\bar{A}CD$		$B\bar{C}$
$\bar{A}B\bar{C}\bar{D}$	✓	$\bar{B}CD$		$B\bar{C}$
$\bar{A}B\bar{C}D$	✓	$\bar{A}B\bar{C}$	✓	
$\bar{A}BCD$	✓	$B\bar{C}\bar{D}$	✓	
$A\bar{B}\bar{C}D$	✓	$\bar{A}BD$		
$A\bar{B}CD$	✓	$B\bar{C}D$	✓	
$AB\bar{C}\bar{D}$	✓	$A\bar{B}D$		
$AB\bar{C}D$	✓	$A\bar{C}D$		
		$AB\bar{C}$	✓	

Note that $B\bar{C}$ can be formed by combining either $\bar{A}B\bar{C}$ and $AB\bar{C}$ or else $B\bar{C}\bar{D}$ and $B\bar{C}D$. The fact that a prime implicant appears twice on the list does not concern us.

c. Table A is now prepared, as illustrated in Table 4-4.

Table 4-4

	$\bar{A}\bar{B}CD$	$\bar{A}B\bar{C}\bar{D}$	$\bar{A}B\bar{C}D$	$\bar{A}BCD$	$A\bar{B}\bar{C}D$	$A\bar{B}CD$	$AB\bar{C}\bar{D}$	$AB\bar{C}D$
$\bar{A}CD$	✓			✓				
$\bar{B}CD$	✓					✓		
$\bar{A}BD$			✓	✓				
$A\bar{B}D$					✓	✓		
$A\bar{C}D$					✓			✓
$B\bar{C}$		✓	✓				✓	✓

d. There is one essential term, corresponding to the single check mark in column $A\bar{B}\bar{C}\bar{D}$.

Table B may now be formed, as in Table 4-5.

Table 4-5

	$\bar{A}\bar{B}CD$	$\bar{A}BCD$	$A\bar{B}\bar{C}D$	$A\bar{B}CD$
$\bar{A}CD$	✓	✓		
$\bar{B}CD$	✓			✓
$\bar{A}BD$		✓		
$A\bar{B}D$			✓	✓
$A\bar{C}D$			✓	

e and f. No application. Table C is the same as Table B.

g. Though there are several possible sets of the remaining prime implicants which will include the four minterms of Table C, careful inspection will show that one of these is simpler than all the rest: namely, the two prime implicants $\bar{A}CD$ and $A\bar{B}D$. Therefore, $f = B\bar{C} + \bar{A}CD + A\bar{B}D$ is the *only* simplest possible expression.

Table

	$ABCDE$	$AB\bar{C}DE$	$AB\bar{C}\bar{D}E$	$ABC\bar{D}E$	$\bar{A}BCDE$	$\bar{A}B\bar{C}DE$	$\bar{A}\bar{B}\bar{C}DE$	$\bar{A}\bar{B}CDE$
$ABCD$	✓							
$A\bar{C}\bar{D}E$			✓					
$\bar{A}\bar{B}CD$								✓
$ACD\bar{E}$								
$A\bar{B}\bar{C}\bar{D}$								
$\bar{A}\bar{C}\bar{D}\bar{E}$								
ABE	✓	✓	✓	✓				
BDE	✓	✓			✓	✓		
$\bar{A}DE$					✓	✓	✓	✓
$\bar{B}C\bar{E}$								
$\bar{B}\bar{D}\bar{E}$								

Example 4-6. Simplify the following five-variable function:

$$f = m_{31} + m_{27} + m_{25} + m_{29} + m_{15} + m_{11} + m_3 + m_7 + m_6 + m_{22} + m_{20} + m_{16} + m_{17} + m_0 + m_4 + m_8 + m_{30}$$

b. The set of prime implicants is now found:

$ABCDE$ ✓
$AB\bar{C}DE$ ✓
$AB\bar{C}\bar{D}E$ ✓
$ABC\bar{D}E$ ✓
$\bar{A}BCDE$ ✓
$\bar{A}B\bar{C}DE$ ✓
$\bar{A}\bar{B}\bar{C}DE$ ✓
$\bar{A}\bar{B}CDE$ ✓
$\bar{A}\bar{B}CD\bar{E}$ ✓

$A\bar{B}CD\bar{E}$ ✓
$A\bar{B}C\bar{D}\bar{E}$ ✓
$A\bar{B}\bar{C}\bar{D}\bar{E}$ ✓
$A\bar{B}\bar{C}\bar{D}E$ ✓
$\bar{A}\bar{B}\bar{C}\bar{D}\bar{E}$ ✓
$\bar{A}\bar{B}C\bar{D}\bar{E}$ ✓
$\bar{A}B\bar{C}\bar{D}\bar{E}$ ✓
$ABCD\bar{E}$ ✓

$ABDE$ ✓
$ABCE$ ✓
$BCDE$ ✓
$ABCD$
$AB\bar{C}E$ ✓
$B\bar{C}DE$ ✓
$AB\bar{D}E$ ✓
$A\bar{C}\bar{D}E$
$\bar{A}BDE$ ✓
$\bar{A}CDE$ ✓
$\bar{A}\bar{C}DE$ ✓

$\bar{A}\bar{B}DE$ ✓
$\bar{A}\bar{B}CD$
$\bar{B}CD\bar{E}$ ✓
$\bar{A}\bar{B}C\bar{E}$ ✓
$A\bar{B}C\bar{E}$ ✓
$ACD\bar{E}$
$A\bar{B}\bar{D}\bar{E}$ ✓
$\bar{B}C\bar{D}\bar{E}$ ✓
$A\bar{B}\bar{C}\bar{D}$
$\bar{B}\bar{C}\bar{D}\bar{E}$ ✓
$\bar{A}\bar{B}\bar{D}\bar{E}$ ✓
$\bar{A}\bar{C}\bar{D}\bar{E}$

ABE
BDE
$\bar{A}DE$
$\bar{B}C\bar{E}$
$\bar{B}\bar{D}\bar{E}$

c. Table A may now be prepared; see Table 4-6.

4-6

$\bar{A}\bar{B}CD\bar{E}$	$A\bar{B}CD\bar{E}$	$A\bar{B}C\bar{D}\bar{E}$	$A\bar{B}\bar{C}\bar{D}\bar{E}$	$A\bar{B}\bar{C}\bar{D}E$	$\bar{A}\bar{B}\bar{C}\bar{D}\bar{E}$	$\bar{A}\bar{B}C\bar{D}\bar{E}$	$\bar{A}B\bar{C}\bar{D}\bar{E}$	$ABCD\bar{E}$
								✓
				✓				
✓								
	✓							✓
			✓	✓				
					(✓)		(✓)	
✓	✓	✓				✓		
		✓	✓		✓	✓		

d. Examining Table A, we find and encircle three essential terms: $\bar{A}\bar{C}\bar{D}\bar{E}$, ABE, and $\bar{A}DE$.

Table B may now be formed; see Table 4-7.

Table 4-7

	$\bar{A}\bar{B}CD\bar{E}$	$A\bar{B}CD\bar{E}$	$A\bar{B}C\bar{D}\bar{E}$	$A\bar{B}\bar{C}\bar{D}\bar{E}$	$A\bar{B}\bar{C}\bar{D}E$	$\bar{A}\bar{B}C\bar{D}\bar{E}$	$ABCD\bar{E}$
$ABCD$							✓
$A\bar{C}\bar{D}E$					✓		
$\bar{A}\bar{B}CD$	✓						
$ACD\bar{E}$		✓					✓
$A\bar{B}\bar{C}\bar{D}$				✓	✓		
BDE							
$\bar{B}C\bar{E}$	✓	✓	✓			✓	
$\bar{B}\bar{D}\bar{E}$			✓	✓		✓	

e. Column $A\bar{B}C\bar{D}\bar{E}$ has a check mark on every row that $\bar{A}\bar{B}C\bar{D}\bar{E}$ has a check mark, and so the former may be eliminated.

f. Table C may now be formed by omitting column $A\bar{B}C\bar{D}\bar{E}$ from Table B, and also omitting row BDE, which contains no check marks; see Table 4-8.

Table 4-8

	$\bar{A}\bar{B}CD\bar{E}$	$A\bar{B}CD\bar{E}$	$A\bar{B}\bar{C}\bar{D}\bar{E}$	$A\bar{B}\bar{C}\bar{D}E$	$\bar{A}\bar{B}C\bar{D}\bar{E}$	$ABCD\bar{E}$
$ABCD$						✓
$A\bar{C}\bar{D}E$				✓		
$\bar{A}\bar{B}CD$	✓					
$ACD\bar{E}$		✓				✓
$A\bar{B}\bar{C}\bar{D}$			✓	✓		
$\bar{B}C\bar{E}$	✓	✓			✓	
$\bar{B}\bar{D}\bar{E}$			✓		✓	

g. Examination of Table C shows that two sets of terms are equally simple and still include all the remaining minterms. These sets are ($\bar{B}C\bar{E}$, $A\bar{B}\bar{C}\bar{D}$, $ABCD$) and ($\bar{B}C\bar{E}$, $A\bar{B}\bar{C}\bar{D}$, $ACD\bar{E}$). Therefore, the two simplest possible expressions are indicated by

$$f = ABE + \bar{A}DE + \bar{B}C\bar{E} + \bar{A}\bar{C}\bar{D}\bar{E} + A\bar{B}\bar{C}\bar{D} + \begin{cases} ABCD \\ ACD\bar{E} \end{cases}$$

(The brace indicates that either of the last two terms may be used in the expression.)

Before leaving the Quine simplification method, it will be worthwhile to discuss another method for finding prime implicants and therefore for setting up Table A. W. V. Quine and R. G. Nelson have each devised alternative methods, but Nelson's is the simpler to understand and to apply. His method requires that the function be expressed first as a Boolean product of Boolean sums. The application of the distributive law, i.e., "multiplying out," will then result in an expression which contains all the prime implicants! Furthermore, it will contain *only* the prime implicants if all subsuming terms are dropped out. One term subsumes another if every letter in the second term also appears in the first and appears in the same form, complemented or uncomplemented as the case may be. Thus the omission of a subsumed term amounts to an application of the theorem $A + AB = A$. For example, we can find the prime implicants in Example 4-5 by first expanding that function into maxterm form and then multiplying out the maxterms as follows:

$$
\begin{aligned}
f &= m_3 + m_4 + m_5 + m_7 + m_9 + m_{11} + m_{12} + m_{13} \\
&= M_{15}M_{14}M_{13}M_9M_7M_5M_1M_0 \\
&= (A + B + C + D)(A + B + C + \bar{D})(A + B + \bar{C} + D) \\
&\quad\; (A + \bar{B} + \bar{C} + D)(\bar{A} + B + C + D)(\bar{A} + B + \bar{C} + D) \\
&\quad\; (\bar{A} + \bar{B} + \bar{C} + D)(\bar{A} + \bar{B} + \bar{C} + \bar{D}) \\
&= (A + B + C)(A + \bar{C} + D)(\bar{A} + B + D)(\bar{A} + \bar{B} + \bar{C}) \\
&= (A + AB + B\bar{C} + BD + AC + CD)(\bar{A} + \bar{A}B + B\bar{C} + \bar{A}D \\
&\quad\; + \bar{B}D + \bar{C}D) \\
&= (A + B\bar{C} + BD + CD)(\bar{A} + B\bar{C} + \bar{B}D + \bar{C}D) \\
&= AB\bar{C} + A\bar{B}D + A\bar{C}D + \bar{A}B\bar{C} + B\bar{C} + B\bar{C}D + \bar{A}BD + B\bar{C}D \\
&\quad\; + B\bar{C}D + \bar{A}CD + \bar{B}CD \\
&= A\bar{B}D + A\bar{C}D + B\bar{C} + \bar{A}BD + \bar{A}CD + \bar{B}CD
\end{aligned}
$$

Of course, once the prime implicants have been found, we can construct Table A and continue with the previously described steps in the Quine method.

The Harvard Simplification Method

The Harvard Computer Group proposed a method of simplification which at first glance looks quite different from the Quine method. However, as we shall see, the Harvard method really does nothing more than provide an alternative way of finding Quine's prime implicants, and so constructing Table A.

The Harvard method employs a chart which contains all the variables under consideration and which displays all $(3^n - 1)$ possible terms which may appear in the result. Examples of the table for two and for three variables appear in Table 4-9. Note that the *rows* this time correspond to minterms (in the Quine method the *columns* did), and there is a column for every letter, for every combination of two letters, for every combination of three letters, etc., until the last column is headed by all the letters. (See Table 4-12, p. 80, for the column headings for four variables.) The entry at any square on the table

Table 4-9

	A	B	AB
m_0	$\bar{A}$	$\bar{B}$	$\bar{A}\bar{B}$
m_1	$\bar{A}$	B	$\bar{A}B$
m_2	A	$\bar{B}$	$A\bar{B}$
m_3	A	B	AB

	A	B	C	AB	AC	BC	ABC
m_0	$\bar{A}$	$\bar{B}$	$\bar{C}$	$\bar{A}\bar{B}$	$\bar{A}\bar{C}$	$\bar{B}\bar{C}$	$\bar{A}\bar{B}\bar{C}$
m_1	$\bar{A}$	$\bar{B}$	C	$\bar{A}\bar{B}$	$\bar{A}C$	$\bar{B}C$	$\bar{A}\bar{B}C$
m_2	$\bar{A}$	B	$\bar{C}$	$\bar{A}B$	$\bar{A}\bar{C}$	$B\bar{C}$	$\bar{A}B\bar{C}$
m_3	$\bar{A}$	B	C	$\bar{A}B$	$\bar{A}C$	BC	$\bar{A}BC$
m_4	A	$\bar{B}$	$\bar{C}$	$A\bar{B}$	$A\bar{C}$	$\bar{B}\bar{C}$	$A\bar{B}\bar{C}$
m_5	A	$\bar{B}$	C	$A\bar{B}$	AC	$\bar{B}C$	$A\bar{B}C$
m_6	A	B	$\bar{C}$	AB	$A\bar{C}$	$B\bar{C}$	$AB\bar{C}$
m_7	A	B	C	AB	AC	BC	ABC

contains the same letters which appear at the top of the corresponding column, and the letters are either complemented or uncomplemented depending on how they appear in the minterm in the corresponding row. Another form for these charts is shown in Table 4-10. Here the

Table 4-10

	A	B	AB
m_0	0	0	0
m_1	0	1	1
m_2	1	0	2
m_3	1	1	3

	A	B	C	AB	AC	BC	ABC
m_0	0	0	0	0	0	0	0
m_1	0	0	1	0	1	1	1
m_2	0	1	0	1	0	2	2
m_3	0	1	1	1	1	3	3
m_4	1	0	0	2	2	0	4
m_5	1	0	1	2	3	1	5
m_6	1	1	0	3	2	2	6
m_7	1	1	1	3	3	3	7

column and row headings are the same, but every entry on the chart has been replaced with a decimal number. The decimal number is formed in exactly the same way a minterm subscript is formed: for each complemented letter a zero is written and for each uncomplemented letter a one, and the resulting binary number is converted to a decimal. For example, the m_4 entry in the AC column of Table 4-9 is $A\bar{C}$. The corresponding entry in Table 4-10 is 2, which is found by replacing $A\bar{C}$ by 10, or binary two.

The first step in using the Harvard simplification scheme is to prepare a Harvard chart having the same number of variables as does the function to be simplified. Notice that for n variables the chart has 2^n rows and $\sum_{i=0}^{n-1} {}_iC_n = 2^n - 1$ columns. The total number of entries is therefore $2^n(2^n - 1)$. This is considerably larger than the number of possible terms of n variables, which we have seen to be $(3^n - 1)$. The reason is, of course, that many of the terms appear in the Harvard chart more than once. Once the chart is prepared, the following rules must be followed:

a. Express the given function as a Boolean sum of minterms. Draw horizontal lines in the Harvard chart through each row which corresponds to a minterm which does *not* appear in the function.

b. Each column now contains some numbers which have been crossed off with horizontal lines. Examine each column in turn, and wherever a number has been crossed out, cross out all other appearances of that number in that column.

c. Now examine the rows one at a time. Starting at the left-hand side of the first row, find the first term which has not been crossed off, and note the letter or letters on the top of that column. Cross off all other terms in that row whose column headings contain those same letters. Now repeat the process for all the remaining terms in that row and in all the other rows.

The entries which still remain are the prime implicants of the Quine method. It would be possible at this point to switch to the Quine method for the remainder of the solution. Alternatively, and more conveniently, the Harvard chart may be used to complete the search for a simplest form by following steps d and e below.

d. We will now select a set of terms from the prime implicants so that the Boolean sum of this set of terms represents the original function. Every time we select a term we will encircle it and all its appearances in the chart, so that when we have finished there will be at least one circle on every row that was not crossed off in step a. We begin by searching for essential terms, which are located by seeking rows which contain only one entry which has not been crossed out. Each such lone entry is encircled, together with all other appearances of the same entry in the table—i.e., together with all other appearances of the same number in the same column. (Of course, every time a term is selected to be part of the desired expression, a decimal number on the chart must be retranslated into a Boolean term. Thus, an eleven in the $ACDE$ column will be written $A\bar{C}DE$.)

e. The rows which neither have lines drawn through them nor contain encircled terms must now be accounted for. Each of these rows contains two or more unmarked terms. Encircle at least one term on each row in such a way as to minimize the number of terms encircled. Remember, of course, that when any decimal number is encircled, all other occurrences of that number in that column must also be encircled. Finally, translate all encircled numbers into Boolean terms. Their Boolean sum is the desired expression.

Note that step e here corresponds to step g in the Quine method. Step f of the Quine method is automatically taken care of in the Harvard plan, and although Quine's step e can be included as a step in the Harvard method, it is a little difficult to carry out and may be eliminated.

Three more examples will illustrate the use of this method of simplification.

Example 4-7. Simplify the function $f = A\bar{B}\bar{C} + ABC + \bar{A}\bar{B}C + \bar{A}B\bar{C}$.

a. The function is already expressed as a Boolean sum of minterms: $f = m_4 + m_7 + m_1 + m_2$. Therefore, m_0, m_3, m_5, and m_6, which are not included in f, are eliminated from the chart of Table 4-11 by drawing horizontal lines through the corresponding rows.

Table 4-11

	A	B	C	AB	AC	BC	ABC
~~m_0~~	~~0~~	~~0~~	~~0~~	~~0~~	~~0~~	~~0~~	~~0~~
m_1	~~0~~	~~0~~	~~1~~	~~0~~	~~1~~	~~1~~	(1)
m_2	~~0~~	~~1~~	~~0~~	~~1~~	~~0~~	~~2~~	(2)
~~m_3~~	~~0~~	~~1~~	~~1~~	~~1~~	~~1~~	~~3~~	~~3~~
m_4	~~1~~	~~0~~	~~0~~	~~2~~	~~2~~	~~0~~	(4)
~~m_5~~	~~1~~	~~0~~	~~1~~	~~2~~	~~3~~	~~1~~	~~5~~
~~m_6~~	~~1~~	~~1~~	~~0~~	~~3~~	~~2~~	~~2~~	~~6~~
m_7	~~1~~	~~1~~	~~1~~	~~3~~	~~3~~	~~3~~	(7)

b. Examining the first (A) column, we find that a zero and a one were crossed off in step a. We therefore cross off all other zeros and ones in that column. Since a zero and a one have also been crossed off in the B and C columns, we cross off all zeros and ones there, too. Going on to the fourth (AB) column, we see that a zero (in m_0 row), a one (in m_3 row), a two (in m_5 row), and a three (in m_6 row) have already been crossed off. We therefore cross off all other zeros, ones, twos, and threes in that column. The same rule applied to columns AC and BC crosses off all entries in those two columns, too. In the last column, however, the rule obviously does not require that any more terms be eliminated.

c and d. Each row now contains only one term which has not been eliminated in step b. Therefore, step d may be applied directly, and the four terms encircled on the chart are all essential terms. Translating the decimal numbers back into Boolean terms, we find $f = \bar{A}\bar{B}C + \bar{A}B\bar{C} + A\bar{B}\bar{C} + ABC$. This is of course the expression we started out with: no simpler second-order expression of this function exists.

Example 4-8. Simplify the function $f = \bar{C}\bar{D} + AB\bar{D} + \bar{A}B\bar{C} + \bar{A}\bar{B}C\bar{D} + A\bar{B}\bar{C}D + \bar{B}CD$.

a. Expanding f, we find

$$f = \bar{A}\bar{B}\bar{C}\bar{D} + \bar{A}B\bar{C}\bar{D} + A\bar{B}\bar{C}\bar{D} + AB\bar{C}\bar{D} + ABC\bar{D} + \bar{A}B\bar{C}D$$

$$+ \bar{A}\bar{B}C\bar{D} + A\bar{B}\bar{C}D + A\bar{B}CD + \bar{A}\bar{B}CD$$

$$= m_0 + m_2 + m_3 + m_4 + m_5 + m_8 + m_9 + m_{11} + m_{12} + m_{14}$$

In Table 4-12, all minterm rows not listed in f are crossed out with horizontal lines.

b. In Table 4-12, small horizontal dashes have been used to carry out step b. For example, consider the CD column. The numbers one, two, and three were crossed out in step a by horizontal lines through rows m_1, m_6, and m_7 respectively. Therefore, the numbers one, two, and three are marked with horizontal dashes wherever they appear in this column. The number zero is nowhere crossed off in this column, and so remains. In the other five two-letter columns all numbers must be crossed off, and the same is true for columns A, B, C, and D. Let us examine one more column before going on to step c. In column BCD, for example, we find horizontal lines through numbers one, six, seven, two, and five. We therefore use horizontal dashes to cross out these five numbers wherever they appear in column BCD. Numbers zero, three, and four remain on the chart. (At this point the *diagonal* marks do not yet appear on the diagram.)

c. Next we examine the rows and apply step c, using diagonal marks to cross off undesired terms. Starting at the left of row m_0, the zero in column CD is the first unmarked term. We therefore look to the right on that row to find other unmarked columns which contain the letters CD; and we put diagonal marks through the numbers in columns ACD, BCD, and $ABCD$. Column ABD does not contain both letters C and D, so it is not affected. Next we examine columns to the right of column ABD, which contain the letters ABD. There are none ($ABCD$ has already been eliminated), and so we go on to the next row. In row m_2 the first unmarked column we find is ABD, and applying the rule of step c, we cross off the entry in column $ABCD$. The

Table 4-12

	A	*B*	*C*	*D*	*AB*	*AC*	*AD*	*BC*	*BD*	*CD*	*ABC*	*ABD*	*ACD*	*BCD*	*ABCD*
m_0	~~0~~	~~0~~	~~0~~	~~0~~	~~0~~	~~0~~	~~0~~	~~0~~	~~0~~	$\boxed{0}$	~~0~~	△0	~~0~~	~~0~~	~~0~~
~~m_1~~	~~0~~	~~0~~	~~0~~	~~1~~	~~0~~	~~0~~	~~1~~	~~0~~	~~1~~	~~1~~	~~0~~	~~1~~	~~1~~	~~1~~	~~1~~
m_2	~~0~~	~~0~~	~~1~~	~~0~~	~~0~~	~~1~~	~~0~~	~~1~~	~~0~~	~~2~~	$\boxed{1}$	△0	~~2~~	~~2~~	~~2~~
m_3	~~0~~	~~0~~	~~1~~	~~1~~	~~0~~	~~1~~	~~1~~	~~1~~	~~1~~	~~3~~	$\boxed{1}$	~~1~~	~~3~~	△3	~~3~~
m_4	~~0~~	~~1~~	~~0~~	~~0~~	~~1~~	~~0~~	~~0~~	~~2~~	~~2~~	$\boxed{0}$	②	~~2~~	~~0~~	~~4~~	~~4~~
m_5	~~0~~	~~1~~	~~0~~	~~1~~	~~1~~	~~0~~	~~1~~	~~2~~	~~3~~	~~1~~	②	~~3~~	~~1~~	~~5~~	~~5~~
~~m_6~~	~~0~~	~~1~~	~~1~~	~~0~~	~~1~~	~~1~~	~~0~~	~~3~~	~~2~~	~~2~~	~~3~~	~~2~~	~~2~~	~~6~~	~~6~~
~~m_7~~	~~0~~	~~1~~	~~1~~	~~1~~	~~1~~	~~1~~	~~1~~	~~3~~	~~3~~	~~3~~	~~3~~	~~3~~	~~3~~	~~7~~	~~7~~
m_8	~~1~~	~~0~~	~~0~~	~~0~~	~~2~~	~~2~~	~~2~~	~~0~~	~~0~~	$\boxed{0}$	△4	~~4~~	~~4~~	~~0~~	~~8~~
m_9	~~1~~	~~0~~	~~0~~	~~1~~	~~2~~	~~2~~	~~3~~	~~0~~	~~1~~	~~1~~	△4	$\boxed{5}$	~~5~~	~~1~~	~~9~~
~~m_{10}~~	~~1~~	~~0~~	~~1~~	~~0~~	~~2~~	~~3~~	~~2~~	~~1~~	~~0~~	~~2~~	~~5~~	~~4~~	~~6~~	~~2~~	~~10~~
m_{11}	~~1~~	~~0~~	~~1~~	~~1~~	~~2~~	~~3~~	~~3~~	~~1~~	~~1~~	~~3~~	~~5~~	$\boxed{5}$	~~7~~	△3	~~11~~
m_{12}	~~1~~	~~1~~	~~0~~	~~0~~	~~3~~	~~2~~	~~2~~	~~2~~	~~2~~	$\boxed{0}$	~~6~~	⑥	~~4~~	~~4~~	~~12~~
~~m_{13}~~	~~1~~	~~1~~	~~0~~	~~1~~	~~3~~	~~2~~	~~3~~	~~2~~	~~3~~	~~1~~	~~6~~	~~7~~	~~5~~	~~5~~	~~13~~
m_{14}	~~1~~	~~1~~	~~1~~	~~0~~	~~3~~	~~3~~	~~2~~	~~3~~	~~2~~	~~2~~	~~7~~	⑥	~~6~~	~~6~~	~~14~~
~~m_{15}~~	~~1~~	~~1~~	~~1~~	~~1~~	~~3~~	~~3~~	~~3~~	~~3~~	~~3~~	~~3~~	~~7~~	~~7~~	~~7~~	~~7~~	~~15~~

process is repeated for all rows, and all the diagonal marks shown in Table 4-12 are inserted.

d. Rows m_5 and m_{14} now contain only one unmarked entry each. These entries, a two in column *ABC* (representing the term $\bar{A}B\bar{C}$), and a six in column *ABD* (representing the term $AB\bar{D}$), are encircled, together with the other two and six in the *ABC* and *ABD* columns respectively. $\bar{A}B\bar{C}$ and $AB\bar{D}$ are essential terms.

e. We must now choose enough unmarked terms to account for the rows m_0, m_2, m_3, m_8, m_9, and m_{11}. Two possibilities seem hopeful: one is indicated by the squares in Table 4-12, the other by the triangles. Notice that (ignoring the triangles) there is either a circle or a square on every row not eliminated in step a; and (ignoring the squares), there is either a circle or a triangle on each such row. Therefore either of these combinations (circle-square or circle-triangle) represents a complete expression for f.

The two expressions are:

$$f = \bar{A}B\bar{C} + AB\bar{D} + \bar{C}\bar{D} + \bar{A}\bar{B}C + A\bar{B}D$$

and

$$f = \bar{A}B\bar{C} + AB\bar{D} + \bar{A}\bar{B}\bar{D} + A\bar{B}\bar{C} + \bar{B}CD$$

The first expression employs one less diode than the second, and is therefore the simplest.

Example 4-9. Simplify the function of Example 4-6, using the Harvard chart method.

The solution is shown in Table 4-13. The essential terms arise from rows m_3, m_8, and m_{29}, each of which contains only one term not crossed out. The essential terms are encircled, and the other terms necessary for the simplest expression are marked with squares. Note that row m_{30} can be included either with $ABCD$ (15) or $ACD\bar{E}$ (14). The latter is indicated by a dotted square. The solution is evidently either

$$f = ABE + \bar{A}DE + \bar{B}C\bar{E} + A\bar{B}\bar{C}\bar{D} + \bar{A}\bar{C}\bar{D}\bar{E} + ABCD$$

or

$$f = ABE + \bar{A}DE + \bar{B}C\bar{E} + A\bar{B}\bar{C}\bar{D} + \bar{A}\bar{C}\bar{D}\bar{E} + ACD\bar{E}$$

and these of course agree with the previous solution.

The Veitch Diagram Simplification Method

A special form of Venn diagram was suggested by Veitch for use in the simplification of Boolean functions, and a modified form was proposed by Karnaugh. These diagrams, which will here be called Veitch diagrams, have already been described in Chapter 3 (Figures 3-15 and 3-16). They provide a very quick and easy way for finding the simplest expression of a function, and will be used throughout this book for that purpose as well as for visualizing some of the complicated features of particular functions.

In Figure 4-2 the Veitch diagrams for functions of two and three variables are repeated, and diagrams for four, five, six, and eight variables are added. In each square is entered the number of the minterm represented by that square. (The minterms are formed, as usual, by writing the letters in alphabetical order and assigning the value "one" to normal letters, and "zero" to complemented letters. In Figure 4-2*e*, for example, the number 45 is binary 101101, and $m_{45} = A\bar{B}CD\bar{E}F$. A glance at the square labeled 45 will show that it lies within the area assigned to variables A, C, D, and F, but outside of B and E.) Veitch diagrams for a number of variables greater than four may be formed by combining smaller diagrams. For example, Figure 4-2*d* for five variables is formed by combining two four-variable diagrams; Figure

Table 4-13

	A	B	C	D	E	AB	AC	AD	AE	BC	BD	BE	CD	CE	DE	ABC	ABD	ABE	ACD	ACE	ADE	BCD	BCE	BDE	CDE	$ABCD$	$ABCE$	$ABDE$	$ACDE$	$BCDE$	$ABCDE$
m_0	0	0	0	0	0	0	0	0	0	0	0	0	0	0	0	0	0	0	0	0	0	0	0	0	0	0	0	0	(0)	0	0
m_1	0	0	0	0	1	0	0	0	1	0	0	1	0	1	1	0	0	1	0	1	1	0	1	1	1	0	1	1	1	1	1
m_2	0	0	0	1	0	0	0	1	0	0	1	0	1	0	2	0	1	0	1	0	2	1	0	2	2	1	0	2	2	2	2
m_3	0	0	0	1	1	0	0	1	1	0	1	1	1	1	3	0	1	1	1	1	(3)	1	1	3	3	1	1	3	3	3	3
m_4	0	0	1	0	0	0	1	0	0	1	0	0	2	2	0	1	0	0	2	2	0	2	[2]	0	4	2	2	0	4	4	4
m_5	0	0	1	0	1	0	1	0	1	1	0	1	2	3	1	1	0	1	2	3	1	2	3	1	5	2	3	1	5	5	5
m_6	0	0	1	1	0	0	1	1	0	1	1	0	3	2	2	1	1	0	3	2	2	3	[2]	2	6	3	2	2	6	6	6
m_7	0	0	1	1	1	0	1	1	1	1	1	1	3	3	3	1	1	1	3	3	(3)	3	3	3	7	3	3	3	7	7	7
m_8	0	1	0	0	0	1	0	0	0	2	2	2	0	0	0	2	2	2	0	0	0	4	4	4	0	4	4	4	(0)	8	8
m_9	0	1	0	0	1	1	0	0	1	2	2	3	0	1	1	2	2	3	0	1	1	4	5	5	1	4	5	5	1	9	9
m_{10}	0	1	0	1	0	1	0	1	0	2	3	2	1	0	2	2	3	2	1	0	2	5	4	6	2	5	4	6	2	10	10
m_{11}	0	1	0	1	1	1	0	1	1	2	3	3	1	1	3	2	3	3	1	1	(3)	5	5	7	3	5	5	7	3	11	11
m_{12}	0	1	1	0	0	1	1	0	0	3	2	2	2	2	0	3	2	2	2	2	0	6	6	4	4	6	6	4	4	12	12
m_{13}	0	1	1	0	1	1	1	0	1	3	2	3	2	3	1	3	2	3	2	3	1	6	7	5	5	6	7	5	5	13	13
m_{14}	0	1	1	1	0	1	1	1	0	3	3	2	3	2	2	3	3	2	3	2	2	7	6	6	6	7	6	6	6	14	14
m_{15}	0	1	1	1	1	1	1	1	1	3	3	3	3	3	3	3	3	3	3	3	(3)	7	7	7	7	7	7	7	7	15	15

Table 4-13 (CONTINUED)

	A	B	C	D	E	AB	AC	AD	AE	BC	BD	BE	CD	CE	DE	ABC	ABD	ABE	ACD	ACE	ADE	BCD	BCE	BDE	CDE	$ABCD$	$ABCE$	$ABDE$	$ACDE$	$BCDE$	$ABCDE$
m_{16}	1	0	0	0	0	2	2	2	2	0	0	0	0	0	0	4	4	4	4	4	4	0	0	0	0	8	8	8	8	0	16
m_{17}	1	0	0	0	1	2	2	2	3	0	0	1	0	1	1	4	4	5	4	5	5	0	1	1	1	8	9	9	9	1	17
m_{18}	1	0	0	1	0	2	2	3	2	0	1	0	1	0	2	4	5	4	5	4	6	1	0	2	2	9	8	10	10	2	18
m_{19}	1	0	0	1	1	2	2	3	3	0	1	1	1	1	3	4	5	5	5	5	7	1	1	3	3	9	9	11	11	3	19
m_{20}	1	0	1	0	0	2	3	2	2	1	0	0	2	2	0	5	4	4	6	6	4	2	2	0	4	10	10	8	12	4	20
m_{21}	1	0	1	0	1	2	3	2	3	1	0	1	2	3	1	5	4	5	6	7	5	2	3	1	5	10	11	9	13	5	21
m_{22}	1	0	1	1	0	2	3	3	2	1	1	0	3	2	2	5	5	4	7	6	6	3	2	2	6	11	10	10	14	6	22
m_{23}	1	0	1	1	1	2	3	3	3	1	1	1	3	3	3	5	5	5	7	7	7	3	3	3	7	11	11	11	15	7	23
m_{24}	1	1	0	0	0	3	2	2	2	2	2	2	0	0	0	6	6	6	4	4	4	4	4	4	0	12	12	12	8	8	24
m_{25}	1	1	0	0	1	3	2	2	3	2	2	3	0	1	1	6	6	7	4	5	5	4	5	5	1	12	13	13	9	9	25
m_{26}	1	1	0	1	0	3	2	3	2	2	3	2	1	0	2	6	7	6	5	4	6	5	4	6	2	13	12	14	10	10	26
m_{27}	1	1	0	1	1	3	2	3	3	2	3	3	1	1	3	6	7	7	5	5	7	5	5	7	3	13	13	15	11	11	27
m_{28}	1	1	1	0	0	3	3	2	2	3	2	2	2	2	0	7	6	6	6	6	4	6	6	4	4	14	14	12	12	12	28
m_{29}	1	1	1	0	1	3	3	2	3	3	2	3	2	3	1	7	6	7	6	7	5	6	7	5	5	14	15	13	13	13	29
m_{30}	1	1	1	1	0	3	3	3	2	3	3	2	3	2	2	7	7	6	7	6	6	7	6	6	6	15	14	14	14	14	30
m_{31}	1	1	1	1	1	3	3	3	3	3	3	3	3	3	3	7	7	7	7	7	7	7	7	7	7	15	15	15	15	15	31

4-2*e* for six variables is formed by combining four four-variable diagrams; Figure 4-2*f*, also for six variables, is formed by combining eight three-variable diagrams. Either Figure 4-2*e* or 4-2*f* may be used for

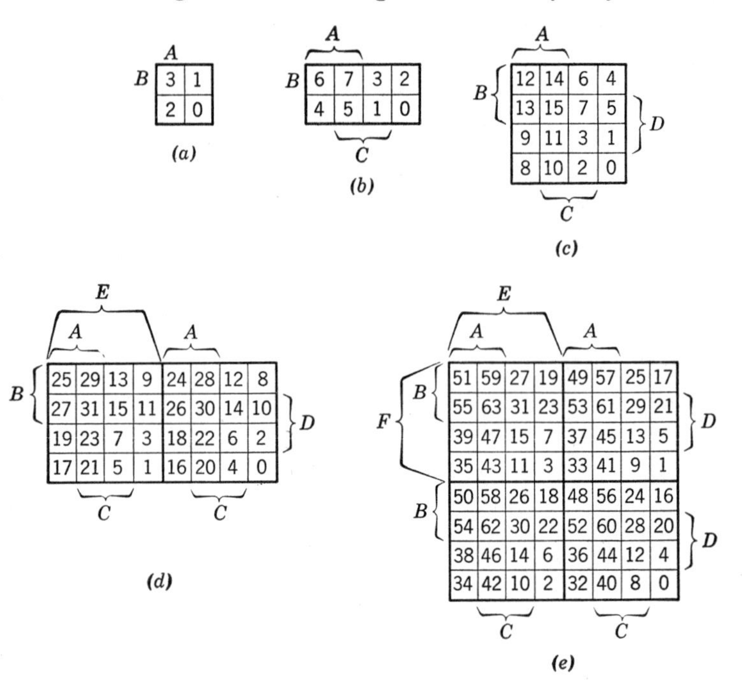

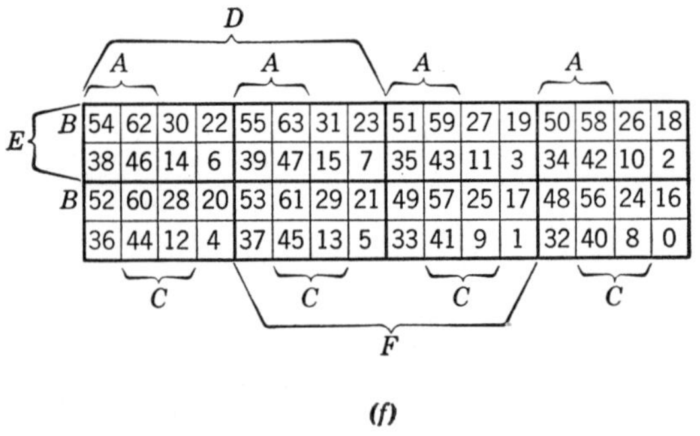

(*f*)

FIGURE 4-2

six-variable simplifications; each, of course, contains $2^6 = 64$ squares, one for each minterm. However, we shall see later that for some functions one of these diagrams may be better than the other. Any number of variables may be plotted on a Veitch diagram, though the diagrams are difficult to use if more than eight variables are involved.

To see how a function may be "plotted" on a Veitch diagram, consider the function of Example 4-8: $f = \bar{C}\bar{D} + AB\bar{D} + \bar{A}B\bar{C} + \bar{A}\bar{B}C\bar{D} + A\bar{B}\bar{C}D + \bar{B}CD$. One way to plot this function of four variables would be to expand it into minterm form and then put a mark in the appropriate square for each minterm. However, one advantage of the Veitch diagram method is that this expansion is unnecessary. Figure

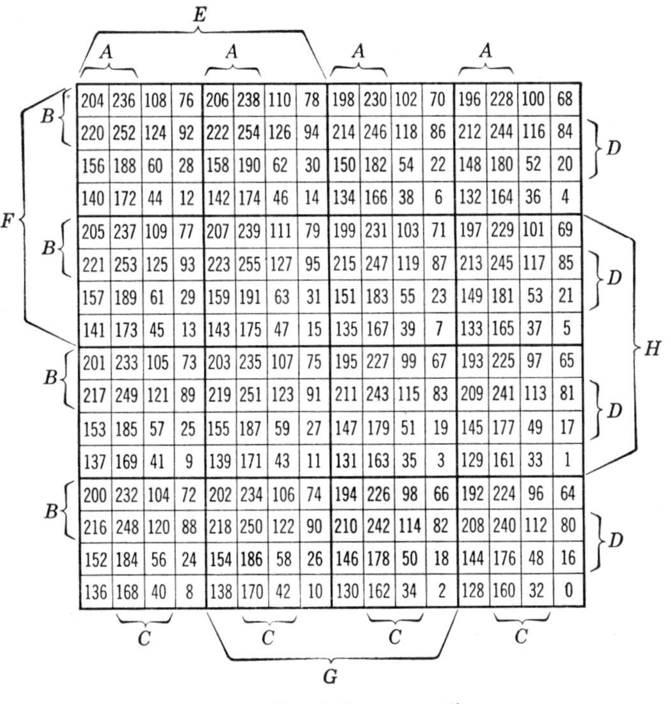

204	236	108	76	206	238	110	78	198	230	102	70	196	228	100	68
220	252	124	92	222	254	126	94	214	246	118	86	212	244	116	84
156	188	60	28	158	190	62	30	150	182	54	22	148	180	52	20
140	172	44	12	142	174	46	14	134	166	38	6	132	164	36	4
205	237	109	77	207	239	111	79	199	231	103	71	197	229	101	69
221	253	125	93	223	255	127	95	215	247	119	87	213	245	117	85
157	189	61	29	159	191	63	31	151	183	55	23	149	181	53	21
141	173	45	13	143	175	47	15	135	167	39	7	133	165	37	5
201	233	105	73	203	235	107	75	195	227	99	67	193	225	97	65
217	249	121	89	219	251	123	91	211	243	115	83	209	241	113	81
153	185	57	25	155	187	59	27	147	179	51	19	145	177	49	17
137	169	41	9	139	171	43	11	131	163	35	3	129	161	33	1
200	232	104	72	202	234	106	74	194	226	98	66	192	224	96	64
216	248	120	88	218	250	122	90	210	242	114	82	208	240	112	80
152	184	56	24	154	186	58	26	146	178	50	18	144	176	48	16
136	168	40	8	138	170	42	10	130	162	34	2	128	160	32	0

FIGURE 4-2 (continued)

4-3 shows the successive steps followed in plotting the function directly. The terms are considered one at a time, and "ones" are put in the proper square for each term. For the term $\bar{C}\bar{D}$, we note that $\bar{C}$ is represented by the four squares at the left-hand side of the diagram, together with the four at the right-hand side. Similarly, $\bar{D}$ is represented by the four squares along the top edge together with the four along the bottom edge. $\bar{C}\bar{D}$ is therefore represented by the squares common to $\bar{C}$ and to $\bar{D}$—the four in the corners. Next, we see that $AB\bar{D}$ must lie in the left-hand side to be in A, in the top left quarter to be in both A and B, and in the top left eighth to be in both A and

B but not D. Of course, the corner square is common both to $\bar{C}\bar{D}$ and to $AB\bar{D}$. This need not concern us. It is simply the Veitch diagram indication of the fact that the expansions of $\bar{C}\bar{D}$ and of $AB\bar{D}$ have one minterm in common: $AB\bar{C}\bar{D}$.

In the same way the remainder of the terms are plotted on the diagram. The completed diagram—the last one of Figure 4-3—is the exact equivalent of the minterm form of the equation for f, and the first step in the Veitch diagram simplification method is to make such

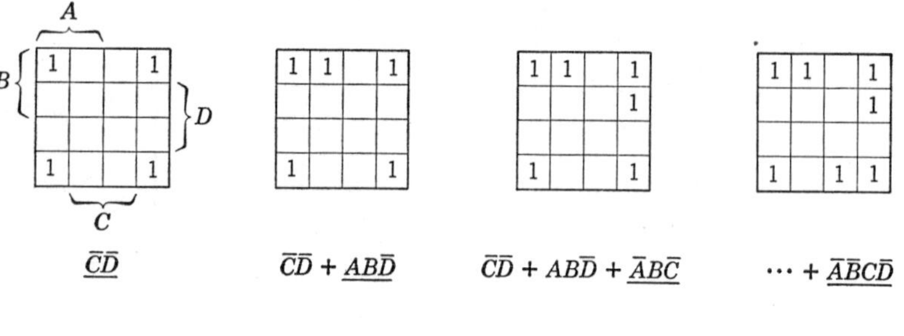

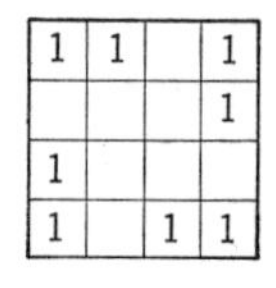

$\cdots + A\bar{B}\bar{C}D$

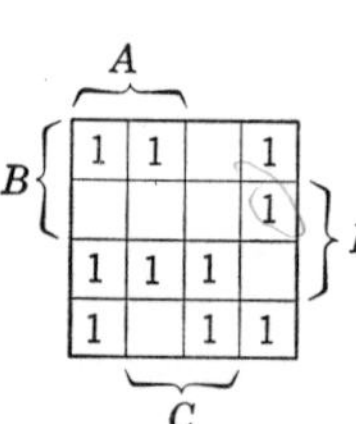

$f = \bar{C}\bar{D} + AB\bar{D} + \bar{A}B\bar{C} + \bar{A}\bar{B}C\bar{D} + A\bar{B}\bar{C}D + \bar{B}CD$

FIGURE 4-3

a chart of the function which is to be minimized. In order to understand the next steps, it will be necessary to learn how to recognize a prime implicant on a Veitch diagram.

It will be remembered that prime implicants were developed in the Quine method by combining minterms to form terms of $(n - 1)$ variables, combining these to form terms of $(n - 2)$ variables, etc., until no further combinations are possible. Each combination was really an application of the theorem $AXXXX + \bar{A}XXXX = XXXX$, and the prime implicants were defined as those terms which were never themselves combined with any other terms. For example, consider the function $f = AB\bar{C}\bar{D} + ABCD + A\bar{B}CD + \bar{A}\bar{B}\bar{C}\bar{D} + AB\bar{C}D + ABC\bar{D} + \bar{A}\bar{B}CD$. Applying the Quine method, we find the prime implicants as follows:

$AB\bar{C}\bar{D}$	✓	$AB\bar{C}$	✓	AB
$ABCD$	✓	$AB\bar{D}$	✓	AB
$A\bar{B}CD$	✓	ACD		
$\bar{A}\bar{B}\bar{C}\bar{D}$		ABD	✓	
$AB\bar{C}D$	✓	ABC	✓	
$ABC\bar{D}$	✓	$\bar{B}CD$		
$\bar{A}\bar{B}CD$	✓			

All the terms listed except $\bar{A}\bar{B}\bar{C}\bar{D}$, ACD, $\bar{B}CD$, and AB could be combined with other terms and therefore simplified. The uncombined (unchecked) terms are the prime implicants.

This function is plotted on a Veitch diagram in Figure 4-4. In order to be able to recognize the prime implicants by a glance at a chart of this kind, it is only necessary to learn the Veitch diagram patterns which represent combinations of terms. Let us therefore leave for a moment the problem represented by Figure 4-4, and examine patterns.

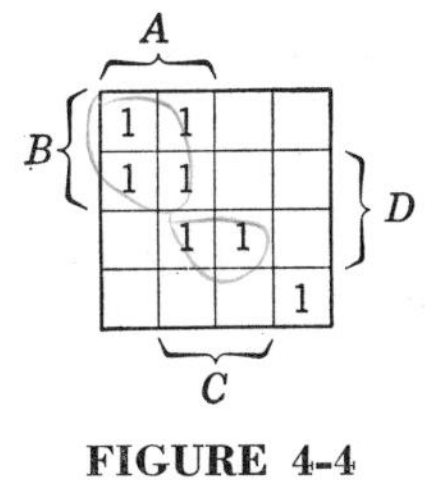

FIGURE 4-4

In Figure 4-5 a number of two-minterm combinations are illustrated. Examining the first three, for example, we see that they can be considered to represent the equations $A\bar{C} = AB\bar{C} + A\bar{B}\bar{C}$, $AB = ABC + AB\bar{C}$, and $B\bar{C} = AB\bar{C} + \bar{A}B\bar{C}$. All equations contain the common minterm $AB\bar{C}$; and $A\bar{B}\bar{C}$, ABC, and $\bar{A}B\bar{C}$ are the *only* minterms that may be combined with $AB\bar{C}$. The other diagrams in Figure 4-5 may be interpreted in the same way. However, the usefulness of these diagrams arises from the fact that combinations can be identified by the patterns they form, and it is therefore as important to learn the patterns as to understand the logic behind the combinations. The following generalization about patterns may be verified by studying Figure 4-5. Two minterms may be combined whenever: they are adjacent to one another in the same row or column; they are at opposite ends of the same row or column; they occupy identical positions in two three-variable or four-variable or multivariable charts which are themselves adjacent to one another or at the opposite ends of row or column. This last rule of the generalization, which is only necessary for charts of many variables, is illustrated in the six-variable chart of Figure 4-6. The minterm labeled a on this chart ($\bar{A}BCDE\bar{F}$) may be combined with b (to form $\bar{A}BCDE$), c (to form $\bar{A}BCE\bar{F}$), or d (to form $\bar{A}BCD\bar{F}$). However, it may not be combined with e because the three-variable chart containing a is not adjacent to the three-variable chart containing e. (In other words, one cannot combine $\bar{A}BCDE\bar{F}$ with

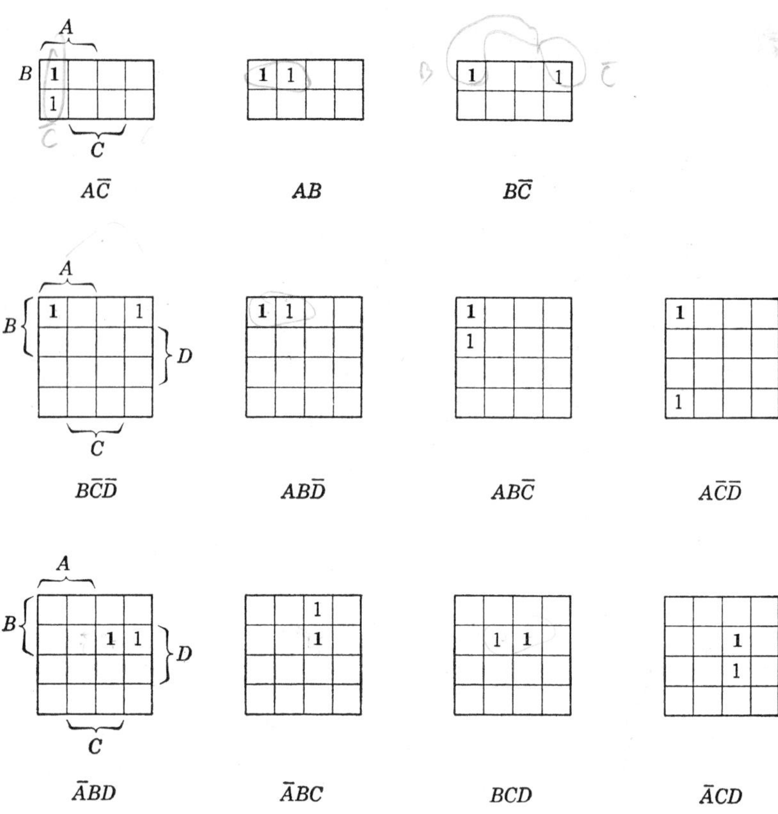

FIGURE 4-5

$\bar{A}BC\bar{D}\bar{E}F$.) Also, a may not be combined with f because a and f do not occupy identical positions on the three-variable charts. (It is of course still possible to combine a with g, h, and i.)

We have seen how it is possible to recognize that two minterms may be combined by looking for certain patterns on the Veitch diagram.

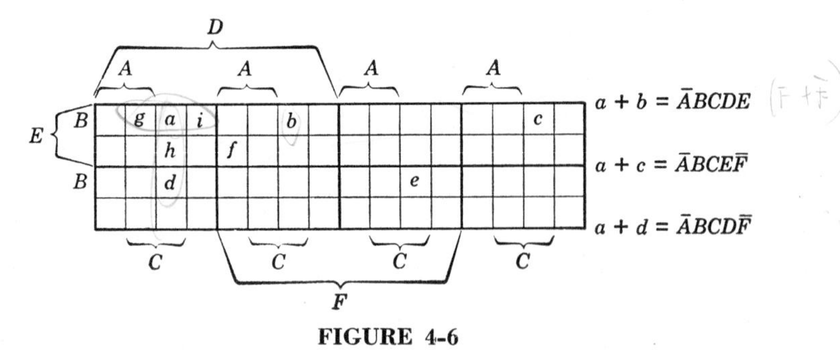

FIGURE 4-6

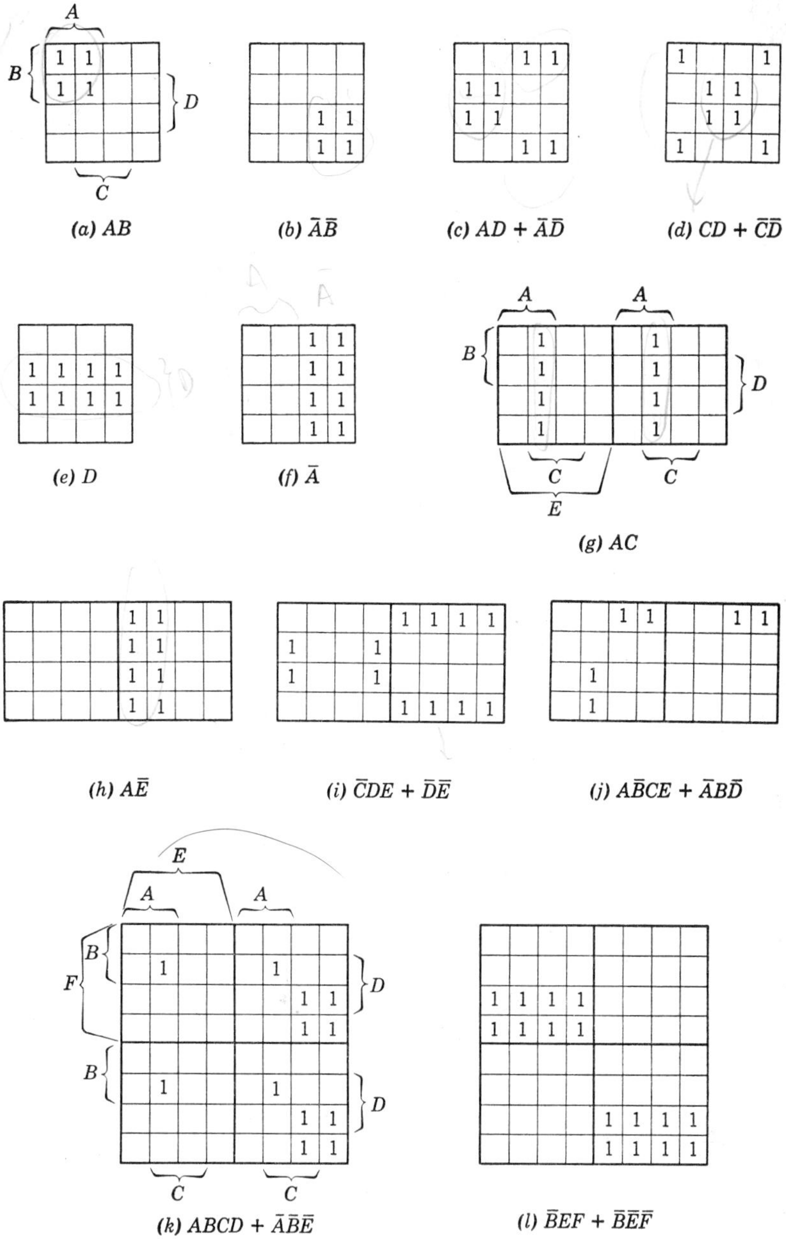

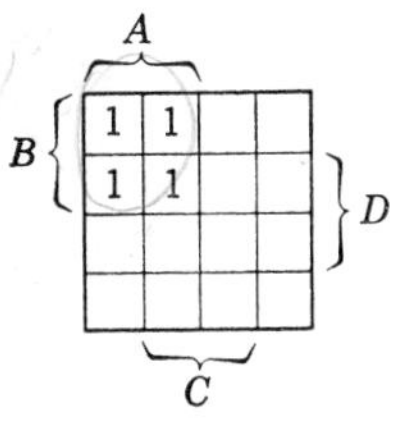

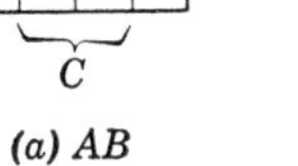

(a) AB

		1	1
		1	1

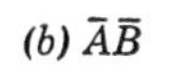

(b) $\bar{A}\bar{B}$

		1	1
1	1		
1	1		
		1	1

(c) $AD + \bar{A}\bar{D}$

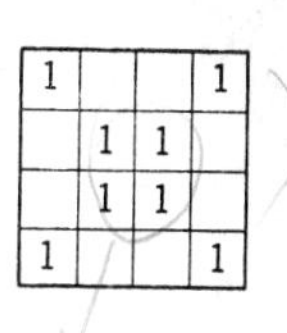

(d) $CD + \bar{C}\bar{D}$

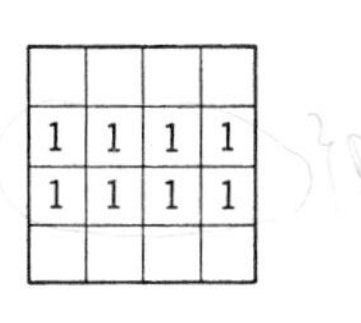

(e) D

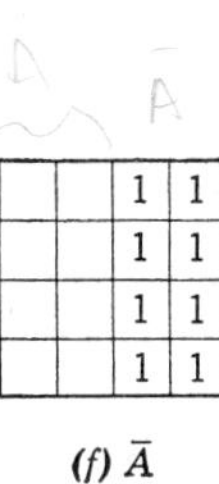

(f) $\bar{A}$

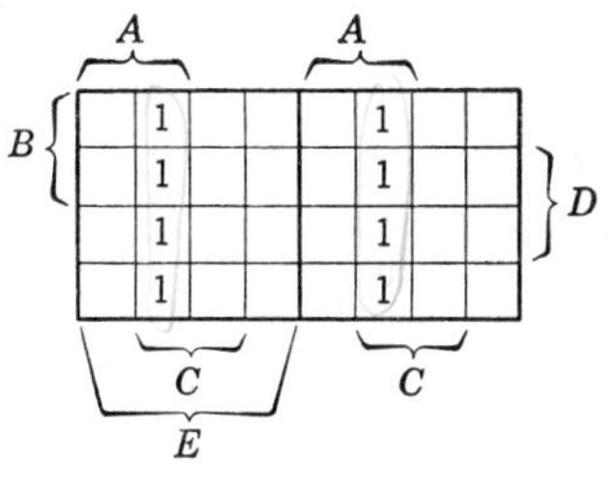

(g) AC

				1	1		
				1	1		
				1	1		
				1	1		

(h) $A\bar{E}$

				1	1	1	1
1			1				
1			1				
				1	1	1	1

(i) $\bar{C}DE + \bar{D}\bar{E}$

		1	1			1	1
	1						
	1						

(j) $A\bar{B}CE + \bar{A}B\bar{D}$

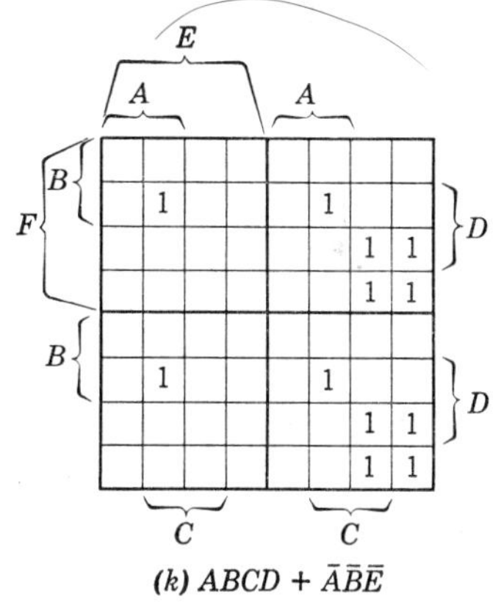

(k) $ABCD + \bar{A}\bar{B}\bar{E}$

1	1	1	1				
1	1	1	1				
				1	1	1	1
				1	1	1	1

(l) $\bar{B}EF + \bar{B}\bar{E}\bar{F}$

FIGURE 4-7

It is also quite easy to recognize patterns which permit the combination of four or eight or sixteen (or, in general, 2^i) minterms. A representative sample of such patterns is shown in Figure 4-7, and the reader is urged to study them carefully. Certain generalizations might be made about these patterns; e.g., squares, double squares, complete rows or columns always combine. However, the diagrams of Figure 4-7 and the examples which will follow should prove more effective than generalizations in teaching patterns.

It is now possible to state the rules for determining the prime implicants and essential terms of a function plotted on a Veitch diagram.

1. Choose a square containing a "one." Note *all* of the possible combinations of this square and any other squares containing "ones."
2. Examine these combinations. If all the "ones" in any combination are contained in some other combination, discard the former. The combinations which are not discarded in this step are the prime implicants. Furthermore, if all combinations but one are discarded, that prime implicant is an essential term (see step d of the Quine method). If the chosen square cannot be combined with any other squares, it is itself a prime implicant and essential term.
3. Repeat steps 1 and 2 for each square containing a "one," i.e., for each minterm.

Stated in this way, these rules may seem tedious and difficult to carry out. Certainly at first it will be necessary for the reader to note and perhaps to write down all combinations of each square and any others in order to determine which can be discarded and which are prime implicants. As he gains experience, however, he will not need to write down the intermediate combinations but will be able to carry out all rules in his head and find the prime implicants immediately.

Let us return now to the function plotted in Figure 4-4 and apply these rules to find the prime implicants. Starting in the upper left-hand corner (with minterm $AB\overline{C}\overline{D}$) we see immediately that it combines with the one to its right (forming $AB\overline{D}$), the one below it ($AB\overline{C}$), and the other three in the corner (AB). However, both the "ones" in $AB\overline{C}$ are also in AB, so that $AB\overline{C}$ is discarded. $AB\overline{D}$ is discarded for the same reason. Since the "ones" in AB are not in any other combination, AB is a prime implicant. And since it is the only prime implicant containing $AB\overline{C}\overline{D}$, it is an essential term.

Application of the same rules to squares $ABC\overline{D}$ and $AB\overline{C}D$ yields the same result: the single prime implicant AB. However, square $ABCD$ combines with the square above it (ABC), the square to its left (ABD), the three other corner squares (AB), and the square below it (ACD). ABC and ABD may be discarded because they are contained in AB, but the two "ones" in ACD are *not* both in AB, and so ACD and AB are both prime implicants.

Square $A\bar{B}CD$ combines in only two ways: with $ABCD$ to form ACD, and with $\bar{A}\bar{B}CD$ to form $\bar{B}CD$. Both these terms are therefore prime implicants. Square $\bar{A}\bar{B}CD$ combines only with $A\bar{B}CD$, and so $\bar{B}CD$ is a prime implicant and an essential term. Finally, square $\bar{A}\bar{B}\bar{C}\bar{D}$ may not be combined with any other square, and so the minterm $\bar{A}\bar{B}\bar{C}\bar{D}$ itself is also a prime implicant and an essential term.

We now have found four prime implicants (which agree with those found using the Quine method) and three essential terms. Since the Boolean sum of these essential terms happens to include all function minterms, the simplest expression for the function is $f = AB + \bar{B}CD + \bar{A}\bar{B}\bar{C}\bar{D}$.

Of course, it is not always true that the Boolean sum of the essential terms is the simplest expression. Some functions have no essential terms. For other functions, the simplest expression is a sum of essential terms and other prime implicants. The complete and general rules for the Veitch diagram method of simplification must take care of all these possibilities, and may be stated as follows:

a. Plot the function on a Veitch diagram.

b. Examine each "one" square for prime implicants, as described above. As they are found, prepare a Table A (see Quine method, step c) listing the prime implicants in rows and the minterms in columns. Whenever an essential term is found, put a small dot in the corner of each square of the Veitch diagram which is part of that essential term. In continuing the search for prime implicants, it is not necessary to look for the prime implicants associated with any square which already contains a dot. For example, in the problem just solved (Figure 4-4), examination of the upper left-hand square revealed an essential term AB. At that point, we could have put dots in the four upper left-hand squares, and then examined only squares $A\bar{B}CD$, $\bar{A}\bar{B}CD$, and $\bar{A}\bar{B}\bar{C}\bar{D}$. Obviously we would have arrived at the same result. Note that it is very definitely to our advantage to try to find squares *in the beginning* which will contribute essential terms, for every essential term reduces the number of squares which must subsequently be examined. Furthermore, it is to our advantage first to find essential terms containing only a few letters, for these terms occupy a large number of squares and therefore reduce the work involved considerably. In first examining a Veitch diagram plot, then, it is wise to look for a pattern in which many "ones" can be combined, and then try to find a square in that large pattern which will make the corresponding combination an essential term.

c. When Table A has been formed and the essential terms discovered and marked on the Veitch diagram, it still may be necessary to choose the smallest set of prime implicants which complete the description of the function. If the function contains a great many minterms, only a few of which are included in essential terms, it may be easiest to follow steps d, e, f, and g of the Quine method. Under most circumstances, however, a study of the Veitch diagram will disclose which prime implicants should be used to include in the function all those "one" squares which contain no dots, and will point out alternative expressions if there are any.

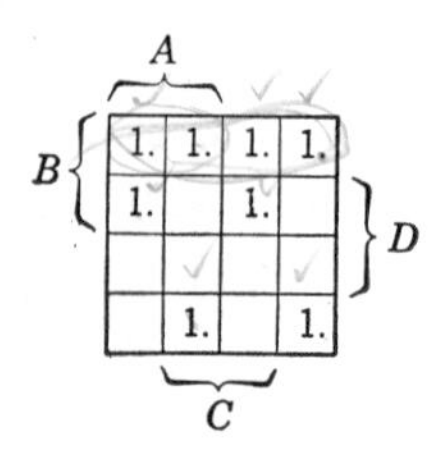

FIGURE 4-8

As usual, the procedure will be illustrated by a series of examples.

Example 4-10. Simplify $f = B\bar{D} + AB\bar{C} + \bar{A}BC + A\bar{B}C\bar{D} + \bar{A}\bar{B}\bar{C}\bar{D}$.

a. The function is plotted in Figure 4-8.

b. The diagram suggests that $B\bar{D}$ may be an essential term. However, careful examination of each of the four squares in $B\bar{D}$ reveals that each of them has two prime implicants, as follows:

minterm	$AB\bar{C}\bar{D}$	$ABC\bar{D}$	$\bar{A}BC\bar{D}$	$\bar{A}B\bar{C}\bar{D}$
prime implicants	$B\bar{D}$	$B\bar{D}$	$B\bar{D}$	$B\bar{D}$
	$AB\bar{C}$	$AC\bar{D}$	$\bar{A}BC$	$\bar{A}\bar{C}\bar{D}$

Since each of the four minterms included in $B\bar{D}$ has two prime implicants, we conclude that $B\bar{D}$ is *not* an essential term. Examination of the remaining four minterms discloses that each of them has only one prime implicant, and that these are therefore essential terms. Furthermore, their sum includes all the minterms in f, so that the simplest minterm-type expression for the function is:

$$f = AB\bar{C} + AC\bar{D} + \bar{A}BC + \bar{A}\bar{C}\bar{D}$$

Example 4-11. Simplify $f = A\bar{B} + B\bar{C} + \bar{B}C + \bar{A}B$ (cf. Examples 4-3 and 4-4).

a. The function is plotted in Figure 4-9.

b. No square can be combined with any three others. Furthermore, every square may be combined with one other in two ways. There are therefore no essential terms, but six prime implicants: $A\bar{C}$, $A\bar{B}$, $\bar{B}C$, $\bar{A}C$, $\bar{A}B$, and $B\bar{C}$. We see that if we choose any one of these, we need choose only two more to complete the expression for the function. Therefore $f = A\bar{C} + \bar{B}C + \bar{A}B$ or $f = A\bar{B} + \bar{A}C + B\bar{C}$.

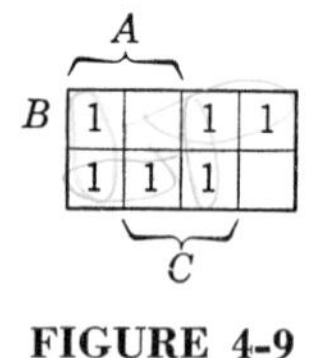

FIGURE 4-9

Example 4-12. Simplify the function of Example 4-6, p. 73.

a. The function is plotted in Figure 4-10.

b. We first note that pattern ABE looks promising as an essential term. Examining the squares one by one, we see that $AB\bar{C}\bar{D}E$ has two prime implicants, ABE and $A\bar{C}\bar{D}E$, and it therefore does not contribute an essential term. Square $ABC\bar{D}E$, however, has only the prime implicant ABE. ABE is therefore an essential term, and we make a note of it and put a dot in each of the four squares. Next, we may try $\bar{A}DE$ to see whether it is an essential term. (We see very quickly that BCE is not one, for two of its four squares share $A\bar{D}E$ as prime implicants, and the other two share ABE.) $\bar{A}DE$ is the sole prime implicant for square $\bar{A}\bar{B}\bar{C}DE$, and so $\bar{A}DE$ is essential and we place dots in its four squares. Only one minterm in the left-hand side of Figure 4-10 has not been accounted for, and we examine $A\bar{B}\bar{C}\bar{D}E$ for prime implicants. We find two ($A\bar{C}\bar{D}E$ and $A\bar{B}\bar{C}\bar{D}$) and since there is no essential term, we go on to the right-hand side. Two combinations immediately catch our eye as prime implicants which may be essential: $\bar{B}C\bar{E}$ and $\bar{B}\bar{D}\bar{E}$. However, close examination reveals that each of the six squares involved has two prime implicants. For example, the prime implicants for $A\bar{B}CD\bar{E}$ are $\bar{B}C\bar{E}$ and $ACD\bar{E}$, whereas those for $\bar{A}\bar{B}\bar{C}\bar{D}\bar{E}$ are $\bar{B}\bar{D}\bar{E}$ and $\bar{A}\bar{C}\bar{D}\bar{E}$. Therefore neither $\bar{B}C\bar{E}$ nor $\bar{B}\bar{D}\bar{E}$ is essential. In fact, careful study of the right-hand half of Figure 4-10 shows that there is only one more essential term, $\bar{A}\bar{C}\bar{D}\bar{E}$, which arises because square $\bar{A}B\bar{C}\bar{D}\bar{E}$ has only one prime implicant. Putting dots in those two squares completes step b.

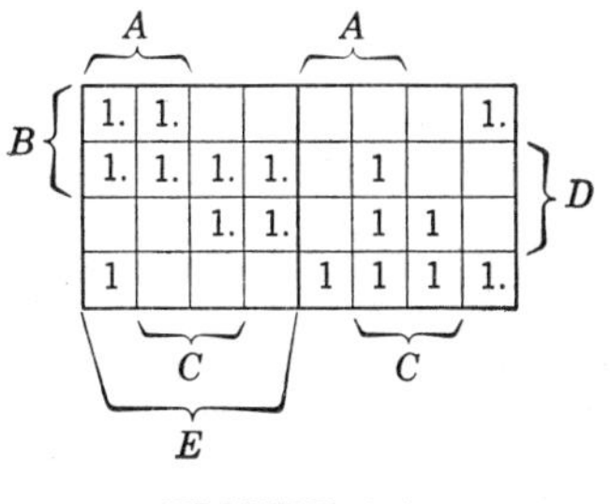

FIGURE 4-10

c. We must now choose a set of prime implicants which will include the remaining seven terms. If we begin with $\bar{B}\bar{D}\bar{E} + \bar{B}C\bar{E}$, we still must add two more prime implicants to include $ABCD\bar{E}$ and $A\bar{B}\bar{C}\bar{D}E$. The fact that $\bar{A}\bar{B}\bar{C}\bar{D}\bar{E}$ already contains a dot suggests that $\bar{B}\bar{D}\bar{E}$ might be omitted; and starting with $\bar{B}C\bar{E}$, we see that two of the three "ones" remaining may be included with $A\bar{B}\bar{C}\bar{D}$. The last minterm, $ABCD\bar{E}$, may be brought in either with $ABCD$ or with $ACD\bar{E}$, and so we are able to write either

$$f = ABE + \bar{A}DE + \bar{A}\bar{C}\bar{D}\bar{E} + \bar{B}C\bar{E} + A\bar{B}\bar{C}\bar{D} + ABCD$$

or

$$f = ABE + \bar{A}DE + \bar{A}\bar{C}\bar{D}\bar{E} + \bar{B}C\bar{E} + A\bar{B}\bar{C}\bar{D} + ACD\bar{E}$$

These, of course, agree with the answers to Example 4-6.

Before going on to some other aspects of simplification, let us briefly summarize the advantages and disadvantages of the four methods just described.

The *cut-and-try method* is useful only for the simplest of functions. Though it is easy to apply, it has two difficulties: there is in general no way of determining whether or when one has obtained the simplest expression for a function; and if several alternative simplest forms exist, it will be difficult to find them all.

The *Quine method* is straightforward but somewhat tedious to apply. However, once Table A has been drawn up, the procedure is fairly straightforward and automatically leads to all possible simplest solutions.

The *Harvard method* is as effective as the Quine method, but has the disadvantage that for a large number of variables the chart grows very large. Although the Harvard chart alone may be used to find the simplest expression, it is perhaps most effective when used in combination with the Quine method. That is, the Harvard chart may be used to find the prime implicants and set up Table A, whereupon the Quine method is used to find the simplest expression. This is an especially effective way to tackle the simplification problem when the function to be simplified is the Boolean sum of more than half of all possible minterms. In such a situation the process of combining terms, as required by the Quine method, is very tedious and grows more tedious as the number of minterms involved increases; whereas the process of eliminating terms, which is the essence of the Harvard method, becomes easier.

A general-purpose computer program for simplifying Boolean equations may be based on either the Harvard or the Quine method.

The *Veitch diagram method* has three very important advantages: the function to be simplified may be attacked directly, without being expanded into minterm form; the process of finding prime implicants is simplified by making them correspond to patterns familiar to the eye; and the ease with which essential terms may be identified makes it possible to reduce the labor involved in seeking prime implicants. For these reasons, and because (as we shall see) the Veitch diagram provides a very convenient way of visualizing various relationships useful in logical design, this method will be the primary one used in this book.

OTHER ASPECTS OF SIMPLIFICATION

Maxterm-Type Expressions

The whole of the last section was devoted to the problem of finding the simplest minterm-type expression for a given function, i.e., a second-order expression in the form of a sum of products. For every function there also exists a simplest maxterm-type expression: a second-order expression in the form of a product of sums. Since both expressions are of the second order, the choice between them should be made on the basis of the number of diodes necessary for each. For some functions (e.g., $A\bar{B}C$, $A + B + C + \bar{D}$) the simplest minterm-type and simplest maxterm-type expressions are identical. For most functions, however, one expression will be simpler than the other.

The rules for finding the simplest maxterm-type expressions of a given function f are as follows:

a. Find the *complement* of the function. This may be done by writing down all minterms which do not appear in the minterm expansion of the function itself. Or, if the Veitch diagram method is used, it may be done by plotting the function on a Veitch diagram, and then plotting on another diagram a function which has "ones" everywhere except where the first function had "ones."

b. Using any of the methods of the previous section, find the simplest *minterm-type* expression for $\bar{f}$.

c. Take the complement of the expression of step b. This new expression will represent the function, since $\overline{(\bar{f})} = f$; it will be a maxterm-type expression, since all "and" circuits have been replaced by "or" circuits and vice versa; and it will be the simplest maxterm-type expression, because if there were a simpler one its complement would be simpler than the minterm-type expression we got for $\bar{f}$ in step b, and this is impossible. Note that, since there may be several alternative simplest minterm-type expressions for $\bar{f}$, there will also be an equivalent number of simplest maxterm-type expressions for f.

Three examples will illustrate the method.

Example 4-13. Find the simplest maxterm-type expression for $f = AB + A\bar{C} + \bar{A}C + \bar{A}\bar{B}\bar{D}$ (cf. Example 4-2).

a. The function and its complement are plotted in Figure 4-11*a* and *b* respectively.

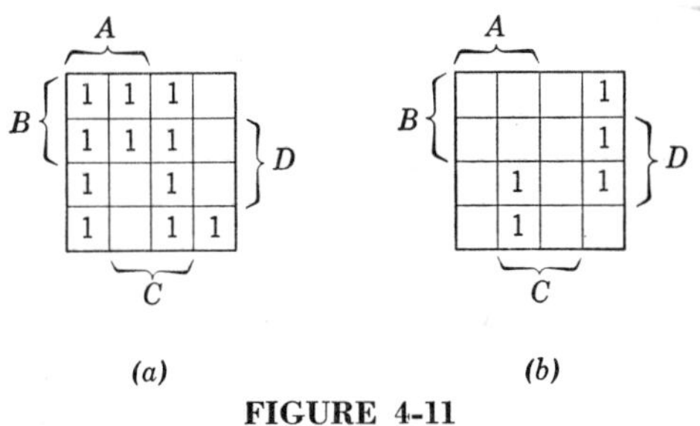

FIGURE 4-11

b. The simplest minterm expression for $\bar{f}$ is $\bar{f} = A\bar{B}C + \bar{A}B\bar{C} + \bar{A}\bar{C}D$.

c. Therefore $f = (\bar{A} + B + \bar{C})(A + \bar{B} + C)(A + C + \bar{D})$. This expression requires twelve diodes, compared with thirteen for the original one.

Example 4-14. Find the simplest maxterm-type expression for the function whose simplest minterm-type expression is $f = BD + AD + \bar{C}DE + CD\bar{E}$ (fourteen diodes).

a. The function f is plotted on the Veitch diagram of Figure 4-12*a*, and $\bar{f}$ in Figure 4-12*b*.

b. Obviously the simplest minterm-type expression for $\bar{f}$ is $\bar{f} = \bar{D} + \bar{A}\bar{B}CE + \bar{A}\bar{B}\bar{C}\bar{E}$.

c. Therefore $f = D(A + B + \bar{C} + \bar{E})(A + B + C + E)$. This expression requires only eleven diodes.

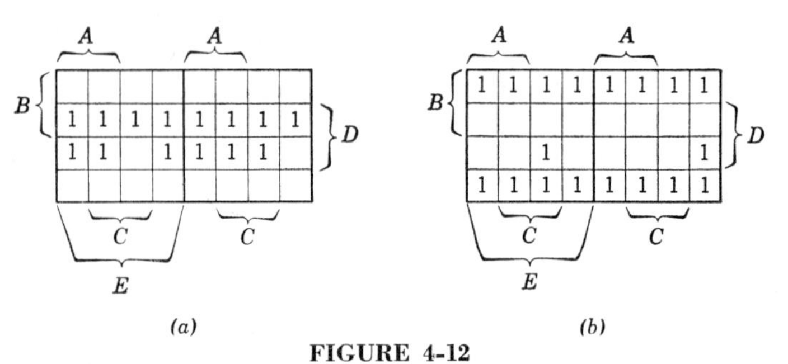

FIGURE 4-12

Example 4-15. Find the simplest maxterm expression for $f = ABE + \bar{A}DE + \bar{B}C\bar{E} + \bar{A}\bar{C}\bar{D}\bar{E} + A\bar{B}\bar{C}\bar{D} + ABCD$ (cf. Examples 4-6 and 4-12).

a. The function itself is plotted in Figure 4-10; its complement is plotted in Figure 4-13.

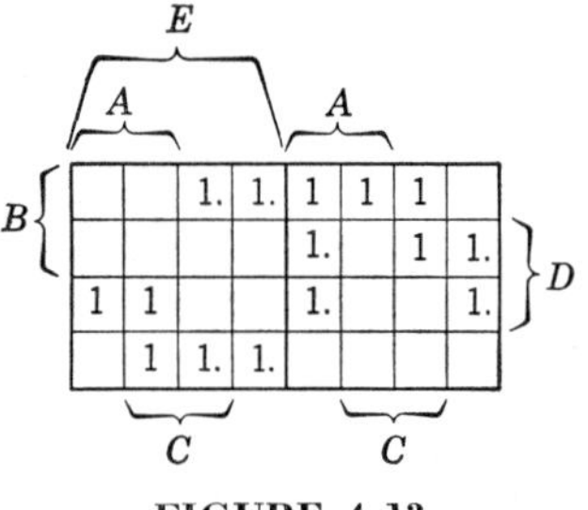

FIGURE 4-13

b. The essential terms for $\bar{f}$ are $\bar{A}\bar{D}E$ and $\bar{C}D\bar{E}$. Further examination of the diagram reveals three simplest minterm-type expressions:

$$\bar{f} = \bar{C}D\bar{E} + \bar{A}\bar{D}E + AB\bar{D}\bar{E} + \bar{A}BC\bar{E} + \begin{cases} A\bar{B}\bar{C}D + A\bar{B}CE \\ \text{or } A\bar{B}DE + A\bar{B}CE \\ \text{or } A\bar{B}DE + \bar{B}C\bar{D}E \end{cases}$$

c. There are therefore three simplest maxterm-type expressions for $\bar{f}$:

$$f = (C + \bar{D} + E)(A + D + \bar{E})(\bar{A} + \bar{B} + D + E)(A + \bar{B} + \bar{C} + E)$$

$$\cdot \begin{cases} (\bar{A} + B + C + \bar{D})(\bar{A} + B + \bar{C} + \bar{E}) \\ \text{or } (\bar{A} + B + \bar{D} + \bar{E})(\bar{A} + B + \bar{C} + \bar{E}) \\ \text{or } (\bar{A} + B + \bar{D} + \bar{E})(B + \bar{C} + D + \bar{E}) \end{cases}$$

Each of these three expressions requires twenty-eight diodes, however, compared with twenty-seven for the original, minterm-type expression.

Redundancies

It very often happens in the design of digital systems that certain combinations of the variables, i.e., certain minterms, are for some reason prohibited. These prohibited combinations will here be called *redundancies* (they have also been called irrelevancies, "don't cares," and forbidden combinations), and they can usually be used to simplify Boolean functions.

Before setting out the rules which make it possible to incorporate redundancies into simplification procedures, it will be well to explain in a little more detail how such redundancies arise. We have already seen that each of the Boolean variables—A, B, C, etc.—which we have been manipulating represents a wire able to assume only one of two voltages, called "zero" and "one." In general, any such wire may come from the output of a flip-flop (or some such one-bit memory device), an amplifier, a diode circuit which implements a Boolean equation, or some kind of mechanical device like a relay or switch. For definiteness, let us assume that each wire is the output line of a flip-flop. If we are using four such flip-flops, A, B, C, and D, for some purpose, the flip-flops may assume $2^4 = 16$ different configurations. Redundancies occur when one or more of those configurations are never allowed. For example, the four flip-flops may be used to represent the numbers zero to nine, as follows:

$$f_0 = \bar{A}\bar{B}\bar{C}\bar{D} \qquad f_5 = \bar{A}B\bar{C}D$$
$$f_1 = \bar{A}\bar{B}\bar{C}D \qquad f_6 = \bar{A}BC\bar{D}$$
$$f_2 = \bar{A}\bar{B}C\bar{D} \qquad f_7 = \bar{A}BCD$$
$$f_3 = \bar{A}\bar{B}CD \qquad f_8 = A\bar{B}\bar{C}\bar{D}$$
$$f_4 = \bar{A}B\bar{C}\bar{D} \qquad f_9 = A\bar{B}\bar{C}D$$

The ten functions f_0 through f_9 can then be used to tell which number is in the flip-flops at any time. If the logic governing the operation of the flip-flops is in some way designed so that the remaining six combinations

A	B	C	D
1	0	1	0
1	0	1	1
1	1	0	0
1	1	0	1
1	1	1	0
1	1	1	1

can never occur, these six combinations (corresponding, of course, to minterms m_{10} through m_{15}) are redundancies.* We shall now show how some of the Boolean equations which define f_0 through f_9 can be simplified because of the presence of these redundancies.

Rules may be devised for incorporating redundancies into any of the simplification schemes described in the last section. However, for brevity's sake only the Quine and Veitch diagram methods will be so treated here. Furthermore, only the Veitch diagram will be used in the remainder of this book.

The modified Quine rules may be stated as follows:

a. Express the function as a sum of minterms. Also, expand all redundancies into minterm form, if they are not already so expressed. Next, examine the two lists of minterms, and eliminate from the function-list any minterms which appear in the redundancy-list. The minterms *remaining* on the function-list will be called function-minterms.

b. Derive a set of "prime implicants" from the *combined* group of function-minterms and redundancy-minterms, exactly as was done in step b of the normal Quine method.

c. Prepare a table (Table A) having as many rows as there are prime implicants, and as many columns as there are function-minterms. Identify each row with a prime implicant, and each column with a function-minterm. Note that the redundant minterms do *not* appear as column headings. Place a check mark in a square on the table wherever the prime implicant to the left of the square was derived, in part, from the minterm above the square. This will be true if the letters in the prime implicant all appear in the same form as they do in the minterm. If any row contains no check marks, it may be crossed off the table. Furthermore, if some row (e.g., row i) contains only one check mark (in column j) and if there is at least one other check mark in column j in a row whose prime implicant contains fewer letters than the prime implicant of row i, row i may be crossed out.

The remainder of the rules (d, e, f, and g) are exactly the same as for the normal Quine method. The procedure will be illustrated by three examples.

* A convenient way of expressing redundancies is to write a Boolean equation setting the redundant combinations equal to zero. For example, we can write: $A\bar{B}C\bar{D} = A\bar{B}CD = AB\bar{C}\bar{D} = AB\bar{C}D = ABC\bar{D} = ABCD = 0$ for the redundancies above. More simply, we can set $AB = AC = 0$, which says exactly the same thing. Care must be taken, however, in interpreting an equation of this kind. Specifically, the equation $AB = 0$ means that a diode "and" circuit with two inputs A and B will always have an output of "zero," since the combination $A = B = 1$ can never occur. We can therefore add the term AB to any equation without affecting the function defined by that equation.

Example 4-16. Simplify $f_6 = \bar{A}BC\bar{D}$ if minterms m_{10} through m_{15} are not allowed.

a. Function and redundancies are already in minterm form. There is of course only one function-minterm: $\bar{A}BC\bar{D}$.

b. The prime implicants are determined as follows:

$\bar{A}BC\bar{D}$	✓	$BC\bar{D}$		AC
$A\bar{B}C\bar{D}$	✓	$A\bar{B}C$	✓	AB
$A\bar{B}CD$	✓	$AC\bar{D}$	✓	
$AB\bar{C}\bar{D}$	✓	ACD	✓	
$AB\bar{C}D$	✓	$AB\bar{C}$	✓	
$ABC\bar{D}$	✓	$AB\bar{D}$	✓	
$ABCD$	✓	ABD	✓	
		ABC	✓	

c. Table A may now be formed, as shown by Table 4-14. It contains only one column.

Table 4-14

	$\bar{A}BC\bar{D}$
$BC\bar{D}$	✓
AC	
AB	

Rows AC and AB may be crossed out.

d. Evidently $BC\bar{D}$ is an essential term and the only one necessary. Therefore, $f_6 = BC\bar{D}$. In other words, to recognize the number "six" in flip-flops A, B, C, and D, one need not examine flip-flop A at all, but only look for the combination $B = C = 1$, $D = 0$ in the three flip-flops B, C, and D. The reason for this should be obvious: if $BC\bar{D} = 1$, then A *must* be zero, because the combination $ABC\bar{D}$ is not permitted.

Example 4-17. Simplify $f = \bar{A}\bar{C} + AB\bar{D} + \bar{A}\bar{B}CD + ABCD$ if the combinations $ABC\bar{D}$, $A\bar{B}\bar{C}\bar{D}$, and $\bar{A}B\bar{C}D$ are redundant.

a. Expanding f, we find

$$f = \bar{A}\bar{B}\bar{C}\bar{D} + \bar{A}\bar{B}\bar{C}D + \bar{A}B\bar{C}\bar{D} + \bar{A}B\bar{C}D + ABC\bar{D} + AB\bar{C}\bar{D} + \bar{A}\bar{B}CD + ABCD$$

Comparing these with the redundancy-minterms, we find the function-minterms to be $\bar{A}\bar{B}\bar{C}\bar{D}$, $\bar{A}\bar{B}\bar{C}D$, $\bar{A}B\bar{C}\bar{D}$, $AB\bar{C}\bar{D}$, $\bar{A}\bar{B}CD$, and $ABCD$.

b. Finding the prime implicants:

$\bar{A}\bar{B}\bar{C}\bar{D}$	✓	$\bar{A}\bar{B}\bar{C}$	✓	$\bar{A}\bar{C}$
$\bar{A}\bar{B}\bar{C}D$	✓	$\bar{A}\bar{C}\bar{D}$	✓	$\bar{C}\bar{D}$
$\bar{A}B\bar{C}\bar{D}$	✓	$\bar{B}\bar{C}\bar{D}$	✓	
$AB\bar{C}\bar{D}$	✓	$\bar{A}\bar{B}D$		
$\bar{A}\bar{B}CD$	✓	$\bar{A}\bar{C}D$	✓	
$ABCD$	✓	$\bar{A}B\bar{C}$	✓	
$ABC\bar{D}$	✓	ABC		
$A\bar{B}\bar{C}\bar{D}$	✓	$AB\bar{D}$	✓	
$\bar{A}B\bar{C}D$	✓	$A\bar{C}\bar{D}$	✓	
		$B\bar{C}\bar{D}$	✓	

c. Table A may now be prepared; see Table 4-15.

Table 4-15

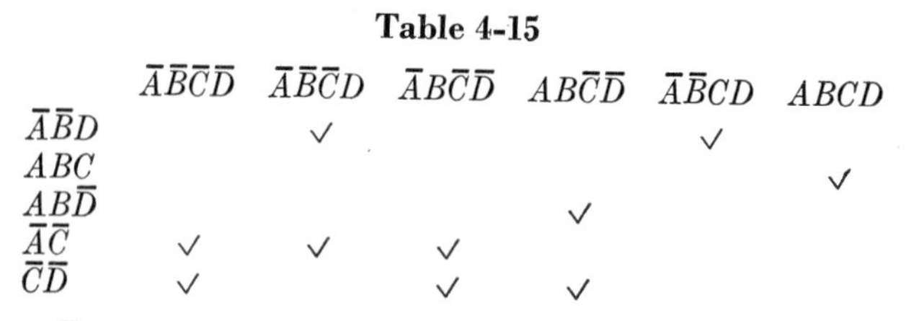

	$\bar{A}\bar{B}\bar{C}\bar{D}$	$\bar{A}\bar{B}\bar{C}D$	$\bar{A}B\bar{C}\bar{D}$	$AB\bar{C}\bar{D}$	$\bar{A}\bar{B}CD$	$ABCD$
$\bar{A}\bar{B}D$		✓			✓	
ABC						✓
$AB\bar{D}$				✓		
$\bar{A}\bar{C}$	✓	✓	✓			
$\bar{C}\bar{D}$	✓		✓	✓		

Row $AB\bar{D}$ may be eliminated because it contains only one check mark (in column $AB\bar{C}\bar{D}$), and the other check mark in that column is opposite prime implicant $\bar{C}\bar{D}$, which contains fewer letters than $AB\bar{D}$.

d. With row $AB\bar{D}$ eliminated, there are three columns—$AB\bar{C}\bar{D}$, $\bar{A}\bar{B}CD$, and $ABCD$—with only one check mark. The corresponding prime implicants are essential terms, and completely characterize the function. Therefore, $f = \bar{C}\bar{D} + ABC + \bar{A}\bar{B}D$.

Example 4-18. Given $f = Ax + By$, where the combination xy is redundant, find the simplest *maxterm-type* expression for this function.

a. To find the simplest maxterm-type expression for f, we first find the simplest minterm-type expression for $\bar{f}$. A convenient way of finding $\bar{f}$ as a sum of minterms is:

$$\begin{aligned}\bar{f} &= \overline{Ax + By} = (\bar{A} + \bar{x})(\bar{B} + \bar{y}) = \bar{A}\bar{B} + \bar{A}\bar{y} + \bar{B}\bar{x} + \bar{x}\bar{y} \\ &= \bar{A}\bar{B}xy + \bar{A}\bar{B}x\bar{y} + \bar{A}\bar{B}\bar{x}y + \bar{A}\bar{B}\bar{x}\bar{y} + \bar{A}Bx\bar{y} + \bar{A}B\bar{x}\bar{y} \\ &\quad + A\bar{B}\bar{x}y + A\bar{B}\bar{x}\bar{y} + AB\bar{x}\bar{y}\end{aligned}$$

Expanding the redundancy, we find $xy = \bar{A}\bar{B}xy + \bar{A}Bxy + A\bar{B}xy + ABxy$. (Note that we do *not* take the complement of the redundancy term. The flip-flop combination $x = 1, y = 1$ is not allowed, and so xy is the redundant combination regardless of how we decide to express f.) Comparing the redundancy-minterms with the expression for $\bar{f}$, we find the function-minterms are $\bar{A}\bar{B}x\bar{y}$, $\bar{A}\bar{B}\bar{x}y$, $\bar{A}\bar{B}\bar{x}\bar{y}$, $\bar{A}Bx\bar{y}$, $\bar{A}B\bar{x}\bar{y}$, $A\bar{B}\bar{x}y$, $A\bar{B}\bar{x}\bar{y}$, and $AB\bar{x}\bar{y}$.

b. Finding the prime implicants:

$\bar{A}\bar{B}x\bar{y}$	✓	$\bar{A}\bar{B}\bar{y}$	✓	$\bar{A}\bar{B}$
$\bar{A}\bar{B}\bar{x}y$	✓	$\bar{A}x\bar{y}$	✓	$\bar{A}\bar{y}$
$\bar{A}\bar{B}\bar{x}\bar{y}$	✓	$\bar{A}\bar{B}x$	✓	$\bar{A}x$
$ABx\bar{y}$	✓	$\bar{B}\bar{x}y$	✓	$\bar{B}\bar{x}$
$\bar{A}B\bar{x}\bar{y}$	✓	$\bar{A}\bar{B}y$	✓	$\bar{B}y$
$A\bar{B}\bar{x}y$	✓	$\bar{A}\bar{x}\bar{y}$	✓	$\bar{x}\bar{y}$
$A\bar{B}\bar{x}\bar{y}$	✓	$B\bar{x}\bar{y}$	✓	xy
$AB\bar{x}\bar{y}$	✓	$\bar{A}B\bar{y}$	✓	
$\bar{A}\bar{B}xy$	✓	$\bar{A}Bx$	✓	
$\bar{A}Bxy$	✓	$B\bar{x}\bar{y}$	✓	
$A\bar{B}xy$	✓	$A\bar{B}\bar{x}$	✓	
$ABxy$	✓	$A\bar{x}\bar{y}$	✓	
		$A\bar{B}y$	✓	
		$\bar{A}xy$	✓	
		$\bar{B}xy$	✓	
		Bxy	✓	
		Axy	✓	
		$\bar{A}\bar{B}x$	✓	

c. Table A may now be prepared; see Table 4-16.

Table 4-16

	$\bar{A}\bar{B}x\bar{y}$	$\bar{A}\bar{B}\bar{x}y$	$\bar{A}\bar{B}\bar{x}\bar{y}$	$\bar{A}Bx\bar{y}$	$\bar{A}B\bar{x}\bar{y}$	$A\bar{B}\bar{x}y$	$A\bar{B}\bar{x}\bar{y}$	$AB\bar{x}\bar{y}$
$\bar{A}\bar{B}$	✓	✓	✓					
$\bar{A}\bar{y}$	✓		✓	✓	✓			
$\bar{A}x$	✓			✓				
$\bar{B}\bar{x}$		✓	✓			✓	✓	
$\bar{B}y$		✓				✓		
$\bar{x}\bar{y}$ (circled)			✓ (circled)		✓ (circled)		✓ (circled)	✓ (circled)
xy								

d. Examining Table A, we find only one essential term: $\bar{x}\bar{y}$. We therefore form Table B, as shown by Table 4-17.

Table 4-17

	$\bar{A}\bar{B}x\bar{y}$	$\bar{A}\bar{B}\bar{x}y$	$\bar{A}Bx\bar{y}$	$A\bar{B}\bar{x}y$
$\bar{A}\bar{B}$	✓	✓		
$\bar{A}\bar{y}$	✓		✓	
$\bar{A}x$	✓		✓	
$\bar{B}\bar{x}$		✓		✓
$\bar{B}y$		✓		✓

e. Column $\bar{A}\bar{B}\bar{x}y$ has a check mark on every row that $A\bar{B}\bar{x}y$ has a check mark, and so the former may be eliminated. For the same reason, column $\bar{A}\bar{B}x\bar{y}$ may be eliminated after comparison with $\bar{A}Bx\bar{y}$.

f. Table C may now be formed by eliminating columns $\bar{A}\bar{B}\bar{x}y$ and $\bar{A}\bar{B}x\bar{y}$; and eliminating row $\bar{A}\bar{B}$, in which no entries remain; see Table 4-18.

Table 4-18

	$\bar{A}Bx\bar{y}$	$A\bar{B}\bar{x}y$
$\bar{A}\bar{y}$	✓	
$\bar{A}x$	✓	
$\bar{B}\bar{x}$		✓
$\bar{B}y$		✓

g. Examination of Table C shows there are four ways of choosing prime implicants so as to include the remaining two minterms. The resulting expressions for $\bar{f}$ are (remembering the essential term $\bar{x}\bar{y}$)

$$\begin{aligned} \bar{f} &= \bar{x}\bar{y} + \bar{A}\bar{y} + \bar{B}\bar{x} \\ &= \bar{x}\bar{y} + \bar{A}\bar{y} + \bar{B}y \\ &= \bar{x}\bar{y} + \bar{A}x + \bar{B}\bar{x} \\ &= \bar{x}\bar{y} + \bar{A}x + \bar{B}y \end{aligned}$$

and the corresponding expression for f may be indicated by

$$f = (x + y) \begin{Bmatrix} (A + y) \\ \text{or } (A + \bar{x}) \end{Bmatrix} \begin{Bmatrix} (B + x) \\ \text{or } (B + \bar{y}) \end{Bmatrix}$$

The modified Veitch diagram rules for incorporating redundancies into simplification may be stated as follows:

a. Plot the redundant terms on a Veitch diagram, placing crosses in the appropriate squares. Next, plot the function to be simplified on the same diagram, putting "ones" in the appropriate squares. Wherever a cross appears in the same square as a "one," omit the "one." The "one" squares remaining are the function-minterms.

b. Examine each "one" square for prime implicants, as was done in the original Veitch diagram method. However, in seeking the prime implicants which are associated with a given "one" square, any and all × squares may be used *as if they contained* "ones." This will ensure that the redundancy-minterms are used wherever they can contribute to simplification.

The succeeding steps are exactly parallel to the corresponding steps of the original Veitch diagram method. As the prime implicants are discovered, they may be set down in a Table A as has just been described for the Quine method, and the Quine redundancy rules may then be carried out. Alternatively, the essential terms may be identified by careful study of the diagram, dots may be put in squares associated with those terms, and then some combination of terms may be sought which will include all the other "one" squares as simply as possible. During this entire procedure, of course, any × square may be interpreted as a "one" wherever it will make possible simpler prime implicants.

Example 4-19. Simplify $f = \bar{A}B\bar{C} + \bar{A}\bar{C}D + \bar{B}\bar{D}$ where the combinations AD, AC, CD, and $\bar{A}\bar{B}\bar{D}$ are redundant.

a. In Figure 4-14*a* the redundancies are plotted, and in Figure 4-14*b* the function-minterms have been added.

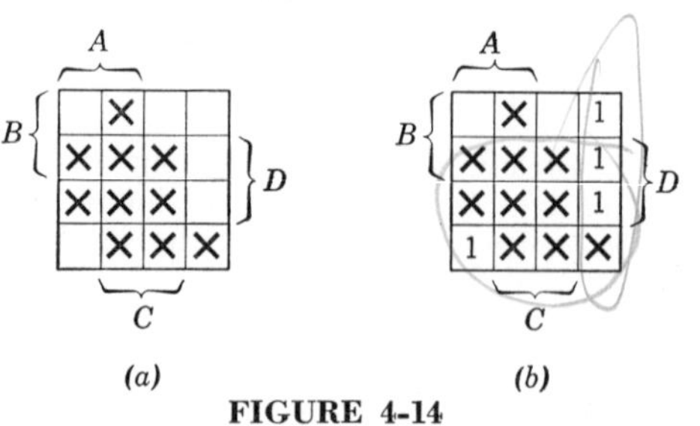

FIGURE 4-14

b. Examination of the four function-minterms reveals the following prime implicants:

Minterm	Prime Implicants
$\bar{A}B\bar{C}\bar{D}$	$\bar{A}\bar{C}$
$\bar{A}B\bar{C}D$	$\bar{A}\bar{C}$, D
$\bar{A}\bar{B}\bar{C}D$	$\bar{A}\bar{C}$, D, $\bar{B}$
$A\bar{B}\bar{C}\bar{D}$	$\bar{B}$

c. $\bar{A}B\bar{C}\bar{D}$ and $A\bar{B}\bar{C}\bar{D}$ each have only a single prime implicant, and therefore $\bar{A}\bar{C}$ and $\bar{B}$ are essential terms. Furthermore, these two essential terms include all function-minterms. Therefore $f = \bar{B} + \bar{A}\bar{C}$ is the simplest expression.

Example 4-20. Find the simplest expression for each of the binary-coded decimal digits contained in flip-flops A, B, C, and D, if minterms m_{10} through m_{15} are not allowed.

a. In Figure 4-15 three of the ten functions are plotted: f_1, f_6, and f_8. On each Veitch diagram the six redundant minterms are indicated by ×'s.

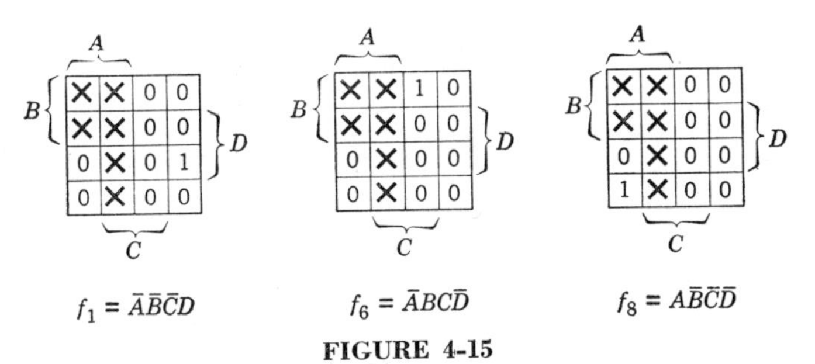

$f_1 = \bar{A}\bar{B}\bar{C}D$ $f_6 = \bar{A}BC\bar{D}$ $f_8 = A\bar{B}\bar{C}\bar{D}$

FIGURE 4-15

b. Consider first $f_1 = \bar{A}\bar{B}\bar{C}D$. Even though all the × squares are considered to be "ones," the $\bar{A}\bar{B}\bar{C}D$ square will not combine with any of them. Therefore $f_1 = \bar{A}\bar{B}\bar{C}D$ is the simplest form for f_1.

Examining f_6 next, we see that $BC\bar{D}$ is a prime implicant and an essential term. Therefore $f_6 = BC\bar{D}$.

We also find an essential term for f_8. The "one" square here may be combined with three × squares to give $f_8 = A\bar{D}$.

In a similar way, the reader may verify that the following expressions are the simplest:

$$
\begin{aligned}
f_0 &= \bar{A}\bar{B}\bar{C}\bar{D} & f_5 &= B\bar{C}D \\
f_1 &= \bar{A}\bar{B}\bar{C}D & f_6 &= BC\bar{D} \\
f_2 &= \bar{B}C\bar{D} & f_7 &= BCD \\
f_3 &= \bar{B}CD & f_8 &= A\bar{D} \\
f_4 &= B\bar{C}\bar{D} & f_9 &= AD
\end{aligned}
$$

Example 4-21. Given $f = Ax + By + Cz$, where the combinations xy, xz, and yz are redundant, find the simplest maxterm-type expression for this function.

a. We are here concerned with a function of six variables: A, B, C, x, y, and z. The function f and the redundancies are plotted in Figure 4-16 in two different but entirely equivalent Veitch diagrams. However, note that Figure 4-16*a* is composed of four four-variable diagrams, and Figure 4-16*b* contains eight three-variable diagrams. Since the given function has a certain three-variable symmetry, the second diagram exhibits a corresponding symmetry

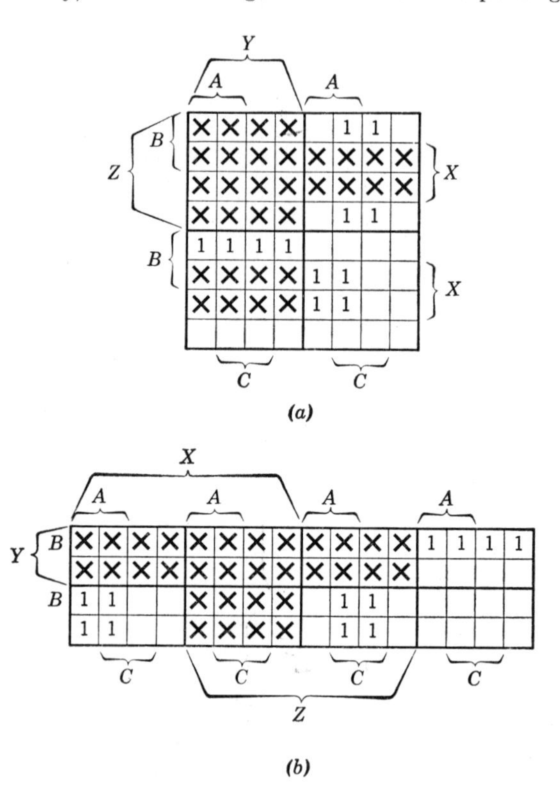

(a)

(b)

FIGURE 4-16

not present in the first. The second diagram will therefore be used in solving this problem.

We are of course interested in a maxterm-type expression for f, and so we must proceed by finding the simplest minterm-type expression for $\bar{f}$. $\bar{f}$ is therefore plotted in Figure 4-17 simply by inserting "ones" in the "zero" squares of Figure 4-16*b*.

b. We begin an immediate search for an essential term, hoping thereby to reduce the labor of seeking prime implicants. The eight "ones" in the lower right-hand corner of the diagram are represented by $\bar{x}\bar{y}\bar{z}$, and we examine the "ones" successively, hoping to find one with but a single prime implicant. Of these eight "ones," the bottom four also have the prime implicant $\bar{B}\bar{x}\bar{z}$; the right-hand four have $\bar{A}\bar{y}\bar{z}$; and four at the outside have $\bar{C}\bar{x}\bar{y}$. However, square $ABC\bar{x}\bar{y}\bar{z}$ has only one prime implicant, and so $\bar{x}\bar{y}\bar{z}$ is an essential term and we can place eight dots in those eight squares.

The other twelve "ones" must now be accounted for. The symmetry of the function leads us to suspect that each of the three groups of four is probably much like the other, so let us focus our attention on, say, the squares $\bar{A}x\bar{y}\bar{z}$.

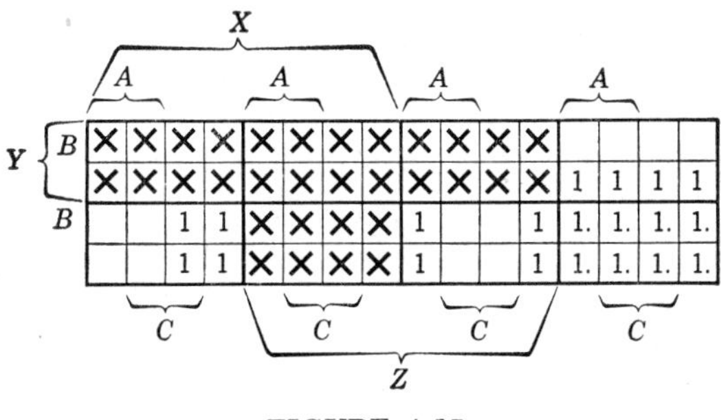

FIGURE 4-17

Drawing up a partial Table A, we write the prime implicants for each function minterm, as in Table 4-19.

Table 4-19

	$x\bar{y}\bar{z}$			
	$\bar{A}BC$	$\bar{A}B\bar{C}$	$\bar{A}\bar{B}C$	$\bar{A}\bar{B}\bar{C}$
$\bar{A}\bar{y}\bar{z}$	✓	✓	✓	✓
$\bar{A}x$	✓	✓	✓	✓
$\bar{A}\bar{C}\bar{y}$		✓		✓
$\bar{A}\bar{B}\bar{z}$			✓	✓
$\bar{A}\bar{B}\bar{C}$				✓

Careful examination of this table reveals that the term $\bar{A}x$ must appear in the simplest expression. Either $\bar{A}x$ or $\bar{A}\bar{y}\bar{z}$ must be present to include $\bar{A}BCx\bar{y}\bar{z}$. Of the twelve "ones" which are not included in the essential term, $\bar{A}x$ and $\bar{A}\bar{y}\bar{z}$ both contribute the same four. $\bar{A}x$ is of course preferred because it is the shorter of the two.

A similar argument applies to the remaining eight "one" squares, and requires that the terms $\bar{B}y$ and $\bar{C}z$ be used. Therefore

$$\bar{f} = \bar{x}\bar{y}\bar{z} + \bar{A}x + \bar{B}y + \bar{C}z$$

and
$$f = (x + y + z)(A + \bar{x})(B + \bar{y})(C + \bar{z})$$

Note that this formula is an extension of *one* of the results of Example 4-18, but that there are no other equally simple solutions here, as there were there.

Third and Higher Order Expressions

At the beginning of this chapter we restricted ourselves by stating that all simplification techniques would be directed toward finding first- and second-order expressions only. Let us now relax that restriction and see what diode savings are made possible by allowing higher order expressions. Consider, for example, the second-order expression

$$f = ABDFH + CDFH + EFH + GH + IJ$$

Constructed this way, f would require twenty-one diodes. If we begin by factoring H from the first four terms, we get the fourth-order expression

$$f = (ABDF + CDF + EF + G)H + IJ$$

requiring nineteen diodes. Next factoring F, we find a sixth-order expression

$$f = [(ABD + CD + E)F + G]H + IJ$$

requiring eighteen diodes. And finally factoring D, we get an eighth-order expression

$$f = \{[(AB + C)D + E]F + G\}H + IJ$$

again requiring eighteen diodes.

For another example, let us examine the function simplified in Example 4-13. We found the simplest minterm- and maxterm-type expressions to be

$$f = AB + A\bar{C} + \bar{A}C + \bar{A}\bar{B}\bar{D} \qquad \text{(thirteen diodes)}$$

$$= (\bar{A} + B + \bar{C})(A + \bar{B} + C)(A + C + D) \qquad \text{(twelve diodes)}$$

Applying P4b to the first expression and P4a to the second, we find

$$f = A(B + \bar{C}) + \bar{A}(C + \bar{B}\bar{D})$$

$$= (\bar{A} + B + \bar{C})(A + C + \bar{B}D)$$

The first of these is a fourth-order expression requiring twelve diodes, and the second a third-order expression requiring ten. It appears that the latter expression is the best possible.

These examples illustrate the very obvious techniques of factoring and multiplying out expressions already in simplest second-order form. As circuit techniques make high-order expressions practical, there will be more and more reason for improving on this method.

Simplified Expressions for Multiple Functions

Each of the digital systems which will be described in later chapters contains many Boolean functions. Whenever diode circuits must be provided for two or more functions, there exists the possibility that some diodes might be eliminated by using part or all of one function in deriving another.

For example, if the two functions

$$f_1 = BC + \bar{A}\bar{B}\bar{C}$$

$$f_2 = C + \bar{A}\bar{B}$$

are constructed from diode circuits in the obvious way, eleven diodes are necessary: seven for f_1 and four for f_2. However, if we notice that $f_2 = f_1 + C$, we can use the output from the f_1 circuit as part of the f_2 circuit, and eliminate two diodes (see Figure 4-18).

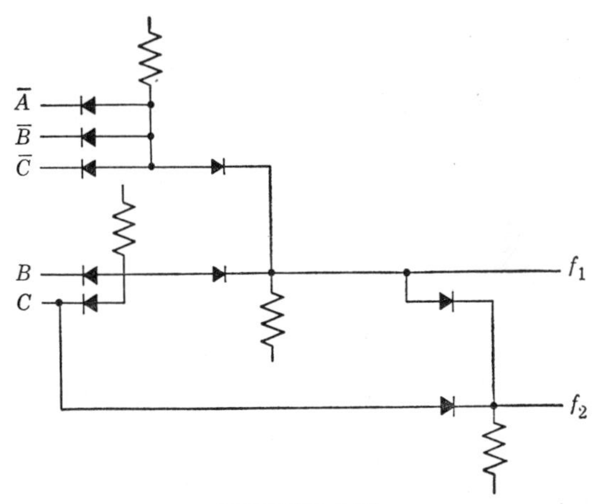

FIGURE 4-18

Aiken and Muller have discussed this problem and have suggested solutions. However, as Veitch himself suggested, the Veitch diagram provides a convenient visual method of combining functions, and its usefulness will be illustrated by an example.

Suppose that three functions are given:

$$f_1 = AB + B\bar{C} + \bar{B}CD + \bar{A}\bar{B}\bar{D}$$

$$f_2 = BD + ACD + \bar{A}\bar{C}\bar{D}$$

$$f_3 = AD + \bar{A}B\bar{C} + \bar{A}\bar{C}\bar{D}$$

(These three expressions are the simplest possible minterm-type expressions for the three functions, and require $14 + 11 + 11 = 36$ diodes altogether.) The three functions are plotted on three Veitch diagrams in Figure 4-19. Comparing the three diagrams, we notice that the three functions have five minterms in common, which may be expressed by the function $Y = ACD + \bar{A}B\bar{C} + \bar{A}\bar{C}\bar{D}$. If we assume that all the functions f_1, f_2, and f_3 are to be expressed as the Boolean sum of

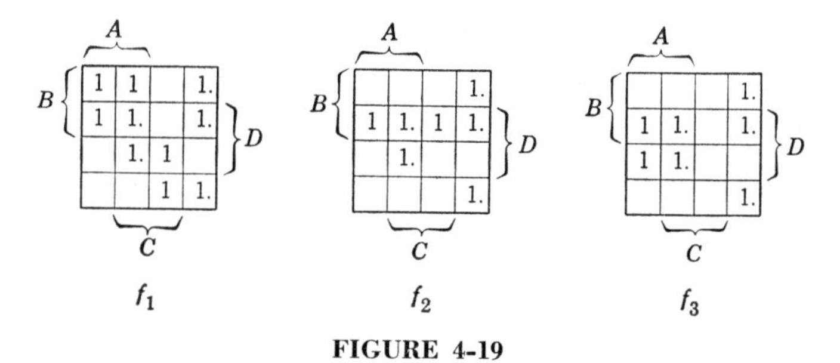

FIGURE 4-19

Y (whose five minterms are indicated by five dots in each Veitch diagram) and some other Boolean terms, we find

$$f_1 = Y + AB + \bar{A}\bar{B}C$$

$$f_2 = Y + BD$$

$$f_3 = Y + AD$$

Twelve diodes are required to form Y, and assuming Y is available, eight more are required for f_1, four more for f_2, and four more for f_3. Thus a total of $12 + 8 + 4 + 4 = 28$ diodes are necessary compared with the thirty-six originally required.

The same technique may of course be applied to maxterm-type functions. For example, it is possible to express f_1, f_2, and f_3 in the following way in terms of three maxterm-type expressions Y_1, Y_2, and Y_3.

$$Y_1 = B + C + \bar{D} \qquad \text{(three diodes)}$$

$$Y_2 = A + \bar{B} + \bar{C} \qquad \text{(three diodes)}$$

$$Y_3 = (\bar{A} + D)(\bar{C} + D)(A + B + \bar{D}) \qquad \text{(ten diodes)}$$

$$f_1 = (\bar{A} + B + D)Y_1Y_2 \qquad \text{(six diodes)}$$

$$f_2 = Y_2Y_3 \qquad \text{(two diodes)}$$

$$f_3 = Y_1Y_3 \qquad \text{(two diodes)}$$

Here a total of only twenty-six diodes are required to form f_1, f_2, and f_3.

BIBLIOGRAPHY

Staff of Harvard Computation Laboratory, *Synthesis of Electronic Computing and Control Circuits*, Harvard University Press, Cambridge, 1951.

E. W. Veitch, "A Chart Method for Simplifying Truth Functions," *Proc., Association for Computing Machinery Conference*, May 2–3, 1952, 127–133.

W. V. Quine, "The Problem of Simplifying Truth-Functions," *American Mathematical Monthly*, **59**, 521–531 (1952).

W. C. Carter and A. S. Rettig, "Analytic Minimization Methods I: Conjunctive Forms," *J. Computing Systems*, **1**, no. 3, 179–195 (1953).

M. Karnaugh, "The Map Method for Synthesis of Combinational Logic Circuits," *Communications and Electronics*, 593–599 (Nov. 1953).

D. E. Muller, "Application of Boolean Algebra to Switching Circuit Design and to Error Detection," *I.R.E. Trans. on Electronic Computers*, **EC-3**, no. 3, 6–11 (1954).

W. V. Quine, "A Way to Simplify Truth Functions," *American Mathematical Monthly*, **62**, 627–631 (1955).

R. J. Nelson, "Simplest Normal Truth Functions," *J. Symbolic Logic*, **20**, no. 2, 105–108 (1955).

R. J. Nelson, "Weak Simplest Normal Truth Functions," *J. Symbolic Logic*, **20**, no. 3, 232–234 (1955).

D. E. Muller, "Complexity in Electronic Switching Circuits," *I.R.E. Trans. on Electronic Computers*, **EC-5**, no. 1, 15–18 (1956).

E. J. McCluskey, Jr., "Minimization of Boolean Functions," *Bell System Technical J.*, **35**, 1417–1444 (1956).

M. J. Ghazala, "Irredundant Disjunctive and Conjunctive Forms of a Boolean Function," *I.B.M. J. of Research and Development*, **1**, no. 2, 171–176 (1957).

EXERCISES

1. State the order of each of the following expressions, and count the diodes required to mechanize each function as given.

a. $\bar{A} + B + \bar{C} + D$
b. $\bar{A} + B[C\bar{D} + E(A + \bar{C})]$
c. $(\bar{A} + BC + \bar{D})(A + \bar{C} + E)$
d. $AB + \{\bar{A}C[D + \bar{E}F + B(GH + I)] + A\bar{C}\}J$

2. Write the simplest maxterm-type and minterm-type expressions for each of the eight functions f_1 through f_8 given in Table 4-20. If more than one simplest expression exists, give them all. State how many diodes are required to mechanize each simplified function.

Table 4-20

A	B	C	D	f_1	f_2	f_3	f_4	f_5	f_6	f_7	f_8
0	0	0	0	0	1	1	0	1	1	0	0
0	0	0	1	0	0	1	0	1	0	1	0
0	0	1	0	1	1	0	1	1	0	0	0
0	0	1	1	1	0	0	1	0	1	1	0
0	1	0	0	1	1	1	0	1	0	0	0
0	1	0	1	1	1	0	1	1	1	0	0
0	1	1	0	1	0	1	0	1	1	0	0
0	1	1	1	1	0	0	1	0	0	0	0
1	0	0	0	0	1	0	0	1	0	0	0
1	0	0	1	0	1	1	1	1	1	0	0
1	0	1	0	0	0	0	0	0	1	0	0
1	0	1	1	0	0	1	1	0	0	0	1
1	1	0	0	1	1	1	0	1	1	0	0
1	1	0	1	0	0	0	0	1	0	0	1
1	1	1	0	1	1	1	1	1	0	0	0
1	1	1	1	0	0	1	1	0	1	0	1

Simplify each of the functions in Table 4-20 using the Quine method; the Harvard method; the Veitch diagram method.

3. Write the simplest maxterm-type and minterm-type expressions for each of the following functions. If more than one simplest expression exists, give them all. State how many diodes are required to mechanize each simplified function.

a. $AB + \bar{A}\bar{B}$
b. $\bar{B}\bar{C} + AB + BC + \bar{A}\bar{C}$
c. $(A + \bar{B} + C)(\bar{A} + B + \bar{C})(\bar{A} + \bar{B} + \bar{C})$
d. $(\bar{B} + \bar{C})(A + B + C)$
e. $AB\bar{C} + \bar{B}\bar{C} + \bar{A}C$
f. $A\bar{C}\bar{D} + \bar{A}\bar{B}\bar{C}\bar{D} + \bar{A}B\bar{C} + A\bar{B}\bar{C}D$
g. $B\bar{C}(A + \bar{D}) + ACD + \bar{A}\bar{B}CD + \bar{A}\bar{B}\bar{D}$
h. $ABCD + \bar{A}\bar{B} + A\bar{C}\bar{D} + B\bar{C}\bar{D}$
i. $\bar{C}\bar{D}\bar{E} + CDE + \bar{A}\bar{B}D + ABCD + BCE + A\bar{B}\bar{C}E$
j. $B\bar{C}\bar{D} + \bar{A}\bar{B}CD + B\bar{C}\bar{E} + \bar{C}\bar{D}\bar{E} + \bar{B}CD\bar{E} + \bar{B}\bar{C}\bar{D}E + \bar{A}\bar{B}C\bar{D}E$
k. $(\bar{A} + C + \bar{D})(A + \bar{B} + \bar{C} + D)(\bar{B} + \bar{C} + \bar{D} + E)(\bar{A} + B + \bar{C} + \bar{D})$
$(A + C + \bar{D} + \bar{E})(A + \bar{B} + \bar{C} + \bar{E})$
l. $\bar{A}C\bar{E} + \bar{A}\bar{C}D + AC\bar{D}\bar{E} + \bar{A}CDE$
m. $BEF + \bar{C}EF + ABDF + \bar{C}\bar{D}\bar{E}F + \bar{B}\bar{C}DF + \bar{B}CD\bar{E}F + ACD\bar{E}$
$+ \bar{C}D\bar{E}\bar{F}$
n. $(\bar{A} + \bar{B} + C + D + \bar{F})(\bar{A} + \bar{B} + C + E + \bar{F})(\bar{A} + \bar{B} + \bar{D} + \bar{E} + \bar{F})$
$(B + \bar{C} + \bar{D} + \bar{E} + \bar{F})(A + C + E + F)(A + B + \bar{C} + D + \bar{E})$
$(\bar{A} + \bar{C} + \bar{D} + \bar{E} + F)(A + B + \bar{C} + \bar{E} + F)(A + B + C + D + F)$

Solve each of the above problems using the Quine method; the Harvard method; the Veitch diagram method.

4. Use Nelson's method to find the prime implicants for the functions of parts *a* through *h* of Exercise 3.

5. Write the simplest maxterm-type and minterm-type expressions for each of the following functions, each having redundancies as specified. If more than one simplest expression exists, give them all. State how many diodes are required to mechanize each simplified function. Use the Quine and Veitch diagram methods for each problem.

a. The eight functions of Exercise 2, with the redundancies $AB = AC = 0$.
b. The four functions of Exercise 3*b*, 3*c*, 3*d*, and 3*e*, each with $AC = 0$.
c. The three functions of Exercise 3*f*, 3*g*, and 3*h*, each with $\bar{A}\bar{B}C = AB\bar{C}D = 0$.
d. The four functions of Exercise 3*i*, 3*j*, 3*k*, and 3*l*, each with $DE = 0$.
e. The two functions of Exercise 3*m* and 3*n*, each with $\bar{B}\bar{D}EF = 0$.
f. $A\bar{B}\bar{C}E + CDE + \bar{A}\bar{C}DE + \bar{A}CE + ABE$, with $BC = BD = 0$.
g. $A\bar{D}E + CDE + \bar{B}\bar{D}\bar{E} + B\bar{C}\bar{D}\bar{E} + \bar{A}\bar{C}\bar{D}E + \bar{A}CD\bar{E}$, with $\bar{A}\bar{C}\bar{E} = 0$.

6. Suppose all restrictions on circuit "order" are removed, so that third and higher order expressions can be mechanized. Under these circumstances, find the simplest expression for each of the following functions:

a. $ABD + CE + CD + ABE$
b. $\bar{A}\bar{C}(\bar{D} + \bar{E}) + \bar{B}(C + \bar{D})(A + \bar{D} + \bar{E})$
c. $AC\bar{D} + AE + \bar{B}C\bar{D} + \bar{B}E + A\bar{C}D$

7. Find the simplest way of expressing each of the following groups of multiple functions:

a. $f_1 = ABD + A\bar{B}C + \bar{C}\bar{D}$
$f_2 = A(B + C)(C + D)$
b. $f_1 = \bar{A}\bar{B} + AB\bar{C}\bar{D}$
$f_2 = A\bar{C}\bar{D} + \bar{A}\bar{B}\bar{C} + \bar{A}\bar{B}D$
$f_3 = AB\bar{D} + C\bar{D} + \bar{A}\bar{B}D$
c. $f_1 = (C + D)(A + \bar{B} + \bar{C} + \bar{D})$
$f_2 = (A + C + D)(\bar{B} + \bar{C} + \bar{D})(\bar{A} + \bar{C} + \bar{D})(\bar{A} + \bar{B} + D)$
$f_3 = (B + D)(A + \bar{B} + \bar{C} + \bar{D})(A + C + D)$

5 / *Memory element input equations*

INTRODUCTION

In Chapter 2 the usefulness and importance of binary memory elements were discussed, and it was stated that a digital computing system of any kind usually consists of a number of such elements connected together by decision circuit elements. In the past two chapters we have become acquainted with decision circuit elements, and have learned how to manipulate them. In this chapter we shall learn how to connect the decision circuits to memory circuits in order to carry out specified operations.

Each memory element must have at least one input line, which determines what state it will assume next, and at least one output line so that the present state of the circuit may be discovered. The voltages on these lines, of course, represent "zeros" and "ones," and the voltages obviously vary with time as the memory elements change state. In Chapter 2 we discussed this situation briefly (see Figures 2-2 and 2-3), and pointed out a difficulty: since only the high "one" voltage and the low "zero" voltage have any meaning in the algebra we have been using, there must be some way of insuring that a voltage level is given a meaning only when it has had time to reach a final value, high or low. Chapter 2 also proposed a solution to this problem in the form of a separate wire upon which appears a steady stream of signals called clock pulses. The memory and decision circuit elements are so designed that all changes take place *between* clock pulses, with the result that only the high and low voltage levels can ever have any effect.

In order to clarify these general remarks, let us discuss a particular example. In Figure 5-1 the output lines of two memory elements A

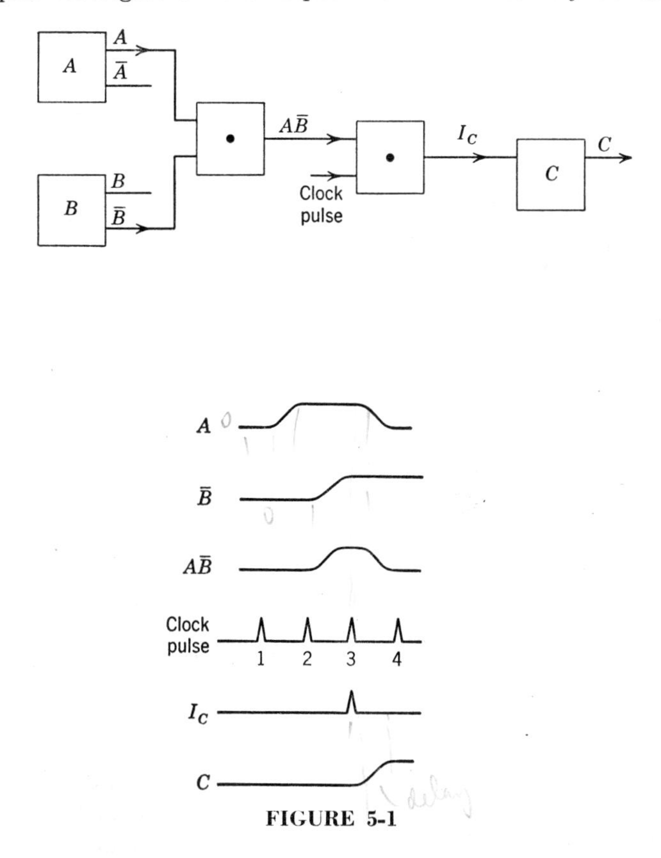

FIGURE 5-1

and B are connected together through an "and" circuit to the input line of memory element C. Line A is high only when element A is in the "one" state; line $\bar{B}$ is high only when element B is in the "zero" state; therefore the line labeled $A\bar{B}$ is high only when A is "one" *and* B is "zero." The $A\bar{B}$ line is next combined with the clock-pulse line through another "and" circuit, so that the input to memory element C (line I_C) is "one" only when $A\bar{B}$ is "one" and a clock pulse appears. The equation $I_C = A\bar{B}\cdot$(clock pulse) may thus be written to define the operation of flip-flop C in this circuit. We shall simplify this equation (and others like it throughout this book) by writing $I_C = A\bar{B}$, with the understanding that a clock-pulse "and" circuit is included somewhere.

Let us now assume that memory element C is designed so that it changes state from "zero" to "one" when I_C is "one." The waveforms of Figure 5-1 then show the sequence of events for four consecutive clock-pulse times, or *bit-times.* Supposing that memory elements A, B, and C are initially in the "zero," "one," and "zero" states respectively, lines A, $\bar{B}$, and C will all be low, line $A\bar{B}$ will be low, and line I_C will be low even when the clock pulse occurs. Memory element C is therefore unaffected. In the interval between clock pulses 1 and 2, memory element A is assumed to change state, but since line $\bar{B}$ is still low, clock pulse 2 has no more effect on element C than did pulse 1. Next, in the interval between pulses 2 and 3, element B is assumed to change state from "one" to "zero," so that line $\bar{B}$ rises. The voltage on line $A\bar{B}$ must therefore rise as well, so that clock pulse 3 appears on line I_C. We have already assumed that a "one" signal on I_C causes element C to change from "zero" to "one," and Figure 5-1 shows C changing in the interval between pulses 3 and 4.

It is particularly important to note the one bit-time delay between the pulse on I_C and the change of element C from "zero" to "one": the I_C pulse occurs at bit-time 3, but not until bit-time 4 can a "one" be observed on line C. In connection with this, note that during the bit-time a memory element receives a signal which will change its state, its output line may be sampled; the output voltage will not begin to change until the sampling time is over. This is illustrated by the waveform on line A at bit-time 3. At that bit-time element A receives an input pulse which will cause it to change state by bit-time 4; nevertheless, at the same bit-time, the fact that A was "one" permitted an input pulse to reach element C.

All of the logic described in this book will be based on circuit elements which have the feature described in the last paragraph. However, the reader should be aware of the fact that computers have been constructed out of components which do not have this characteristic. A whole class of machines, called Princeton machines because of their common background in computer development at the Institute for Advanced Study, contain decision and memory circuits having no "clock pulses" at all. The design of such machines is a problem which will not be discussed here.

The foregoing remarks explain how the element of time will be introduced here into the memory and decision circuit elements. Notice that the detailed circuit design does not concern us. We are interested in the logical properties of the elements, and do not care whether the circuits contain tubes, diodes, transistors, relays, magnetic cores, or what. All of the work of this chapter, and therefore of the remainder of the book, has that characteristic: it may be applied to any set of circuit elements whose logical properties are clearly defined. However, in order to be specific, most of this book will refer to flip-flops and diode circuits as the memory and decision elements respectively.

The important characteristics of memory elements may now be stated in a few sentences:

1. The state of a memory device is defined by the state (i.e., the voltage level) of its output line or lines.
2. The output lines from a number of memory elements may be combined together in logical circuits to form Boolean functions.
3. The state of any memory element or any combination of memory elements is defined only at a set of discrete time intervals known as clock-pulse times.
4. The state of a particular memory element at a particular bit time is a function of (a) the state of the memory element, and of its input line or lines, during the previous bit-time, and (b) the logical properties of the memory element, which express how the state of the memory element is related to its input lines. (These logical properties are built into the element by the circuit designer.)

We shall now proceed to show how memory elements having given logical properties may be connected together to carry out any given operation.

MEMORY ELEMENT INPUTS: THE PROBLEM

In this section the problem will be defined in a little more detail, first for a particular example and then in general. The example to be considered is the design of a certain counter, using a familiar kind of two-input flip-flop.

Flip-Flop Logical Properties

Several flip-flops of a type which will here be called R-S will be used in the design of the counter, and it is therefore first necessary to examine their logical characteristics. The flip-flop has two inputs. Input S, the *set* line, puts the flip-flop in the "one" state, and input R, the *reset* line, puts it into the "zero" state. (The words "set" and "reset" will be used throughout the book for "turn on" and "turn off" respectively.) We can express this in a more precise way by means of the following table.

Bit-Time n		Bit-Time $(n + 1)$
R^n	S^n	Q^{n+1}
0	0	Q^n
0	1	1
1	0	0
1	1	?

Because of the delay between the occurrence of an input signal and the effect on the output, the table is divided into two parts. On the left are the four possible states of the R and S input lines at time n. On the right are indicated the resulting states of the flip-flop (flip-flop Q) at the next bit-time, $(n + 1)$. If both R and S are "zero" at time n, the flip-flop will do nothing and its state at time $(n + 1)$ will be the same as its state at time n. That is, $Q^{n+1} = Q^n$. If R is "zero" and S "one," the flip-flop will be set to "one" at time $(n + 1)$ regardless of its state at time n. Similarly, a "one" on R and a "zero" on S resets the flip-flop to "zero" regardless of its previous state. Finally, if both R and S are "one" at time n, the circuit designer indicates (by the question mark) that the action of the flip-flop is indeterminate. In the example we are discussing, we will therefore make sure that R and S can never be "one" at the same time.

If we impose the restriction that either R or S must always be "zero," we can express the information given in the table by two equations:

$$Q^{n+1} = \bar{R}^n\bar{S}^nQ^n + \bar{R}^nS^n \tag{5-1}$$

$$R^nS^n = 0 \tag{5-2}$$

Changing our notation a bit to avoid the necessity for so many superscripts, we may write

$$Q^{n+1} = (\bar{R}\bar{S}Q + \bar{R}S)^n$$

Furthermore, equations 5-1 and 5-2 may be combined and simplified as follows:

$$Q^{n+1} = (\bar{R}\bar{S}Q + \bar{R}S + RS)^n$$

$$= (S + \bar{R}\bar{S}Q)^n$$

$$Q^{n+1} = (S + \bar{R}Q)^n \tag{5-3}$$

Equations 5-2 and 5-3 completely specify the logical characteristics of the R-S flip-flop. Equation 5-3, which expresses the value of Q at time $(n + 1)$ as a function (among other things) of its value at time n, will be called a *difference equation*. Furthermore, because this difference equation defines the operating characteristics of a particular kind of memory element, it will be called the *characteristic equation* of that memory element.

Later in the chapter we shall analyze other memory elements, following the same procedure used above for the R-S flip-flop. In every instance we will arrive at a difference equation expressing the output of the memory element at time $(n + 1)$ as a function of the state of the element and its input lines at time n. This general relationship may be expressed mathematically for a memory element Q having input lines $I_1, I_2, \cdots, I_m$, as follows:

$$Q^{n+1} = f_1(Q^n, I_1{}^n, I_2{}^n, \cdots, I_m{}^n) \tag{5-4}$$

where f_1 represents a Boolean function of the terms in parenthesis. There may also be some restriction on the allowable combination of input lines, expressed by

$$f_2(Q^n, I_1{}^n, I_2{}^n, \cdots, I_m{}^n) = 0 \tag{5-5}$$

where f_2 is another Boolean function. Equations 5-4 and 5-5 are thus the general characteristic equations defining the logical response of a memory element to signals on its input lines.

Application Equations

Turning now to the example which will illustrate the design problem, suppose the R-S flip-flops described above are to be used in a four-bit counter which counts in the following sequence: 0, 10, 9, 8, 4, 14, 15, 1, 7, 13, 5, 12, 6, 2, 3, 11, 0, etc. Four R-S flip-flops, labeled A, B, C, and D, are to be employed in the counter, and each bit-time the number in this four-bit register is to change to the next number on the above list. When the counter contains the digit eleven, the register should, at the next bit-time, return to zero and start all over again. (This count sequence has no particular merit over other sequences. It was chosen for illustrative purposes.)

Let us now see how we can express the action of this counter by a set of Boolean equations. First, let us copy down the count sequence in a way which will point up the bit-time by bit-time operation of the counter. We do this by noting that if flip-flops A, B, C, and D are all in the "zero" state at bit-time n, they must be in the 1, 0, 1, 0 (ten) state at time $(n + 1)$. Similarly, if they are in the 1, 0, 1, 0 state at time n, they must change to the 1, 0, 0, 1 (nine) state by time $(n + 1)$. These changes, together with all the other changes which define the operation of the counter, are shown in Table 5-1. Note that the last entry in this new table provides for the counter's return to zero after eleven: if it is in the 1, 0, 1, 1 state at time n, it must change to the 0, 0, 0, 0 state by time $(n + 1)$.

Table 5-1

Time n				Time $(n+1)$			
A	B	C	D	A	B	C	D
0	0	0	0	1	0	1	0
1	0	1	0	1	0	0	1
1	0	0	1	1	0	0	0
1	0	0	0	0	1	0	0
0	1	0	0	1	1	1	0
1	1	1	0	1	1	1	1
1	1	1	1	0	0	0	1
0	0	0	1	0	1	1	1
0	1	1	1	1	1	0	1
1	1	0	1	0	1	0	1
0	1	0	1	1	1	0	0
1	1	0	0	0	1	1	0
0	1	1	0	0	0	1	0
0	0	1	0	0	0	1	1
0	0	1	1	1	0	1	1
1	0	1	1	0	0	0	0

We are now in a position to write equations defining counter operation. The first column on the right-hand half of the table shows the state of flip-flop A at bit-time $(n + 1)$. Let us express this state as a function of the states of A, B, C, and D at time n—in terms of the left-hand half of the table. Using the basic theorem, we arrive at the difference equation:

$$A^{n+1} = (\bar{A}\bar{B}\bar{C}\bar{D} + A\bar{B}C\bar{D} + A\bar{B}\bar{C}D + \bar{A}B\bar{C}\bar{D} + ABC\bar{D} + \bar{A}BCD + \bar{A}B\bar{C}D + \bar{A}\bar{B}CD)^n \tag{5-6}$$

Each of the other three columns to the right of Table 5-1 may be treated in the same way, with the following results:

$$B^{n+1} = (A\bar{B}\bar{C}\bar{D} + \bar{A}B\bar{C}\bar{D} + ABC\bar{D} + \bar{A}\bar{B}\bar{C}D + \bar{A}BCD + AB\bar{C}D + \bar{A}B\bar{C}D + AB\bar{C}\bar{D})^n \tag{5-7}$$

$$C^{n+1} = (\bar{A}\bar{B}\bar{C}\bar{D} + \bar{A}B\bar{C}\bar{D} + ABC\bar{D} + \bar{A}\bar{B}\bar{C}D + AB\bar{C}\bar{D} + \bar{A}BC\bar{D} + \bar{A}\bar{B}C\bar{D} + \bar{A}\bar{B}CD)^n \tag{5-8}$$

$$D^{n+1} = (A\bar{B}C\bar{D} + ABC\bar{D} + ABCD + \bar{A}\bar{B}\bar{C}D + \bar{A}BCD + AB\bar{C}D + \bar{A}\bar{B}C\bar{D} + \bar{A}\bar{B}CD)^n \tag{5-9}$$

For reasons which will become clear later on, it is convenient to rearrange each of these expressions by factoring from the right-hand side the same letter which appears on the left-hand side, so that the equations explicitly show the state of the flip-flop at time $(n + 1)$ as a function of its state at time n, as well as of the states of other flip-flops at time n. Difference equations 5-6 through 5-9 then become

$$\begin{aligned} A^{n+1} &= [A(\bar{B}C\bar{D} + \bar{B}\bar{C}D + BC\bar{D}) + \bar{A}(\bar{B}\bar{C}\bar{D} + B\bar{C}D + BCD \\ &\quad + B\bar{C}\bar{D} + \bar{B}CD)]^n \\ &= [A(C\bar{D} + \bar{B}\bar{C}D) + \bar{A}(\bar{C}\bar{D} + CD + BD)]^n \end{aligned} \tag{5-10}$$

$$\begin{aligned} B^{n+1} &= [B(\bar{A}\bar{C}\bar{D} + AC\bar{D} + \bar{A}CD + A\bar{C}D + \bar{A}\bar{C}D + A\bar{C}\bar{D}) \\ &\quad + \bar{B}(A\bar{C}\bar{D} + \bar{A}\bar{C}D)]^n \\ &= [B(\bar{C} + A\bar{D} + \bar{A}D) + \bar{B}(A\bar{C}\bar{D} + \bar{A}\bar{C}D)]^n \end{aligned} \tag{5-11}$$

$$\begin{aligned} C^{n+1} &= [C(AB\bar{D} + \bar{A}B\bar{D} + \bar{A}\bar{B}\bar{D} + \bar{A}\bar{B}D) + \bar{C}(\bar{A}\bar{B}\bar{D} + \bar{A}B\bar{D} \\ &\quad + \bar{A}\bar{B}D + AB\bar{D})]^n \\ &= [C(B\bar{D} + \bar{A}\bar{B}) + \bar{C}(B\bar{D} + \bar{A}\bar{B})]^n \end{aligned} \tag{5-12}$$

$$\begin{aligned} D^{n+1} &= [D(ABC + \bar{A}\bar{B}\bar{C} + \bar{A}BC + AB\bar{C} + \bar{A}\bar{B}C) + \bar{D}(A\bar{B}C \\ &\quad + ABC + \bar{A}\bar{B}C)]^n \\ &= [D(AB + \bar{A}\bar{B} + BC) + \bar{D}(AC + \bar{B}C)]^n \end{aligned} \tag{5-13}$$

Each of these equations describes the application of a flip-flop to a particular design problem, and they will therefore be called *application equations*. Furthermore, we can generalize and say that whenever some design problem is to be solved it will be possible to set up a table describing the operation of the flip-flops bit-time by bit-time; and to derive application equations from that table. The application equations, being difference equations, may always be written in the factored form exemplified by equations 5-10 through 5-13, or more generally in the form

$$Q^{n+1} = (g_1Q + g_2\bar{Q})^n \tag{5-14}$$

In equation 5-14, g_1 and g_2 represent Boolean functions of whatever variables determine the state of Q, although of course neither g_1 nor

g_2 contains Q itself. For example, examining equation 5-13, we see that g_1 and g_2 for flip-flop D are given by

$$g_1 = AB + \bar{A}\bar{B} + BC \qquad g_2 = AC + \bar{B}C$$

Input Equations. The Complete Problem

We are now in a position to understand exactly what the design problem is. We have seen how it is possible, on the one hand, to write a Boolean difference equation describing how the flip-flop responds to signals on its input lines. For the R-S flip-flop the characteristic equation is

$$Q^{n+1} = (S + \bar{R}Q)^n \tag{5-3}$$

The equivalent general equation is

$$Q^{n+1} = f_1(Q^n, I_1{}^n, I_2{}^n, \cdots, I_m{}^n) \tag{5-4}$$

Certain subsidiary equations may also be necessary to describe restrictions applicable to the input lines. The response of the R-S flip-flop to simultaneous signals on both input lines is ambiguous, and so we impose the restriction

$$R^nS^n = 0 \tag{5-2}$$

Again, the equivalent general expression is

$$f_2(Q^n, I_1{}^n, I_2{}^n, \cdots, I_m{}^n) = 0 \tag{5-5}$$

Having described the memory element by a pair of equations, we have seen that it is possible to write another difference equation called the application equation which defines the part a memory element is to play in some desired configuration of equipment. For example, we saw that the application equation for the fourth flip-flop in a particular counter was

$$D^{n+1} = [D(AB + \bar{A}\bar{B} + BC) + \bar{D}(AC + \bar{B}C)]^n \tag{5-13}$$

The equivalent general expression is

$$Q^{n+1} = (g_1Q + g_2\bar{Q})^n \tag{5-14}$$

Note that the application equation is entirely independent of the logical properties of the flip-flops which are to be used, and that the characteristic equations have nothing to do with the eventual application of the flip-flop.

The problem we are faced with is that of writing the Boolean equation for each of the input lines for each of the flip-flops involved. If we want flip-flop D of equation 5-13 to be an R-S flip-flop, for example, we combine equations 5-3 and 5-13 to get

$$\begin{aligned} D^{n+1} &= (S + \bar{R}D)^n \\ &= [D(AB + \bar{A}\bar{B} + BC) + \bar{D}(AC + \bar{B}C)]^n \end{aligned} \tag{5-15}$$

The right-hand part of this double equation is an equation in R, S, A, B, C, and D, and the design problem is solved if we can solve this equation for R and S, subject to the subsidiary condition given by $R^nS^n = 0$. The equivalent general equations are

$$\begin{aligned} Q^{n+1} &= (S + \bar{R}Q)^n = (g_1Q + g_2\bar{Q})^n \\ R^nS^n &= 0 \end{aligned} \tag{5-16}$$

Obviously, if we can solve equations 5-16 for R and S in terms of g_1, g_2, and Q, we can find a solution for equation 5-15 simply by substituting $g_1 = AB + \bar{A}\bar{B} + BC$ and $g_2 = AC + \bar{B}C$ into the general expressions for R and S. In the next section we shall show how equations 5-16 may be solved; and we shall derive equations for other memory elements and write down their solutions. In other words, we shall in general solve the equations

$$\begin{aligned} Q^{n+1} &= f_1(Q^n, I_1{}^n, I_2{}^n, \cdots, I_m{}^n) = (g_1Q + g_2\bar{Q})^n \\ &f_2(Q^n, I_1{}^n, I_2{}^n, \cdots, I_m{}^n) = 0 \end{aligned} \tag{5-17}$$

for I_1, I_2, $\cdots$, and I_m in terms of g_1, g_2, and Q, where f_1 and f_2 are expressions describing the operation of a number of common memory elements.

MEMORY ELEMENT INPUTS: TYPICAL SOLUTIONS

The R-S Flip-Flop

The first flip-flop to be analyzed is the one introduced in the example of the previous section. We have derived from the characteristic and application expressions the equations

$$\begin{aligned} S + \bar{R}Q &= g_1Q + g_2\bar{Q} \\ RS &= 0 \end{aligned} \tag{5-16}$$

and we now proceed to solve these for R and S. Because they are Boolean equations we cannot of course employ subtraction or division in arriving at a solution. Furthermore, we shall find that the solutions have unexpected properties: for example, a completely general solution may involve arbitrary Boolean "constants" which may be assigned any value—analogous, in a way, to the arbitrary constants which arise in the solution of differential equations in calculus.

An algebraic method of solution is indicated in Appendix I at the end of this book. However, Boolean equations may also be solved by applying truth table techniques, and it is this approach which will be used throughout this chapter. It has the advantage that it is easy to understand and to follow, and the fact that it is not an obviously general technique need not concern us here.

Examining equations 5-16 once more, we see that there are three independent variables, g_1, g_2, and Q, each of which may be "zero" or "one." We must therefore consider eight possible input configurations, as shown in Table 5-2. The fourth column in Table 5-2 shows the "value" of the application equation for each input configuration. We

Table 5-2

g_1	g_2	Q	$g_1Q + g_2\overline{Q}$
0	0	0	0
0	0	1	0
0	1	0	1
0	1	1	0
1	0	0	0
1	0	1	1
1	1	0	1
1	1	1	1

shall now consider each of the eight rows of this table in turn, deducing what R and S must be in order for equations 5-16 to hold. Inasmuch as this is the first solution to be determined, we shall set down in detail the arguments which lead to the solution.

In Table 5-3 the two right-hand columns of Table 5-2 are repeated, and equations 5-16 are used to replace $(g_1Q + g_2\overline{Q})$ by $(S + \overline{R}Q)$. Let us first examine the rows for which $S + \overline{R}Q = 0$, i.e., rows 0, 1, 3, and 4. We can immediately write down $S = 0$ for each of these rows, for obviously if $S = 1$, $S + \overline{R}Q = 1$. Furthermore, in rows 1 and 3, $Q = 1$. Therefore if $R = 0$, $S + \overline{R}Q$ would also be "one." We deduce that $R = 1$ in rows 1 and 3. Examining rows 0 and 4 now, we see that because $Q = 0$, R may be either "zero" or "one." In fact, in either row 0 or row 4, R may have any value at all, and this may be indicated by writing two arbitrary Boolean terms, a_0 and a_4, in those rows.

Now we may turn our attention to rows 2, 5, 6, and 7, where $S + \bar{R}Q = 1$. We see at once that in rows 2 and 6, where $Q = 0$, the term $\bar{R}Q = 0$ and so S *must* be "one" if $S + \bar{R}Q$ is to be "one." Because

Table 5-3

Row No.	Q	$S + \bar{R}Q$	R	S
0	0	0	a_0	0
1	1	0	1	0
2	0	1	0	1
3	1	0	1	0
4	0	0	a_4	0
5	1	1	0	a_5
6	0	1	0	1
7	1	1	0	a_7

R and S cannot be "one" at the same time ($RS = 0$, equations 5-16), we must enter "zero" in the R column opposite these two rows. The rows remaining, 5 and 7, must have $R = 0$, for if $R = 1$ then $S = 0$ (since $RS = 0$), and of course $\bar{R}Q = 0$ so $S + \bar{R}Q = 0$. If R is "zero," then S may be either "zero" or "one," and we can again assign arbitrary Boolean quantities, a_5 and a_7, to S in those two rows.

Table 5-4

g_1	g_2	Q	R	S
0	0	0	a_0	0
0	0	1	1	0
0	1	0	0	1
0	1	1	1	0
1	0	0	a_4	0
1	0	1	0	a_5
1	1	0	0	1
1	1	1	0	a_7

The completed truth table is shown in Table 5-4, and from it we can write R and S in terms of g_1, g_2, and Q:

$$
\begin{aligned}
R &= a_0\bar{g}_1\bar{g}_2\bar{Q} + \bar{g}_1\bar{g}_2Q + \bar{g}_1g_2Q + a_4g_1\bar{g}_2\bar{Q} \\
S &= \bar{g}_1g_2\bar{Q} + a_5g_1\bar{g}_2Q + g_1g_2\bar{Q} + a_7g_1g_2Q
\end{aligned}
\qquad (5\text{-}18)
$$

This is the perfectly general solution for R and S in terms of g_1, g_2, and Q. Before assigning definite values to a_0, a_4, a_5, and a_7, let us prove that equations 5-18 are a solution to equations 5-16 irrespective of the values given the a's. We do this simply by substituting 5-18 in 5-16.

$$S + \bar{R}Q = \bar{g}_1g_2\bar{Q} + a_5g_1\bar{g}_2Q + g_1g_2\bar{Q} + a_7g_1g_2Q$$
$$+ Q(g_1 + g_2 + Q + a_0)(g_1 + g_2 + \bar{Q})$$
$$(g_1 + \bar{g}_2 + \bar{Q})(\bar{g}_1 + g_2 + Q + \bar{a}_4)$$

Since

$$Q(Q + x) = Q$$

and

$$Q(g_1 + g_2 + \bar{Q})(g_1 + \bar{g}_2 + \bar{Q}) = Q(g_1 + g_2)(g_1 + \bar{g}_2) = g_1Q$$

we must have

$$S + \bar{R}Q = \bar{g}_1g_2\bar{Q} + a_5g_1\bar{g}_2Q + g_1g_2\bar{Q} + a_7g_1g_2Q + g_1Q$$
$$= g_1Q + g_2\bar{Q}$$

Note that all a's disappear, and we have verified the first part of equations 5-16. We prove that $RS = 0$ as follows:

$$RS = (a_0\bar{g}_1\bar{g}_2\bar{Q} + \bar{g}_1\bar{g}_2Q + \bar{g}_1g_2Q + a_4g_1\bar{g}_2\bar{Q})$$
$$(\bar{g}_1g_2\bar{Q} + a_5g_1\bar{g}_2Q + g_1g_2\bar{Q} + a_7g_1g_2Q)$$

Comparison of R and S here shows that the Boolean product of any term of R with all terms of S is "zero", regardless of the values of the a's. Therefore $RS = 0$.

Equations 5-18 are thus quite correct and are plotted in Figure 5-2. They may easily be simplified by assigning values to the a's. The simplest forms for R and S are obtained by setting $a_0 = a_4 = a_5 = a_7 = 0$, whereupon

$$R = \bar{g}_1Q \qquad S = g_2\bar{Q} \tag{5-19}$$

Although these are the general expressions for R and S, a certain simplification is possible when the application equation has the property that $\bar{g}_1g_2 = 0$. Setting $a_0 = a_7 = 1$ and $a_4 = a_5 = 0$ in equations 5-18, we get

$$R = \bar{g}_1\bar{g}_2\bar{Q} + \bar{g}_1\bar{g}_2Q + \bar{g}_1g_2Q$$
$$S = g_1g_2\bar{Q} + g_1g_2Q + \bar{g}_1g_2\bar{Q}$$

g_2	0	0	1	0
	a_4	0	1	a_0

(columns g_1; Q) R

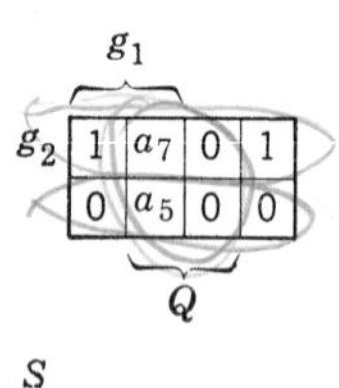

FIGURE 5-2

In each of these equations, the first two terms may

be combined, and the third eliminated because $\bar{g}_1 g_2 = 0$

$$R = \bar{g}_1 \bar{g}_2 \qquad S = g_1 g_2$$

The relationship $\bar{g}_1 g_2 = 0$ may now be introduced, and R and S simplified to

$$R = \bar{g}_1 \qquad S = g_2 \qquad \text{if } \bar{g}_1 g_2 = 0 \tag{5-20}$$

It must now be evident that the Q and $\bar{Q}$ in equations 5-19 are only necessary to insure that $RS = 0$, and both may be omitted if $\bar{g}_1 g_2 = 0$.

Furthermore, it may easily be shown that if we find in some particular example that $\bar{g}_1 = X_1 + X_2$ and $g_2 = Y_1 + Y_2$, we would normally write the complete solution

$$R = X_1 Q + X_2 Q \qquad S = Y_1 \bar{Q} + Y_2 \bar{Q}$$

However, if $X_1 Y_1 = X_1 Y_2 = 0$, we may simplify R by writing $R = X_1 + X_2 Q$. Similarly, if $X_1 Y_1 = X_2 Y_1 = 0$, we may write $S = Y_1 + Y_2 \bar{Q}$. In other words, we may examine $\bar{g}_1$ and g_2 term by term and if the product of any term in one with all terms in the other is zero, we may omit a Q (or $\bar{Q}$) in the corresponding input equation.

We can now complete the design of the counter described by equations 5-10 through 5-13. For flip-flop A (equation 5-10)

$$\bar{g}_1 = g_2 = \bar{C}\bar{D} + CD + BD \tag{5-21}$$

Since $\bar{g}_1 g_2 \neq 0$, we must use equations 5-19, and find

$$R_A = A\bar{C}\bar{D} + ACD + ABD \qquad S_A = \bar{A}\bar{C}\bar{D} + \bar{A}CD + \bar{A}BD \tag{5-22}$$

For flip-flop B (equation 5-11)

$$\bar{g}_1 = ACD + \bar{A}C\bar{D} \qquad g_2 = A\bar{C}\bar{D} + \bar{A}\bar{C}D$$

Here, $\bar{g}_1 g_2 = 0$, and so we can apply equations 5-20

$$R_B = ACD + \bar{A}C\bar{D} \qquad S_B = A\bar{C}\bar{D} + \bar{A}\bar{C}D \tag{5-23}$$

For flip-flop C (equation 5-12), $g_1 = g_2$, and so $\bar{g}_1 g_2 = \bar{g}_1 g_1 = 0$. We can again apply equations 5-20, with the result

$$R_C = BD + A\bar{B} \qquad S_C = B\bar{D} + \bar{A}\bar{B} \tag{5-24}$$

Finally, for flip-flop D (equation 5-13), $\bar{g}_1 g_2 \neq 0$. Examining $\bar{g}_1$ and g_2 term by term, we note that

$$\bar{A}B\bar{C}(g_2) = \bar{A}B\bar{C}(AC + \bar{B}C) = 0$$

We can therefore omit the D from one term in R_D, and write

$$R_D = A\bar{B}D + \bar{A}B\bar{C} \qquad S_D = AC\bar{D} + \bar{B}C\bar{D} \tag{5-25}$$

Equations 5-22 through 5-25 represent the complete design for the counter of the previous section. These input equations (and others throughout this book) are written as minterm-type expressions. In practice, it would be desirable to compare them with the equivalent maxterm-type expressions to determine which is simpler in each case.

The D Flip-Flop

The next flip-flop to be discussed is the *delay* memory element, a circuit having a single input and an output equal to the input one bit-time earlier. The truth table for this memory element showing its state at time $(n + 1)$, (Q^{n+1}), as a function of its input line at time n, (D^n), is now given:

D^n	Q^{n+1}
0	0
1	1

The corresponding characteristic equation for the flip-flop is, of course,

$$Q^{n+1} = D^n \tag{5-26}$$

Since we know from the application equation that

$$Q^{n+1} = (g_1Q + g_2\overline{Q})^n \tag{5-14}$$

we can immediately write D in terms of g_1, g_2, and Q:

$$D = g_1Q + g_2\overline{Q} \tag{5-27}$$

Note that no arbitrary "constant" appears in this expression, of the kind which appeared in the solution to the R-S flip-flop. Also, note that when $g_1 = g_2$,

$$D = g_1Q + g_1\overline{Q} = g_1, \qquad \text{if } g_1 = g_2 \tag{5-28}$$

If the counter of the previous section were to be constructed from D flip-flops, the input equations would be

$$\begin{aligned} D_A &= \overline{A}\overline{C}\overline{D} + \overline{A}CD + AC\overline{D} + A\overline{B}\overline{C}D + \overline{A}BD \\ D_B &= B\overline{C} + A\overline{C}\overline{D} + AB\overline{D} + \overline{A}BD + \overline{A}\overline{C}D \\ D_C &= B\overline{D} + \overline{A}\overline{B} \\ D_D &= ABD + \overline{A}\overline{B}D + AC\overline{D} + \overline{A}\overline{B}C + \overline{A}CD \end{aligned} \tag{5-29}$$

The T Flip-Flop

The T or *trigger* memory element has a single input line which causes the memory element to change state when it is "one," but leaves it in its former state otherwise. We may therefore draw up the following truth table.

T^n	Q^{n+1}
0	Q^n
1	$\overline{Q}^n$

and write the following characteristic equation:

$$Q^{n+1} = (\overline{T}Q + T\overline{Q})^n \tag{5-30}$$

The derivation of the general equation for T is shown in Table 5-5, where the same procedure is followed which was used in deriving the

Table 5-5

g_1	g_2	Q	$g_1Q + g_2\overline{Q} = \overline{T}Q + T\overline{Q}$	T
0	0	0	0	0
0	0	1	0	1
0	1	0	1	1
0	1	1	0	1
1	0	0	0	0
1	0	1	1	0
1	1	0	1	1
1	1	1	1	0

R-S input equations. The result, however, contains no arbitrary constants, and may be written from the truth table as follows,

$$\begin{aligned} T &= \bar{g}_1\bar{g}_2Q + \bar{g}_1g_2\overline{Q} + \bar{g}_1g_2Q + g_1g_2\overline{Q} \\ &= \bar{g}_1Q + g_2\overline{Q} \end{aligned} \tag{5-31}$$

As usual, the equation may be verified by substituting it into equation 5-30, with the result

$$\begin{aligned} Q^{n+1} &= \overline{(\bar{g}_1Q + g_2\overline{Q})}Q + (\bar{g}_1Q + g_2\overline{Q})\overline{Q} \\ &= (g_1Q + \bar{g}_2\overline{Q})Q + g_2\overline{Q} \\ &= g_1Q + g_2\overline{Q} \end{aligned} \tag{5-32}$$

Equation 5-32 is the generalized application equation, and so equation 5-31 must be correct.

If $\bar{g}_1 = g_2$, i.e., if g_1 and g_2 are the complements of one another, we may simplify equation 5-31 to

$$T = \bar{g}_1, \qquad \text{if } \bar{g}_1 = g_2 \tag{5-33}$$

If the counter of the previous section were to be constructed from T flip-flops, the input equations would be

$$\begin{aligned} T_A &= CD + \bar{C}\bar{D} + BD \\ T_B &= ABCD + \bar{A}BC\bar{D} + \bar{A}\bar{B}\bar{C}D + A\bar{B}\bar{C}\bar{D} \\ T_C &= A\bar{B}C + BCD + \bar{A}B\bar{C} + B\bar{C}\bar{D} \\ T_D &= A\bar{B}D + \bar{A}B\bar{C}D + AC\bar{D} + \bar{B}C\bar{D} \end{aligned} \tag{5-34}$$

The J-K Flip-Flop

The J-K memory element has the properties of an R-S memory element, except that the combination $J = K = 1$ is allowed, and causes the circuit to change state. The following truth table therefore applies.

J^n	K^n	Q^{n+1}
0	0	Q^n
0	1	0
1	0	1
1	1	$\bar{Q}^n$

(Note that K, like R, resets the element to the "zero" state, whereas J, like S, sets it to the "one" state.)

From the truth table we see that the characteristic equation is:

$$\begin{aligned} Q^{n+1} &= (\bar{J}\bar{K}Q + J\bar{K} + JK\bar{Q})^n \\ &= (\bar{K}Q + J\bar{Q})^n \end{aligned} \tag{5-35}$$

The derivation of the general expression for inputs J and K in terms of g_1, g_2, and Q is shown in Table 5-6, where we see eight arbitrary constants are necessary. The result may be written

$$\begin{aligned} J &= a_1\bar{g}_1\bar{g}_2Q + \bar{g}_1g_2\bar{Q} + a_3\bar{g}_1g_2Q + a_5g_1\bar{g}_2Q + g_1g_2\bar{Q} + a_7g_1g_2Q \\ K &= a_0\bar{g}_1\bar{g}_2\bar{Q} + \bar{g}_1\bar{g}_2Q + a_2\bar{g}_1g_2\bar{Q} + \bar{g}_1g_2Q + a_4g_1\bar{g}_2\bar{Q} + a_6g_1g_2\bar{Q} \end{aligned} \tag{5-36}$$

Table 5-6

g_1	g_2	Q	$g_1Q + g_2\bar{Q} = \bar{K}Q + J\bar{Q}$	J	K
0	0	0	0	0	a_0
0	0	1	0	a_1	1
0	1	0	1	1	a_2
0	1	1	0	a_3	1
1	0	0	0	0	a_4
1	0	1	1	a_5	0
1	1	0	1	1	a_6
1	1	1	1	a_7	0

Equations 5-36 are illustrated by Veitch diagrams in Figure 5-3, where the arbitrary constants are plotted, together with the minterms for which J and K must be "one" and "zero." A glance at these two charts shows that the simplest expressions for J and K occur when $a_0 = a_2 = a_3 = a_7 = 1$ and $a_1 = a_4 = a_5 = a_6 = 0$, whereupon

$$J = g_2 \qquad \text{and} \qquad K = \bar{g}_1 \tag{5-37}$$

Note that these equations are exactly the same as equations 5-20 for the R-S flip-flop, but that there are no restrictions on g_1 and g_2 here, whereas 5-20 only holds if $\bar{g}_1 g_2 = 0$.

J (g_1 over the middle two columns, g_2 on the top row, Q under the middle two columns):

g_2	1	a_7	a_3	1
	0	a_5	a_1	0

K (g_1 over the middle two columns, g_2 on the top row, Q under the middle two columns):

g_2	a_6	0	1	a_2
	a_4	0	1	a_0

FIGURE 5-3

Applying this result to the design of the counter, we find:

$$
\begin{aligned}
J_A &= K_A = \bar{C}\bar{D} + CD + BD \\
J_B &= \bar{A}\bar{C}D + A\bar{C}\bar{D} \qquad K_B = ACD + \bar{A}C\bar{D} \\
J_C &= \bar{A}\bar{B} + B\bar{D} \qquad K_C = A\bar{B} + BD \\
J_D &= AC + \bar{B}C \qquad K_D = A\bar{B} + \bar{A}B\bar{C}
\end{aligned}
\tag{5-38}
$$

The R-S-T Flip-Flop

The last memory element to be described has three inputs: the R and S input lines have the same effect here that they had in the R-S memory element; the T input line "triggers" the memory element, just as the T input does in the T flip-flop; and the operation of the circuit is undefined when any two inputs are "one" simultaneously. These features are indicated in the truth table, Table 5-7.

Table 5-7

R^n	S^n	T^n	Q^{n+1}
0	0	0	Q^n
0	0	1	$\bar{Q}^n$
0	1	0	1
1	0	0	0
0	1	1	?
1	0	1	?
1	1	0	?
1	1	1	?

From the truth table the characteristic equation may be written

$$Q^{n+1} = (\bar{R}\bar{S}\bar{T}Q + \bar{R}\bar{S}T\bar{Q} + \bar{R}S\bar{T})^n \tag{5-39}$$

with the understanding that $RS = ST = RT = 0$. Making use of these redundancies, we can simplify equation 5-39 to

$$Q^{n+1} = (S + T\bar{Q} + \bar{R}\bar{T}Q)^n \tag{5-40}$$

The derivation of the general expressions for R, S, and T is indicated in Table 5-8. The arguments justifying two typical lines of that table are as follows:

For $g_1 = 1$, $g_2 = 0$, $Q = 1$,

$$\begin{aligned} g_1Q + g_2\bar{Q} &= 1 = S + T\bar{Q} + \bar{R}\bar{T}Q \\ &= S + \bar{R}\bar{T} \end{aligned}$$

Table 5-8

g_1	g_2	Q	$g_1Q + g_2\bar{Q} = S + T\bar{Q} + \bar{R}\bar{T}Q$	R	S	T
0	0	0	0	a_0	0	0
0	0	1	0	a_1	0	$\bar{a}_1$
0	1	0	1	0	a_2	$\bar{a}_2$
0	1	1	0	a_3	0	$\bar{a}_3$
1	0	0	0	a_4	0	0
1	0	1	1	0	a_5	0
1	1	0	1	0	a_6	$\bar{a}_6$
1	1	1	1	0	a_7	0

If S is "one" the equation holds, and since $RS = ST = RT = 0$, we must have $R = T = 0$. On the other hand, if $S = 0$, then both R and T must be "zero" in order for $S + \bar{R}\bar{T}$ to be equal to "one." Thus S may be "zero" or "one" or any arbitrary function, say a_5, but R and T must both be "zero."

For $g_1 = 1$, $g_2 = 1$, $Q = 0$,

$$\begin{aligned} g_1Q + g_2\bar{Q} &= 1 = S + T\bar{Q} + \bar{R}\bar{T}Q \\ &= S + T \end{aligned}$$

Therefore obviously either S or T must be "one." However, since $RS = ST = RT = 0$, S and T cannot both be "one" at the same time, and R must be "zero." S and T must be complementary quantities, and this is indicated by putting $S = a_6$, $T = \bar{a}_6$.

The same conclusions may be reached, perhaps more easily and quickly, by ignoring the equations and observing the desired operation of the circuit. When Q^{n+1} and Q^n are both "one," it is evident that

R^n and T^n must be "zero," for otherwise Q would be "zero" at $(n + 1)$ time. Furthermore, S^n will have no effect on Q and may be either "zero" or "one." On the other hand, if $Q^n = 0$ and $Q^{n+1} = 1$, then R^n must be "zero," and either S^n or T^n *must* be "one."

The general equations for R, S, and T may be written down from Table 5-8.

$$
\begin{aligned}
R &= a_0\bar{g}_1\bar{g}_2\overline{Q} + a_1\bar{g}_1\bar{g}_2Q + a_3\bar{g}_1g_2Q + a_4g_1\bar{g}_2\overline{Q} \\
S &= a_2\bar{g}_1g_2\overline{Q} + a_5g_1\bar{g}_2Q + a_6g_1g_2\overline{Q} + a_7g_1g_2Q \qquad (5\text{-}41) \\
T &= \bar{a}_1\bar{g}_1\bar{g}_2Q + \bar{a}_2\bar{g}_1g_2\overline{Q} + \bar{a}_3\bar{g}_1g_2Q + \bar{a}_6g_1g_2\overline{Q}
\end{aligned}
$$

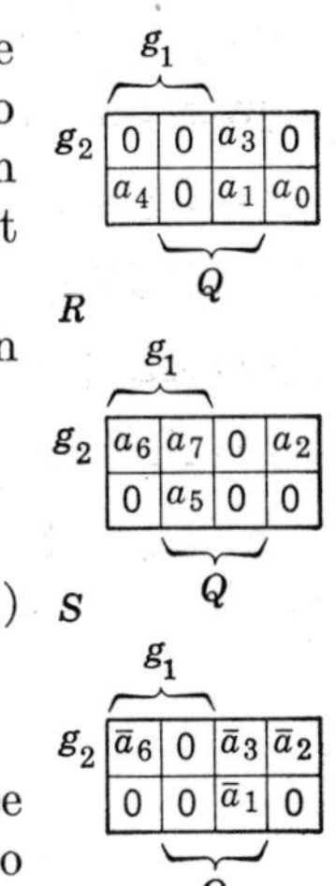

FIGURE 5-4

In Figure 5-4 these three equations are plotted on three Veitch diagrams to make it easier to supply values to the various constants. Of the many possible simplifications to equations 5-41, the most interesting is perhaps the one in which R, S, and T are independent of Q. Letting $a_0 = a_1 = a_6 = a_7 = 1$ and $a_2 = a_3 = a_4 = a_5 = 0$, we find:

$$
\begin{aligned}
R &= \bar{g}_1\bar{g}_2 \\
S &= g_1g_2 \qquad (5\text{-}42) \\
T &= \bar{g}_1g_2
\end{aligned}
$$

In Table 5-9 this solution is repeated, together with four others, each obtained by assigning particular values to the various a's in equations

Table 5-9

Solution Number	R	S	T
1 (equation 5-42)	$\bar{g}_1\bar{g}_2$	g_1g_2	$\bar{g}_1g_2$
2 (R-S)	$\bar{g}_1Q$	$g_2\overline{Q}$	0
3 (T)	0	0	$\bar{g}_1Q + g_2\overline{Q}$
4	$\bar{g}_1\bar{g}_2$	$g_2\overline{Q}$	$\bar{g}_1g_2Q$
5	$\bar{g}_1Q$	g_1g_2	$\bar{g}_1g_2\overline{Q}$

5-41. Solutions 2 and 3 employ the R-S-T flip-flop as an R-S and as a T circuit respectively, but solutions 4 and 5 are distinctively new and different solutions. For any given application equation, all of these forms should be investigated, for one may require fewer diodes than any of the others.

The simplest set of solutions for the counter of the previous section is given below:

$$\begin{aligned} &R_A = S_A = 0 && && T_A = BD + CD + \bar{C}\bar{D} \\ &R_B = ACD + \bar{A}C\bar{D} && S_B = A\bar{C}\bar{D} + \bar{A}\bar{C}D && T_B = 0 \\ &R_C = A\bar{B} + BD && S_C = \bar{A}\bar{B} + B\bar{D} && T_C = 0 \\ &R_D = A\bar{B}D + \bar{A}B\bar{C}D && S_D = AC\bar{D} + \bar{B}C\bar{D} && T_D = 0 \end{aligned} \tag{5-43}$$

The fact that one of the other solutions may, in some instances, be simpler than solution 2 or 3 is illustrated in the following example. Suppose for some flip-flop Q

$$g_1 = B\bar{C} + \bar{A}B \qquad g_2 = B\bar{C} + A\bar{B}C \tag{5-44}$$

Then

$$\begin{aligned} \bar{g}_1 &= \bar{B} + AC \\ \bar{g}_1\bar{g}_2 &= \bar{A}\bar{B} + \bar{B}\bar{C} + ABC \\ g_1g_2 &= B\bar{C} \\ \bar{g}_1g_2 &= A\bar{B}C \end{aligned}$$

and we may write

Solution 1 (fifteen diodes):

$$R_Q = \bar{A}\bar{B} + \bar{B}\bar{C} + ABC, \qquad S_Q = B\bar{C}, \qquad T_Q = A\bar{B}C \tag{5-45}$$

Solution 2 (sixteen diodes):

$$R_Q = \bar{B}Q + ACQ, \qquad S_Q = B\bar{C}\bar{Q} + A\bar{B}C\bar{Q}, \qquad T_Q = 0 \tag{5-46}$$

Solution 3 (sixteen diodes):

$$R_Q = S_Q = 0, \qquad T_Q = \bar{B}Q + ACQ + B\bar{C}\bar{Q} + A\bar{B}C\bar{Q} \tag{5-47}$$

Solution 4 (twenty-three diodes):

$$R_Q = \bar{A}\bar{B} + \bar{B}\bar{C} + ABC,\ S_Q = B\bar{C}\bar{Q} + A\bar{B}C\bar{Q},\ T_Q = A\bar{B}CQ \tag{5-48}$$

Solution 5 (thirteen diodes):

$$R_Q = \bar{B}Q + ACQ, \qquad S_Q = B\bar{C}, \qquad T_Q = A\bar{B}C\bar{Q} \tag{5-49}$$

INPUT EQUATIONS WITH REDUNDANCIES: THE COMPLETE SOLUTION

The introduction of forbidden combinations into the algebra of input equations is complicated by the fact that some of the redundant combinations depend on the memory element whose input equations are being written. Suppose, for example, we are given the application equation:

$$Q^{n+1} = [(A\bar{C} + BC + \bar{B}\bar{C})Q + (A\bar{B}\bar{C} + \bar{A}B)\bar{Q}]^n \qquad (5\text{-}50)$$

together with the additional knowledge that

$$A\bar{B}CQ = A\bar{B}C\bar{Q} = \bar{A}\bar{B}C\bar{Q} = 0 \qquad (5\text{-}51)$$

Here, g_1 and g_2 are of course functions of A, B, and C only, whereas the redundancies are functions of Q as well. How can these redundancies be used to simplify input equations, which are in general functions of g_1 and g_2?

In providing an answer to this question, we will again employ the Veitch diagram, and will develop a simplification procedure which

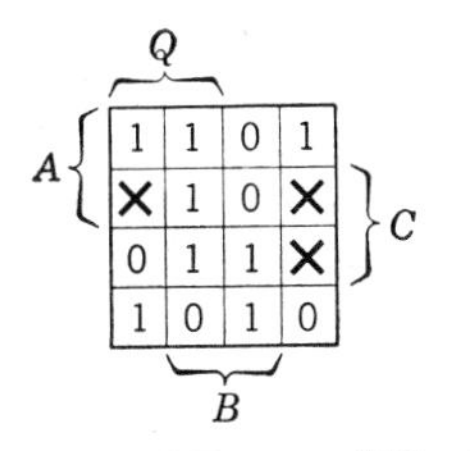

$Q^{n+1} = [(A\bar{C} + BC + \bar{B}\bar{C})Q + (A\bar{B}\bar{C} + \bar{A}B)\bar{Q}]^n$

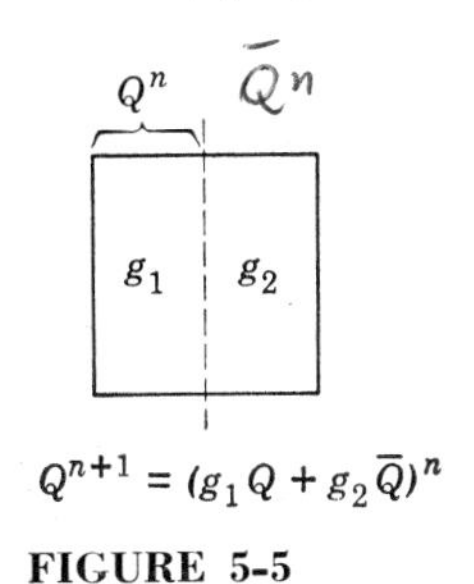

$Q^{n+1} = (g_1 Q + g_2 \bar{Q})^n$

FIGURE 5-5

enables us to go directly from a memory element truth table to a corresponding Veitch diagram and from there directly to the input equations.* It will therefore never be necessary to write the application equation explicitly, nor to write out expressions for g_1 and g_2.

We begin the simplification procedure by plotting the application equation and redundancies on a Veitch diagram. Equations 5-50 and 5-51 are plotted in Figure 5-5. Also plotted in Figure 5-5 is the generalized application equation $Q^{n+1} = (g_1Q + g_2\bar{Q})^n$. Clearly, the func-

* This approach was worked out jointly by Mr. John V. Blankenbaker and the author.

tion g_1 is plotted in the Q half of any application equation Veitch diagram, and g_2 in the $\overline{Q}$ half.

If we refer now to the Veitch diagram plots of the generalized input equations to a flip-flop, we can immediately see how to solve our problem. From Figure 5-2, for example, we find that the Veitch diagram for R in the R-S flip-flop must contain "ones" wherever $\bar{g}_1 Q = 1$, and (since a_0 and a_4 may be either "zero" or "one") crosses wherever $\bar{g}_2\overline{Q} = 1$. That is, we can plot R_Q by placing "ones" where $g_1 = 0$

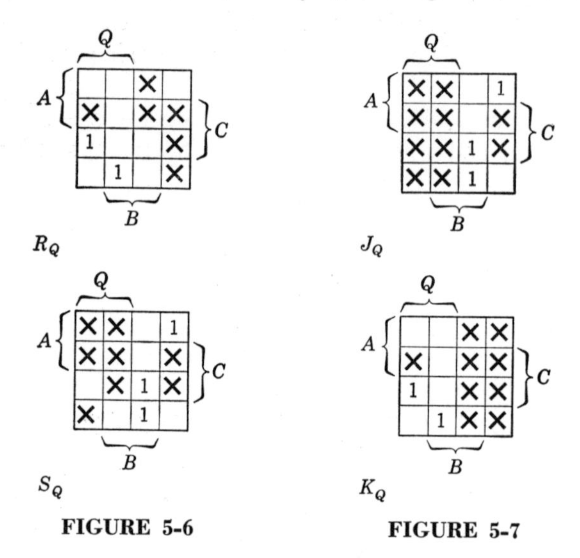

FIGURE 5-6

FIGURE 5-7

on the Q half of Figure 5-5, and ×'s where $g_2 = 0$ on the $\overline{Q}$ half. In Figure 5-6 these rules are employed to plot R_Q (the redundancies of equation 5-51 still appear as ×'s, of course), and the simplest minterm-type expression for R_Q is

$$R_Q = \bar{B}C + \bar{A}B\bar{C}Q \tag{5-52}$$

In a similar way, we can draw a Veitch diagram for S by inserting "ones" wherever $g_2 = 1$ in $\overline{Q}$, and ×'s where $g_1 = 1$ in Q. This is also shown in Figure 5-6, and the simplest S_Q is

$$S_Q = A\bar{B} + \bar{A}B\overline{Q} \tag{5-53}$$

The simplest form for the J-K flip-flop can be found using an exactly analogous scheme. Referring to Figure 5-3, we see that J should be plotted by placing ×'s in the Q half of the Veitch diagram, and "ones" wherever $g_2 = 1$ in $\overline{Q}$; and that K should be plotted by putting ×'s in the $\overline{Q}$ half and "ones" wherever $g_1 = 0$ in Q. The two Veitch diagrams are shown in Figure 5-7, and the simplest minterm-type equations are

$$J_Q = A\bar{B} + \bar{A}B \qquad K_Q = \bar{B}C + \bar{A}B\bar{C} \tag{5-54}$$

This method of writing flip-flop input equations directly from Veitch diagrams will be employed throughout this text. In fact, because the rules are so simple, the input logic can be and will be written directly from the plot of the application equation (Figure 5-5) without the need for drawing special Veitch diagrams for each input (Figures 5-6 and 5-7).

The method described here can obviously be extended to other memory devices.

OTHER METHODS OF DERIVING INPUT EQUATIONS

Until now, our discussion of memory element logic and of the derivation of input equations has been carried out on a rather formal level. The results have invariably been expressed in the form of general Boolean expressions (e.g., $R_Q = \bar{g}_1 Q$, $S_Q = g_2\bar{Q}$) from which the particular input equations for some specific application may be found by substituting particular expressions for g_1 and g_2. We have also seen how the input equations may be derived directly from the Veitch diagram plot of the application equation. We shall now show that it is possible, and sometimes very convenient, to derive the input equations simply by observation, without going through all the formal steps we have been using.

Table 5-10

Time n				Time $n+1$	
U	D	A	B	A	B
0	0	0	0	0	0
0	0	0	1	0	1
0	0	1	0	1	0
0	0	1	1	1	1
0	1	0	0	1	1
0	1	0	1	0	0
0	1	1	0	0	1
0	1	1	1	1	0
1	0	0	0	0	1
1	0	0	1	1	0
1	0	1	0	1	1
1	0	1	1	0	0

Suppose, for example, we want to design a two-flip-flop counter which is to count in the sequence 0, 1, 2, 3, 0, 1, 2, 3, 0, etc., whenever the signal U is "one," and is to count 3, 2, 1, 0, 3, 2, 1, 0, 3, etc., whenever the signal D is "one." The formal approach to the problem of writing input equations for the two counter flip-flops (let us call them A and B) requires that we write down a truth table containing all four variables—A, B, U, and D—in all combinations which can occur at time n, and that we put opposite each entry the desired state of A and B at time $(n + 1)$, as in Table 5-10. (Note that the combination $UD = 1$ is not allowed: we cannot count both up and down at the same time.) The difference equations for A and B may be derived from this table, and the resulting input equations written as follows:

$$\begin{aligned}
A^{n+1} &= [A(\bar{U}\bar{D} + DB + U\bar{B}) + \bar{A}(D\bar{B} + UB)]^n \\
B^{n+1} &= [B(\bar{U}\bar{D}) + \bar{B}(D + U)]^n \\
R_A &= ABU + A\bar{B}D \\
S_A &= \bar{A}BU + \bar{A}\bar{B}D \\
R_B &= BD + BU \\
S_B &= \bar{B}D + \bar{B}U
\end{aligned} \qquad (5\text{-}55)$$

The same equations may be obtained by the following line of reasoning. Let us first design an "up" counter. The truth table is simply

n		$n + 1$	
A	B	A	B
0	0	0	1
0	1	1	0
1	0	1	1
1	1	0	0

Examining the A columns of the table, we observe that the only time flip-flop A is reset from the "one" to the "zero" state is when A and B are both "one." The reset line for the A flip-flop in an up counter must therefore be written $R_A = AB$. Furthermore, the only time flip-flop A is set from "zero" to "one" is when A is "zero" and B, "one." Therefore, $S_A = \bar{A}B$. Following an exactly similar line of reasoning for the B flip-flop, we find

$$R_B = \bar{A}B + AB = B \qquad \text{and} \qquad S_B = \bar{A}\bar{B} + A\bar{B} = \bar{B}$$

We next write the truth table for the "down" counter

n		$n+1$	
A	B	A	B
0	0	1	1
0	1	0	0
1	0	0	1
1	1	1	0

and from it deduce that $R_A = A\bar{B}$, $S_A = \bar{A}\bar{B}$, $R_B = B$, and $S_B = \bar{B}$. We have thus derived the complete input equations for two counters without having written or plotted a single difference equation and without once having had to complement a Boolean function.

We now have two equations for each input line, and must find a way of combining them. The combination is achieved easily by noting that the count up logic should be effective only when $U = 1$, and the count down logic only when $D = 1$. In other words, a given input line should be "one" only when $U = 1$ *and* the count up logic is true, *or* when $D = 1$ *and* the count down logic is true. Written in the form of a Boolean equation, this sentence is

$$\text{Input} = U\ (\text{count up}) + D\ (\text{count down}) \tag{5-56}$$

and we can therefore derive

$$\begin{aligned} R_A &= UAB + DA\bar{B} \\ S_A &= U\bar{A}B + D\bar{A}\bar{B} \\ R_B &= UB + DB \\ S_B &= U\bar{B} + D\bar{B} \end{aligned} \tag{5-57}$$

These equations of course agree with equations 5-55, which we found using the more formal approach.

SUMMARY

Several remarks must be made about the techniques and results developed in this chapter. In the first place, the development of the input equations for a given type of memory element from the characteristic equations of that memory element (and from the general application equation) is a development which may be adapted to new memory circuits as they are invented. Unfortunately, the technique is not so general that it can be applied to any memory element. For example, it will not be of any help in analyzing asynchronous circuits—circuits which operate without regard to any clock pulses. The analysis of such circuits, and the logical design of systems using such circuits, is a problem which will not be considered here.

Second, it is worthwhile to say something about the informal derivation of input equations which was mentioned in the last section. It should be noted that two ideas were presented there: that it is possible to derive input equations directly from truth tables without recourse to difference equations or Veitch diagrams; and that it is possible to combine the logic relating two functions carried out at different times by employing a very simple rule of combination (equation 5-56). The rule of combination is so useful it will be used throughout this book. On the other hand, we shall continue to use the formal difference-equation derivation rather than the direct translation from truth tables, simply because it can be carried out quickly and formally. A practicing designer who must write many equations for a single kind of memory element and who is interested in quick results may develop short cuts of the kind mentioned.

Third and finally, it is interesting to speculate on the relative advantages and disadvantages of the memory elements here described. The question should be asked: For general use in logical design, which memory element will on the average require the least number of diodes in decision circuits? In Table 5-11, for example, are listed the number of diodes required for each flip-flop in the counter which was used for

Table 5-11

Memory Element	Diode Count				
	A	*B*	*C*	*D*	Total
R-S	24	16	12	16	68
D	21	19	6	20	66
T	9	20	16	17	62
J-K	14	16	12	13	55
R-S-T	9	16	12	16	53

an example in this chapter. It would be foolish to deduce from this table that the *R-S-T* flip-flop is to be preferred above all others, for no one example can possibly represent the variety of applications found in practice. However, study of the logic of the memory elements suggests a few generalizations: an *R-S-T* flip-flop will always require *at most* the same number of diodes that an *R-S* flip-flop requires for the same application (obviously the same remark applies to the *T* flip-flop); a *D* memory element is to be preferred in applications in which

$g_1 = g_2$, i.e., where the state of a memory element at each bit-time is independent of its state at the last bit-time; and a *J-K* memory element is generally to be preferred in applications where the state of a memory element at a given bit-time *is* dependent on its state at the previous bit-time. Usually, however, considerations of circuit cost and reliability determine which flip-flop is adopted by a particular group. There is not enough difference between them on the basis of a count of logical diodes to warrant a choice of one over another.

BIBLIOGRAPHY

R. C. Jeffrey and I. S. Reed, "The Use of Boolean Algebra in Logical Design," *M.I.T. Report E-458-1*, Apr. 15, 1952.

E. C. Nelson, "An Algebraic Theory for Use in Digital Computer Design," *I.R.E. Trans. on Electronic Computers*, **EC-3,** no. 3, 12–21 (Sept. 1954).

G. W. Arant, "A Time-Sequential Tabular Analysis of Flip-Flop Logical Operation," *I.R.E. Trans. on Electronic Computers*, **EC-6,** no. 2, 72–74 (June 1957).

EXERCISES

1. Design a four flip-flop counter which counts in the ordinary binary sequence, as shown in Table 5-12. That is, write the input equations for flip-flops *A*, *B*, *C*, and *D*. Use *R-S* flip-flops, and simplify the input equations as

Table 5-12

A	*B*	*C*	*D*
0	0	0	0
0	0	0	1
0	0	1	0
0	0	1	1
0	1	0	0
0	1	0	1
0	1	1	0
0	1	1	1
1	0	0	0
1	0	0	1
1	0	1	0
1	0	1	1
1	1	0	0
1	1	0	1
1	1	1	0
1	1	1	1
0	0	0	0

etc.

much as possible. Repeat the exercise, using *J-K* flip-flops; using *D* flip-flops; using *T* flip-flops.

2. Draw a schematic circuit diagram (cf. Figure 5-1) for (*R-S*) flip-flops *C* and *D* of Exercise 1. Sketch the waveforms for the voltages on wires *C*, *D*, $\overline{C}$, $\overline{D}$, R_C, S_C, R_D, and S_D during six consecutive clock-pulse times.

3. Each of the truth tables shown in Table 5-13 defines the logical properties of a particular kind of memory element. Each of these elements has two

Table 5-13

I_1^n	I_2^n	Q^{n+1}	I_1^n	I_2^n	Q^{n+1}	I_1^n	I_2^n	Q^{n+1}	I_1^n	I_2^n	Q^{n+1}
0	0	Q^n	0	0	Q^n	0	0	Q^n	0	0	0
0	1	Q^n	0	1	$\overline{Q}^n$	0	1	0	0	1	$\overline{Q}^n$
1	0	0	1	0	0	1	0	1	1	0	$\overline{Q}^n$
1	1	1	1	1	1	1	1	1	1	1	1

inputs, I_1 and I_2. Write the characteristic equation for each memory element. Combine each characteristic equation with the general application equation ($Q^{n+1} = g_1Q + g_2\overline{Q}$), and find the most general expression for each pair of inputs as a function of g_1, g_2, and Q. Find the *simplest* expression for each pair of inputs as a function of g_1, g_2, and Q.

4. Design the binary counter of Exercise 1 using each of the memory elements of Exercise 3 in turn. Find the simplest set of input equations for each counter.

Q_1	Q_2
0	0
0	1
1	0
0	0
etc.	

5. An inventor produces a single-input memory device which has the following logical properties: when the input line (I) is "zero" at time n, the device will be in the "zero" state at time $(n + 1)$; when the input line is "one" at time n, the device will change state by bit-time $(n + 1)$. He wishes to build a three-counter from two of these devices (Q_1 and Q_2) which counts in the sequence shown to the left. Write the input logic (I_{Q1} and I_{Q2}) for the inventor. What would you advise him about the general usefulness of his device, and why?

6. By the addition of suitable logic, it is possible to make one kind of memory element look like another one. For example, a *D* flip-flop may be converted

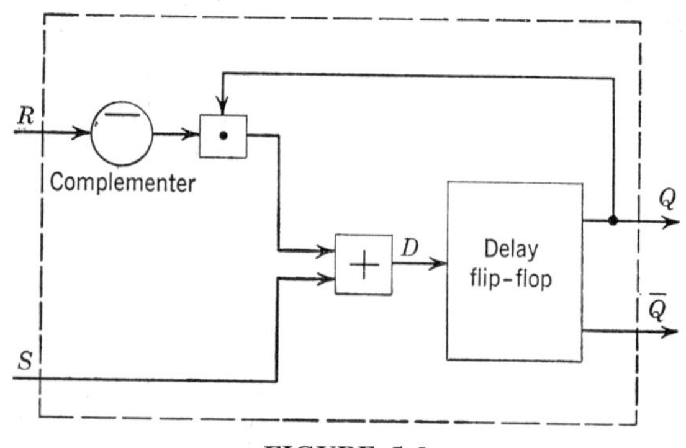

FIGURE 5-8

to an *R-S* flip-flop by the addition of the logic shown in Figure 5-8. Draw similar diagrams for each of the following transformations.

a. Convert a *D* flip-flop into a *J-K* flip-flop.
b. Convert a *D* flip-flop into an *R-S-T* flip-flop.
c. Convert a *T* flip-flop into a *J-K* flip-flop.
d. Convert a *J-K* flip-flop into a *T* flip-flop.
e. Convert an *R-S-T* flip-flop into a *D* flip-flop.

7. Write the simplest possible logic for each of the counters shown in Table 5-14.

Table 5-14

A	*B*	*C*
0	0	0
0	0	1
1	0	1
1	0	0
1	1	0
1	1	1
0	1	1
0	1	0
0	0	0

etc.

A	*B*	*C*
0	0	0
0	1	1
1	0	1
1	1	0
0	0	1
0	1	0
1	0	0
1	1	1
0	0	0

etc.

A	*B*	*C*	*D*
0	0	0	0
0	0	0	1
0	0	1	0
0	1	0	0
1	0	0	0
1	0	0	1
1	0	1	0
1	1	0	0
1	1	0	1
1	1	1	0
1	1	1	1
1	0	1	1
0	1	1	1
0	0	1	1
0	1	1	0
0	1	0	1
0	0	0	0

etc.

Design each counter using:

a. *D* flip-flops.
b. *T* flip-flops.
c. *J-K* flip-flops.
d. *R-S* flip-flops.
e. *R-S-T* flip-flops.

8. Write the simplest possible logic for each of the counters given in Table 5-15. (Note that, in each counter, certain configurations never occur, and these redundancies may be used in simplification. For example, in the first counter $\bar{A}\bar{B}\bar{C} = 0$.)

Design each counter using:

a. *D* flip-flops.
b. *T* flip-flops.
c. *J-K* flip-flops.

Table 5-15

A	B	C
1	0	0
1	1	0
1	1	1
0	1	1
1	0	1
0	1	0
0	0	1
1	0	0
etc.		

A	B	C
0	0	0
0	1	1
1	1	0
1	0	1
0	1	0
0	0	0
etc.		

A	B	C	D
0	0	0	0
1	0	0	0
1	1	0	0
1	0	1	0
1	1	1	0
0	0	0	1
1	0	0	1
1	1	0	1
1	0	1	1
1	1	1	1
0	0	0	0
etc.			

d. *R-S* flip-flops.
e. *R-S-T* flip-flops.

9. The simplified input logic for a T flip-flop (Q) is $T_Q = \bar{B}\bar{C} + BC + \bar{A}Q$, where $ABC\bar{Q} = 0$. Write the simplest input logic for flip-flop Q if it is to be

a. A D flip-flop.
b. A *J-K* flip-flop.
c. An *R-S* flip-flop.
d. An *R-S-T* flip-flop.

6 / *The derivation of application equations*

INTRODUCTION

In the last chapter, we learned how to make the transition from an application equation, which describes precisely the function a flip-flop must perform, to the input equations for that flip-flop. We also saw in a very simple example how to derive an application equation. It is the job of the logical designer to provide input equations for appropriate memory elements, given only a general system description of the operation of the computer. The system description is often specified in somewhat vague terms so that the logical designer has a great deal of freedom in determining on a specific design. It is the purpose of this chapter to describe some general methods for getting from a system description to the application equations or their equivalent.

Because of the nature of this problem and more specifically because it is difficult to describe precisely the data with which the logical designer begins, it is not possible to set down an invariant list of rules or procedures for the logical designer to follow. It is, nevertheless, possible and desirable to describe certain methods of design which have been used and can be used in synthesizing computers. For the most part, the methods will be *described* here and nothing more. The techniques will be applied to particular design problems in later chapters.

A GENERAL MODEL FOR A DIGITAL SYSTEM

Huffman and Moore have proposed a very general model for a digital system (or what they call a *sequential switching circuit*) which can serve as a starting point for our discussion. The model is illustrated in Figure 6-1. It consists of a black box having a certain number of inputs

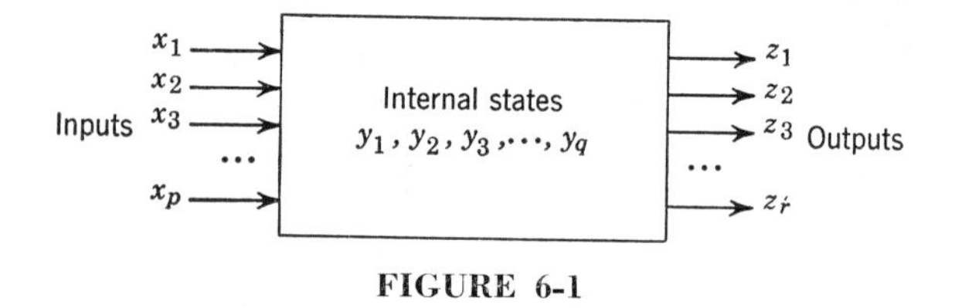

FIGURE 6-1

x_i, a certain number of outputs z_i, and a certain number of internal states y_i. At any bit-time, the state of each of the output lines z_i is a function of the state of all the input lines at that time and of all the internal states at that time. This general relationship is indicated in equation 6-1.

$$z_i^n = f_i(x_1, x_2, \cdots, x_p; y_1, y_2, \cdots, y_q)^n \qquad i = 1, 2, 3, \cdots, r \quad (6\text{-}1)$$

Furthermore, the internal state of the machine—which depends on the state of the memory devices inside the box—is determined by the relationship indicated in equation 6-2.

$$y_i^{n+1} = g_i(x_1, x_2, \cdots, x_p; y_1, y_2, \cdots, y_q)^n \qquad i = 1, 2, 3, \cdots, q \quad (6\text{-}2)$$

This will be recognized as a set of difference equations determining the state of every memory element inside the box at one bit-time as a function of the state of all of the input lines and all of the memory units at the previous bit-time.

This model is general enough that it can represent the operation of any digital system. For example, the box may be a general-purpose digital computer, in which case the input lines are wires from a paper tape reader, say, the output lines are wires to the solenoids of a printer, and the internal memory devices include flip-flops, magnetic cores, magnetic tapes and drums, etc. There are certain minor difficulties with the model (e.g., the synchronizing of slow input and output devices with fast internal memory devices raises some difficulties) but theoretically all of these can be overcome.

The real difficulty with the practical application of this model is that it is too general. It recognizes and allows for the possibility of various permutations and combinations of internal states which in practice are not used, and it therefore introduces complications where none exist. Furthermore, it does not of itself suggest a procedure for the logical designer to follow in translating a system description of a computer into a description in terms of this model. If it did, all the designer's problems would be solved, for equations 6-2 are the application equations for all memory elements in the computer and equations 6-1 determine the state of all the output lines.

Because this model is conceptually too difficult to handle, it is often broken down into at least five parts, as illustrated in Figure 6-2. Each

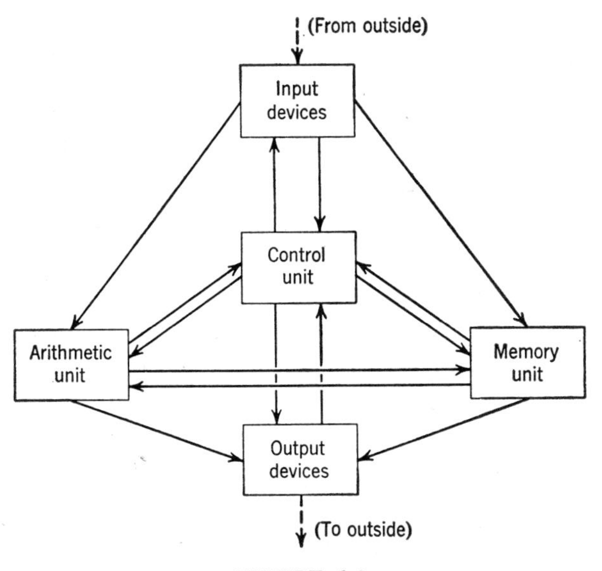

FIGURE 6-2

of these five parts can itself be considered to be a black box having the characteristics of the Huffman-Moore model, and in subdividing the computer in this way we make five simpler problems out of one very difficult one. (Note the relationship between these five parts and the functional description of a digital computer given in Chapter 1.)

Applying this technique of "divide and conquer" to the computer synthesis problem, we settle on a method which may be described as follows. The logical designer begins with a system study which specifies computer inputs, outputs, word lengths, operations to be performed, speeds required, etc. From these specifications he postulates a set of five models as shown in Figure 6-2: one for the input devices, one for the memory unit, one for the arithmetic unit, one for the output devices, and one for the control unit. For each of these models he then describes in words the input and output lines required to and from each, and he writes a description setting out what the output lines should do as a function of the input lines and of time. From this description, the bit-time by bit-time operation of each of the boxes may be derived, a suitable number of memory elements may be determined upon, and the application equations for these memory devices may be written.

Subsequent chapters of this book will describe in some detail the design of each of these five parts of the computer. However, before going any further, it will be useful to discuss each of these parts in a little more detail in order to fix clearly in our minds the meanings of the terms here used.

The memory unit is illustrated in Figure 6-3. For our purposes, a memory unit will consist of a device containing a very large number

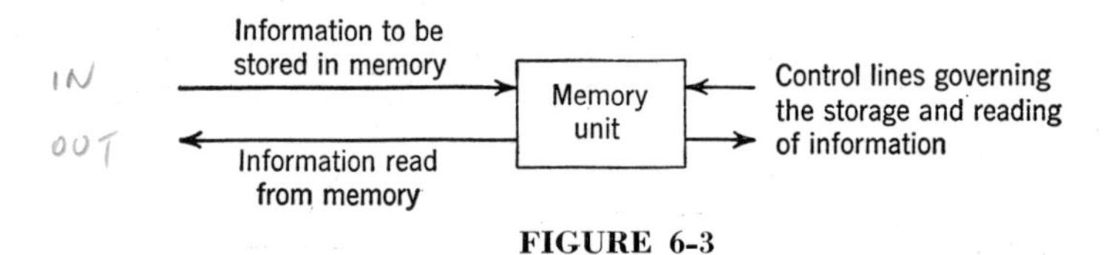

FIGURE 6-3

of bits of information (usually 1000 bits or more) whose primary purpose it is to store a large volume of information without modifying it. Storage of information implies both that the computer can write information in the memory unit and can read information from it without intervention by a human operator. As is indicated in Figure 6-3, the memory unit must be provided with input and output lines for transferring information to and from the memory, and with control lines which govern the storage and reading of information. Examples of memory devices are magnetic core memories, magnetic drums, magnetic tapes (all three described in Chapter 7), mercury delay lines, electrostatic storage tubes, etc.

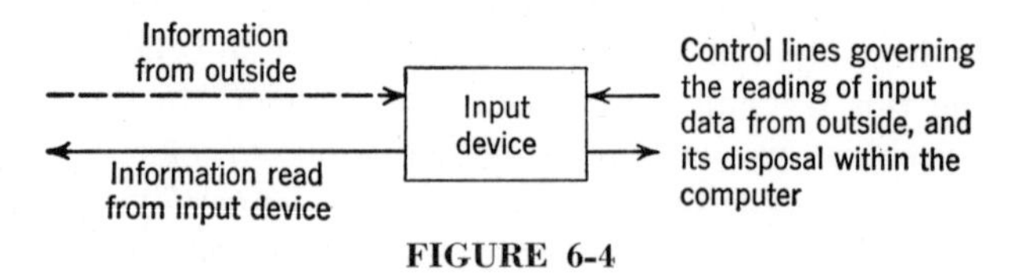

FIGURE 6-4

A generalized input device is illustrated in Figure 6-4. An input device makes it possible for the computer system to accept information from outside itself and to read that information into the computer.

It serves as a translator from the language of the outside world into computer language. Figure 6-4 therefore shows a dotted line representing information which comes from outside and a solid line which represents information read from the input device to the digital system. Control lines are also necessary to govern the reading of the input data. Examples of input devices are punched card readers, punched tape readers, print readers, and analog-to-digital converters.

An output device is illustrated in Figure 6-5. Just as an input device supplied information to the computer from the outside world, so the

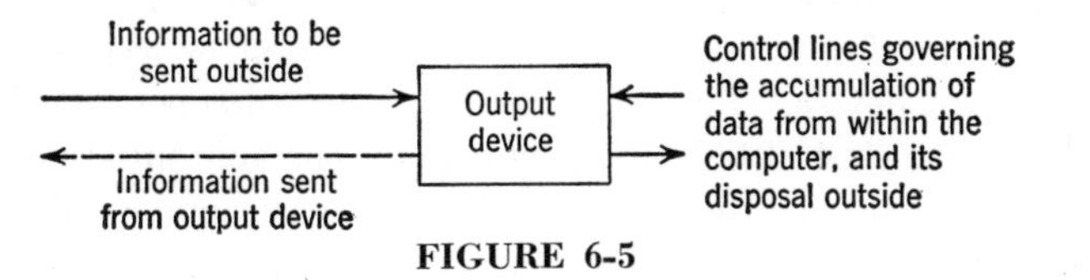

FIGURE 6-5

output device translates and displays information sent from the computer in computer language. Like the input device, the output device must be supplied with control lines which govern the writing of output data. Examples of output devices are printers, paper tape and paper card punches, and digital-to-analog converters. Input and output devices will be described in Chapter 8.

An arithmetic unit is illustrated in Figure 6-6. The principal function of the arithmetic unit is to execute operations on information sent to it from other parts of the computer. The operations executed may

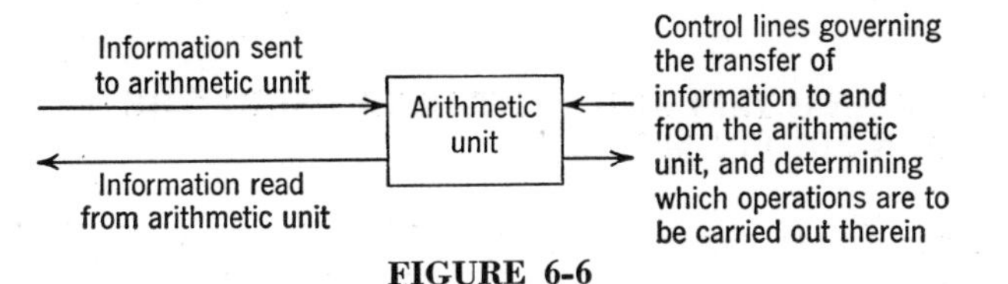

FIGURE 6-6

be of any kind as long as each operation is defined by a rule or a set of rules which tell exactly how the result of the operation is produced. The arithmetic unit may contain enough storage capacity to hold temporarily one or more of the operands and results involved in the various operations. The common operations of addition, subtraction, multiplication, and division are found in most machines, but an arithmetic unit might also be required to extract roots, compare numbers, shift them to the left and to the right, or perform still other operations not usually thought of as arithmetic. Arithmetic unit design will be discussed in Chapter 9.

The last and most difficult unit to describe is the control unit. Referring again to Figure 6-2, we see that the control unit receives information from each of the other units and in turn transmits information back to each of the others. It is the function of the control unit to organize and to co-ordinate the operation of the rest of the computer. When it is time for some input information to be read to the memory, the control unit stimulates the input device appropriately and sets up a path from the input device to the proper place in the arithmetic or memory unit. Similarly, when an output of information is necessary, the control unit must stimulate the output device and arrange that the proper information be sent from the memory unit or arithmetic unit to the output device. Finally, when an arithmetic operation is to be performed, the control unit must arrange for the appropriate operands to be transferred from the memory unit to the arithmetic unit, for the proper operation to be carried out, and for the result to be stored back in the memory unit if that is required. The control unit is generally the last part of a computer to be designed. Each of the other computer subdivisions is detailed and the signals which must later be supplied to it from the control unit are specified. The control unit is then finally synthesized and made to supply all the appropriate signals at the appropriate times. In Chapter 11 we shall show how a control unit may be designed.

The remainder of this chapter will illustrate in considerably more detail the techniques which can be used in arriving at application equations and flip-flop input equations, using the model of Figure 6-1.

CIRCUIT ANALYSIS AND SYNTHESIS USING THE HUFFMAN-MOORE MODEL

Because memory elements are expensive, it is always the logical designer's objective to use as few as possible in design. However, the total cost of a computer depends on the number of *decision* elements used as well as the number of memory elements. In any computer design organization, the circuit designer must supply the logical designer with data about the relative costs of various circuit components so that, for example, it will be possible to decide which is cheaper—a circuit containing two flip-flops and fifty diodes or another equivalent one requiring three flip-flops but only thirty diodes.

Huffman and Mealy have devised a method which will, under certain circumstances, make it possible for the designer to minimize the number of memory elements necessary in a black box which is to perform a given function without regard to the number of decision elements necessary. Although this simplification procedure thus ignores a very important aspect of simplification, it is a useful tool for the logical designer to have at his disposal. We will therefore first describe the method devised by Huffman and Mealy for simplification and will then show how the techniques they use can also be employed in circuit synthesis.

Analysis

Suppose we are given a circuit containing three flip-flops, A, B, and C, and an input wire, x, from some other switching circuit. At each bit-time there may be either a "zero" or a "one" on the input line x, and the next state of flip-flops A, B, and C will depend on their previous state and upon the state of the input line. This relationship is shown in Table 6-1, which also displays the state of an output line, z,

Table 6-1

	Time n			Time $(n+1)$			Time n
x	A	B	C	A	B	C	z
0	0	0	0	0	0	0	0
0	0	0	1	0	1	1	1
0	0	1	0	1	0	0	1
0	0	1	1	1	1	0	0
0	1	0	0	0	0	0	0
0	1	0	1	0	1	1	1
0	1	1	0	1	0	0	1
0	1	1	1		not allowed		
1	0	0	0	0	0	1	0
1	0	0	1	0	1	0	1
1	0	1	0	0	0	1	1
1	0	1	1	1	0	1	0
1	1	0	0	0	0	1	0
1	1	0	1	1	1	0	1
1	1	1	0	1	0	1	1
1	1	1	1		not allowed		

at each of these bit-times. From this table, it is obviously possible to draw up difference equations for flip-flops A, B, and C, and to derive the input equations for any type of flip-flops desired. The logic defining the state of output wire z at each bit-time may also be written. However, we may reasonably ask the questions: Are all three flip-flops really necessary to carry out the function indicated by this truth table? Could the same function be carried out with fewer flip-flops?

The methods derived by Huffman and Mealy to answer these questions first require that we rearrange the truth table slightly as shown in Table 6-2. This has the effect of separating the internal states from

Table 6-2

Internal State at Time n	Internal State at Time $(n+1)$		Output at Time n	
	for $x^n = 0$	for $x^n = 1$	for $x^n = 0$	for $x^n = 1$
$(A\ B\ C)^n$	$(A\ B\ C)^{n+1}$	$(A\ B\ C)^{n+1}$	z^n	z^n
0 0 0	0 0 0	0 0 1	0	0
0 0 1	0 1 1	0 1 0	1	1
0 1 0	1 0 0	0 0 1	1	1
0 1 1	1 1 0	1 0 1	0	0
1 0 0	0 0 0	0 0 1	0	0
1 0 1	0 1 1	1 1 0	1	1
1 1 0	1 0 0	1 0 1	1	1

the inputs and outputs. Next, we note that the various states of A, B, and C are quite arbitrarily chosen. Our real interest in this table, however, is not the state of individual flip-flops within the circuit, but the state of the circuit as a whole. Since there are seven possible combinations of A, B, and C allowed in the circuit, we can say it may take on seven different states; and instead of identifying these with combinations of flip-flop states, we could identify them by giving each of them a name. Let us name them by using the letters a, b, c, d, e, f, and g. We can then form Table 6-3 from Table 6-2 by replacing the combination 000 by a, 001 by b, 010 by c, etc. The information on this table can also be represented diagrammatically as shown in Figure 6-7.

Table 6-3

Internal State at Time n	Internal State at Time $(n+1)$		Output z at Time n	
	for $x^n = 0$	for $x^n = 1$	for $x^n = 0$	for $x^n = 1$
a	a	b	0	0
b	d	c	1	1
c	e	b	1	1
d	g	f	0	0
e	a	b	0	0
f	d	g	1	1
g	e	f	1	1

Here, each of the seven states is indicated by a circle. The state to which the circuit goes upon receipt of a particular input at a given bit-time is indicated by two lines drawn from each circle: one to the new state entered when $x = 1$, the other to that entered for $x = 0$. Furthermore, the output associated with each state is indicated by the numeral 0 or 1 written in the circle associated with that state. Note that the truth table, Table 6-3, and Figure 6-7 are entirely equivalent to one another and are perfectly general systems for representing the operation of any sequential switching unit. Note also that Table 6-3

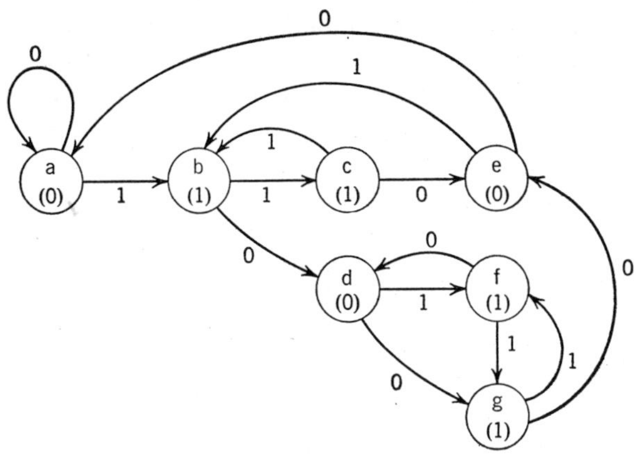

FIGURE 6-7

may be extended to handle circuits for which there are many input lines and many output lines. For example, in Table 6-4 there is shown a part of a table describing a system having two input lines and three output lines. The meaning of the table may be illustrated by examin-

Table 6-4

Internal State at Time n	Internal State at Time $(n+1)$				Outputs (z_1, z_2, z_3) at Time n			
	$(x_1x_2)^n$ 0 0	$(x_1x_2)^n$ 0 1	$(x_1x_2)^n$ 1 0	$(x_1x_2)^n$ 1 1	$(x_1x_2)^n$ 0 0	$(x_1x_2)^n$ 0 1	$(x_1x_2)^n$ 1 0	$(x_1x_2)^n$ 1 1
a	g	f	h	e	001	101	011	111
b	b	c	d	a	010	010	000	000
c	c	d	e	f	001	001	111	101
d	–	–	–	–	–	–	–	–
.	–	–	–	–	–	–	–	–
.	–	–	–	–	–	–	–	–
.	–	–	–	–	–	–	–	–

ing the first two sets of entries on the first row. They state that, first, if the circuit is in state a at bit-time n and the input lines x_1 and x_2 are 0 0 at that bit-time, it will next go into state g, and the three output lines will have values, 001. The second entry states that if the circuit is initially in state a and inputs are 0 1 for x_1 and x_2, the outputs at that bit-time will be 101 for z_1, z_2, and z_3 and the circuit will next go into state f.

Returning now to the example of Table 6-3, we seek the answer to the question posed at the beginning of this section: Can we carry out the function here specified with fewer than seven states? Huffman and Mealy devised two rules for combining states with one another and thus eliminating unnecessary states. We first note that if we wish to combine state i with state j we must proceed by eliminating one of these states from the table entirely, say state i, and by replacing the letter i by the letter j wherever i appears on the chart. Next we observe that it is impossible to combine or to merge two states which have different output configurations, for each output combination represents a different action to be taken by the circuit and these actions cannot be changed without completely changing the function of the circuit. Referring to Table 6-3, we see therefore that it will be impossible to merge states a and b, for example, but that it may be possible to merge states a, d, and e or some pair of them and that it likewise may be possible to combine states b, c, f, and g, or some pair of them We can, therefore, write down the states which are candidates for com bination in the following way.

(a, d, e)   (b, c, f, g)

Once we have arranged the original states into sets whose members have identical outputs, we can attempt to reduce the number of states necessary by applying two rules. Rule I may be stated as follows: Two states may be combined if they are in the same set and if each combination of inputs causes the two states to go to the same new state. Putting it another way, two rows may be combined if all of the entries on the "next state" columns and the output columns are exactly the same for both rows. Applying this to the example of Table 6-3, we see at once that states a and e may be combined, for an input of "zero" takes both of them into state a and an input of "one" takes them both into state b. Eliminating state e and replacing it wherever it appears in the table by state a, we arrive at Table 6-5. The states which remain and are candidates for combination are indicated as follows:

(a, d)   (b, c, f, g)

Table 6-5

Present State	Next State		Output	
	$x = 0$	$x = 1$	$x = 0$	$x = 1$
a	a	b	0	0
b	d	c	1	1
c	a	b	1	1
d	g	f	0	0
f	d	g	1	1
g	a	f	1	1

Examining these combinations, we see there is no further opportunity for the application of rule I and we are therefore forced to turn to the second rule.

Rule II may be applied by carrying out a sequence of steps, as follows:

1. Group the states into sets (as described above) where all members of each set have the same output combination.
2. Assign an arbitrary number to each set. For example, in Table 6-5 we may identify the sets as follows:

1	2
(a, d)	(b, c, f, g)

3. Examine the truth table which defines the operation of the circuit and note down, under each state, the numbers of the *sets* into which that state will change for each of the various input combinations. For example, referring to Table 6-5, row a, we see that a "zero" input will leave the circuit in state a (set 1), and a "one" input will take it to state b (set 2). We therefore write the numerals 1, 2 beneath state a. Next, we examine state d. A "zero" or a "one" input occurring when the circuit is in state d will take the circuit into states g and f respectively. Since both g and f are in set 2, we write 2, 2 beneath d. The same procedure applied to the other states results in the configuration shown below:

1		2			
(a,	d)	(b,	c,	f,	g)
1, 2	2, 2	1, 2	1, 2	1, 2	1, 2

4. We can now examine each set to determine whether it must in fact be divided into two or more new sets. Each new set is formed (*a*) from members of only *one* of the old sets and (*b*) from only the

members of that set which have the same numbers written beneath them. Examining set 1 above, we see that states a and d have two different pairs of numbers written beneath them, and we therefore conclude that this set must be broken into two new sets, one containing a and the other d. The second set, however, need not be broken up because the same pair of numbers, the combination 1, 2, appears under each pair of letters.

5. We now have a new group of sets to which we must again apply steps 2, 3, and 4 above. When this cycle of steps has been applied until it can be applied no longer, the *sets* which still remain correspond to the different *states* that are required in the circuit being analyzed. If any set contains more than one state, all these states may be combined into one.

Applying this series of steps to the example above, we carry out the following sequence of operations.

1	2	3			
(a)	(d)	(b,	c,	f,	g)
		2, 3	1, 3	2, 3	1, 3

1	2	3		4	
(a)	(d)	(b,	f)	(c,	g)
		2, 4	2, 4	1, 3	1, 3

We now see that the third and fourth sets need not be broken down any further. States b and f may be combined, for they provide the same output *and* they give rise to the same state-changes for every input combination. Similarly, states c and g may be combined. The result, Table 6-6, describes a circuit which has the same function as

Table 6-6

Present State	Next State		Output	
	$x = 0$	$x = 1$	$x = 0$	$x = 1$
a	a	b	0	0
b	d	c	1	1
c	a	b	1	1
d	c	b	0	0

that of Table 6-3, but requires four states (and therefore two memory elements) instead of seven (and three). The change-of-state diagram corresponding to this new configuration is given in Figure 6-8, and in

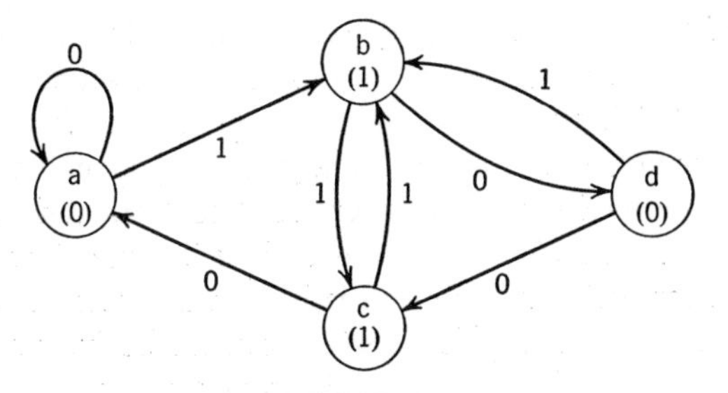

FIGURE 6-8

Table 6-7 are shown the truth table and input equations for two flip-flops, A and B, which may be used to carry out this function.

Table 6-7

Time n			Time $(n+1)$		Time n
x	A	B	A	B	z
0	0	0	0	0	0
0	0	1	1	1	1
0	1	0	0	0	1
0	1	1	1	0	0
1	0	0	0	1	0
1	0	1	1	0	1
1	1	0	0	1	1
1	1	1	0	1	0

$$z = A\bar{B} + \bar{A}B$$
$$J_A = B$$
$$K_A = x + \bar{B}$$
$$J_B = x$$
$$K_B = A\bar{x} + \bar{A}x$$

The two rules described above are perfectly general and may be used to simplify any sequential switching circuit. (In fact, rule II alone may be used, though rule I is so easy to apply that it is often used first as a means of eliminating superfluous states quickly.) Furthermore, the rules may easily be extended to the more practical case where forbidden combinations are present by remembering that these redundancies may be given *any* value whenever two rows are combined together. We shall see some later examples of simplifications involving redundancies.

Before going on to the problem of circuit synthesis, it is desirable to point out some of the problems which are *not* solved by the Huffman-Mealy procedure. We have already mentioned that the procedure does not necessarily give the simplest total system for carrying out a given operation. It only permits the designer to find the fewest number of necessary states under certain circumstances. It is definitely not possible to say that the cheapest configuration is always the one with the least number of memory elements. It is quite easy to construct examples where an extra flip-flop or two contribute simplifications in logic which save diodes worth much more than the cost of the flip-flops. Second, the Huffman-Mealy method gives no help on the problem of how to choose flip-flops to represent the different states in such a way that the logic is minimized. This is a very important problem and one for which no solution is known at present. Finally, the Huffman-Mealy method unfortunately cannot be *guaranteed* to give the fewest number of states possible except when there are no redundancies present in the circuit. We can be certain that the example represented by Table 6-6 has the fewest possible states because there were no redundancies present. It is, however, possible to construct examples including redundancies in which the sequence of combinations carried out influences and determines the minimum number of states found. (See Exercise 6-3.) Notwithstanding these difficulties and disadvantages, the Huffman-Mealy method provides a useful tool for circuit simplification.

Synthesis

The method of analysis described in the last section enables us to start with any circuit which carries out a prescribed function and to reduce the number of states required to carry out that function, if such reduction is possible. Furthermore, as we have seen, the tables employed to carry out this reduction are difference-equation tables from which the application equations for particular flip-flops may be derived. We shall now discuss the *synthesis* of sequential switching circuits using the Huffman-Moore model, and illustrate the method with two examples.

The general method of synthesis may be stated as follows: First, assume the circuit being designed exists in some internal state a. Then for each of the possible combinations of input variables, write down a state to which the circuit should go from state a. At the same time, specify what the outputs should be for each of these input combinations. Now for each of the states into which the circuit could have gone from this first state, repeat the process of defining new states which correspond to different input configurations. Always remember that some input combinations occurring when the circuit is in a given state may return that circuit to some state which has already been defined, so that it is unnecessary to continue adding new states indefinitely. Because all practical problems involve only a finite number of internal states, it will always be possible eventually to end the process by returning all states to states which have already been defined.

As soon as enough states have been defined completely to describe the operation of the circuit, the Huffman-Mealy rules described in the last section may be applied to reduce the number of internal states required. When this reduction has been carried out, the resulting table may be employed to draw up a difference-equation truth table and the input equations may be derived therefrom.

To indicate how this general method can be applied in practice, let us carry out the complete design of a very simple sequential circuit. This circuit is illustrated in Figure 6-9 and may be described as follows: A

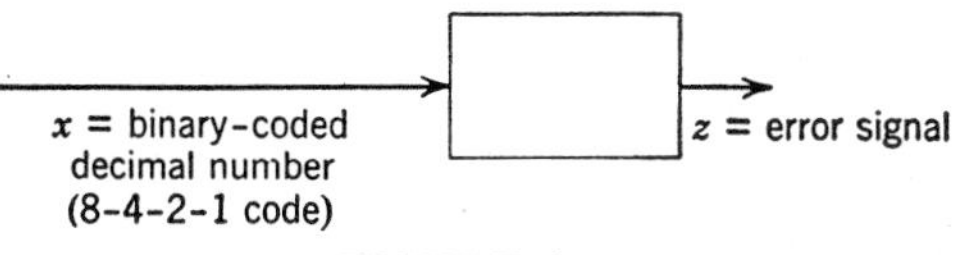

FIGURE 6-9

series of binary-coded decimal digits (8-4-2-1 code) appear, least significant bit first, on input line x. Each decimal digit requires four bit-times, and at the fourth bit-time it is desired that the output from the circuit, z, be "zero" if the preceding four bits actually represented a decimal digit between zero and nine and be "one" if the preceding four bits represented a not-allowed digit—10, 11, 12, 13, 14, or 15. The circuit is to repeat this error-detecting function indefinitely.

We begin the synthesis by drawing up the table shown as Table 6-8. Internal state a is postulated as the state of the circuit at t_1 time, i.e.,

Table 6-8

Present State	Next State		Output		(Notes)
	$x = 0$	$x = 1$	$x = 0$	$x = 1$	
a	b	c	0	0	t_1 time (start)
b	d	e	0	0	t_2
c	f	g	0	0	
d	h	i	0	0	t_3
e	j	k	0	0	
f	l	m	0	0	
g	n	o	0	0	
h	a	a	0	0	t_4
i	a	a	0	1	
j	a	a	0	1	
k	a	a	0	1	
l	a	a	0	0	
m	a	a	0	1	
n	a	a	0	1	
o	a	a	0	1	

at the time the least significant bit of a given decimal digit occurs. If this least significant bit x is "zero," the circuit next goes to state b. If it is "one," the circuit next goes to state c, and these states are indicated in Table 6-8. We now have to show what happens to the circuit for each of the two input possibilities when the circuits start in state b or in state c. We arbitrarily say that a "zero" in state b or state c takes the circuit to state d and state f respectively, and a "one" in those two states takes the circuit to either state e or state g. At the third bit-time, then, the circuit will be in either state d, e, f, or g, and depending on the state of the input line we can define eight more states from h to o in one of which the circuit must be at t_4 time. Whatever state the circuit is in at t_4 time, it will return to state a at t_1 time and so a is indicated as the next state for all possible input combinations in states h through o.

We must also define the output signal z for each state and for each of the input combinations. We of course want no output at all until t_4 time, so all outputs are "zero" for states a through g. Furthermore, at bit-time t_4 the circuit will be in one of eight different states determined by what the three previous digits were. At that same time, a new input will be either a "zero" or a "one" and these two possibilities from each of eight states give the sixteen possible binary numbers having four binary digits. Six of those sixteen are not proper binary-coded decimal digits and must therefore cause a "one" on the output line. To determine which six they are, we need only identify each of the states h through o with the binary code from which it arises and put a "one" in the appropriate output column for each of the six forbidden combinations. This is done in Table 6-8 where, for example, the "one" occurring in state i when $x = 1$ corresponds to the code 1100 or 12. Each of the other "ones" in the z column may be identified with one of the forbidden combinations.

Table 6-9

Present State	Next State		Output		(Notes)
	$x = 0$	$x = 1$	$x = 0$	$x = 1$	
a	b	b	0	0	t_1
b	d	e	0	0	t_2
d	h	i	0	0	t_3
e	i	i	0	0	
h	a	a	0	0	t_4
i	a	a	0	1	

We are now ready to apply the Huffman-Mealy rules for simplification. The resulting simplification is shown in Table 6-9 where we have applied rule I as follows: lines j, k, m, n, and o may be combined with line i. Line g may then be combined with line e and line f with line d. Finally, line c may be combined with line b. In Table 6-10, rule II is

Table 6-10

1					2
(a	b	d	e	h)	(i)
1	1	1	2	1	
1	1	2	2	1	

1			2	3	4
(a	b	h)	(d)	(e)	(i)
1	2	1			
1	3	1			

1		2	3	4	5
(a	h)	(b)	(d)	(e)	(i)
2	1				
2	1				

applied to Table 6-9 with the result that we decide no further simplification is possible.

It is now possible to assign flip-flop states to each of the internal states required in Table 6-9. Since there are six states there must be at least three flip-flops. In Table 6-11 we postulate three flip-flops, A,

Table 6-11

	Present State			Next State						Output $= z$	
				$x = 0$			$x = 1$			$x = 0$	$x = 1$
	A	B	C	A	B	C	A	B	C		
a	0	0	1	1	0	1	1	0	1	0	0
b	1	0	1	1	1	0	1	0	0	0	0
d	1	1	0	1	1	1	0	1	1	0	0
e	1	0	0	0	1	1	0	1	1	0	0
h	1	1	1	0	0	1	0	0	1	0	0
i	0	1	1	0	0	1	0	0	1	0	1

B, and C, and assign states to these flip-flops corresponding to states a, b, d, e, h, and i. From this table we can write the difference equations for flip-flops A, B, and C and can derive the corresponding input equations. The difference equations, input equations for J-K flip-flops, and the equation for the output line z are given below.

$$\begin{aligned} A^{n+1} &= (\bar{B}C + B\bar{C}\bar{x})^n \\ B^{n+1} &= (\bar{C} + A\bar{B}\bar{x})^n \\ C^{n+1} &= (\bar{A} + B + \bar{C})^n \end{aligned} \tag{6-3}$$

$$z^n = (\bar{A}Bx)^n \tag{6-4}$$

$$\begin{aligned} J_A &= \bar{B} \qquad & K_A &= BC + \bar{B}\bar{C} + Bx \\ J_B &= \bar{C} + A\bar{x} \qquad & K_B &= C \\ J_C &= 1 \qquad & K_C &= A\bar{B} \end{aligned} \tag{6-5}$$

Two comments are in order about this solution. First, a little study of Table 6-9 should convince the student that he could have written that table down in the beginning rather than starting with Table 6-8. It should be evident, for example, that a forbidden code may be recognized by the fact that a "one" occurs at t_4 time and at t_3 time, or at t_4 time and at t_2 time, or both. The state of the input at t_1 time is immaterial and so state b may be entered from state a regardless of what x was in state a. It is then necessary to distinguish whether a "zero" or a "one" occurred at t_2 time, so both states d and e are required. A "zero" occurring in state d corresponds to a "zero" both at t_2 and t_3 time, and therefore means that whatever the input is at t_4, the code is a permissible one representing either the digits zero, one, eight, or nine. However, a "one" occurring when the circuit is in state d and a "zero" or a "one" occurring in state e mean that the incoming code represents a binary number of two or greater, and that the final question as to whether the code is allowed or not will be determined by whether or not a "one" occurs at t_4 time. State i is provided to account for these contingencies. Very often when a circuit has been reduced to its simplest possible form using the Huffman-Mealy method, it will be obvious that this reduced state could have been arrived at in the beginning, had the designer been more clever. In designing complicated circuits, however, it is difficult to be clever.

The second comment to be made about the solution of this problem is that the particular mechanization described by equations 6-3, 6-4, and 6-5, although it certainly represents the minimum number of internal states necessary for a solution to the problem, does not necessarily represent the solution which gives the cheapest possible mechanization for this function. Other arrangements of flip-flops and other kinds of flip-flops may give a solution which is better by some measure of value.

Before leaving the subject of circuit synthesis, let us solve this same problem with slightly different boundary conditions. Let us suppose, referring to Figure 6-10, that we still have an input line x containing a sequence of serial binary-coded decimal digits. However, suppose we also have four timing lines, t_1, t_2, t_3, and t_4, which identify the four bit-times of each decimal digit. Each of these four lines contains a

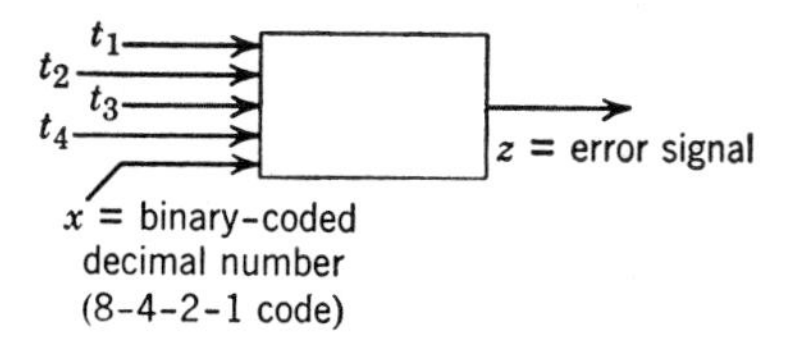

FIGURE 6-10

"one" at the appropriate bit-time and no two lines are "one" at the same time. For example, if the decimal number 943 appeared on line x, the states of lines x, t_1, t_2, t_3, and t_4 would be as shown in Table 6-12.

Table 6-12

	x	t_1	t_2	t_3	t_4
	1	1	0	0	0
3	1	0	1	0	0
	0	0	0	1	0
	0	0	0	0	1
	0	1	0	0	0
4	0	0	1	0	0
	1	0	0	1	0
	0	0	0	0	1
	1	1	0	0	0
9	0	0	1	0	0
	0	0	0	1	0
	1	0	0	0	1

The circuit of Figure 6-10 thus has five input lines having a total of eight possible input combinations at any bit-time. The problem as before is to design a detection circuit which will make z a "one" at t_4 time whenever a digit greater than nine has been read from line x.

We can begin exactly as we began in the previous synthesis problem with a table listing exhaustively all of the possible input combinations. Table 6-13 then corresponds exactly to Table 6-8. However, it contains a number of dashes, each of which indicates that the corresponding configuration is not allowed. For example, in Table 6-13, it is not

Table 6-13

Present State	Next State								Output							
Time →	t_1		t_2		t_3		t_4		t_1		t_2		t_3		t_4	
x →	0	1	0	1	0	1	0	1	0	1	0	1	0	1	0	1
a	b	c	–	–	–	–	–	–	0	0	–	–	–	–	–	–
b	–	–	d	e	–	–	–	–	–	–	0	0	–	–	–	–
c	–	–	f	g	–	–	–	–	–	–	0	0	–	–	–	–
d	–	–	–	–	h	i	–	–	–	–	–	–	0	0	–	–
e	–	–	–	–	j	k	–	–	–	–	–	–	0	0	–	–
f	–	–	–	–	l	m	–	–	–	–	–	–	0	0	–	–
g	–	–	–	–	n	o	–	–	–	–	–	–	0	0	–	–
h	–	–	–	–	–	–	a	a	–	–	–	–	–	–	0	0
i	–	–	–	–	–	–	a	a	–	–	–	–	–	–	0	1
j	–	–	–	–	–	–	a	a	–	–	–	–	–	–	0	1
k	–	–	–	–	–	–	a	a	–	–	–	–	–	–	0	1
l	–	–	–	–	–	–	a	a	–	–	–	–	–	–	0	0
m	–	–	–	–	–	–	a	a	–	–	–	–	–	–	0	1
n	–	–	–	–	–	–	a	a	–	–	–	–	–	–	0	1
o	–	–	–	–	–	–	a	a	–	–	–	–	–	–	0	1

possible to be in state a at t_2 time and there are therefore dashes in the t_2 column opposite row a. Rule I of the Huffman-Mealy simplification procedure may now be applied to Table 6-13 with the result that Table 6-14 is formed corresponding exactly to Table 6-9. However, because

Table 6-14

Present State	Next State								Output							
Time →	t_1		t_2		t_3		t_4		t_1		t_2		t_3		t_4	
x →	0	1	0	1	0	1	0	1	0	1	0	1	0	1	0	1
a	b	b	–	–	–	–	–	–	0	0	–	–	–	–	–	–
b	–	–	d	e	–	–	–	–	–	–	0	0	–	–	–	–
d	–	–	–	–	h	i	–	–	–	–	–	–	0	0	–	–
e	–	–	–	–	i	i	–	–	–	–	–	–	0	0	–	–
h	–	–	–	–	–	–	a	a	–	–	–	–	–	–	0	0
i	–	–	–	–	–	–	a	a	–	–	–	–	–	–	0	1

of the redundancies which exist in this new problem, further simplifications are still possible here which were not possible in Table 6-9. We can, for example, apply rule I to row b and combine it with row a, for rows a and b do not contain any contradictory entries. If we do this, however, we must not only replace all the b's in the table by a's, but we must also replace the "don't care" entries in row a by the corresponding entries in row b. The result is shown in Table 6-15. By

Table 6-15

Present State	Next State								Output							
Time →	t_1		t_2		t_3		t_4		t_1		t_2		t_3		t_4	
x →	0	1	0	1	0	1	0	1	0	1	0	1	0	1	0	1
a	a	a	d	e	–	–	–	–	0	0	0	0	–	–	–	–
d	–	–	–	–	h	i	–	–	–	–	–	–	0	0	–	–
e	–	–	–	–	i	i	–	–	–	–	–	–	0	0	–	–
h	–	–	–	–	–	–	a	a	–	–	–	–	–	–	0	0
i	–	–	–	–	–	–	a	a	–	–	–	–	–	–	0	1

continuing with this same procedure it is possible to combine rows d and h with a, and row i with e to obtain Table 6-16 which is obviously the simplest configuration and which requires only two states. The

Table 6-16

Present State	Next State								Output							
Time →	t_1		t_2		t_3		t_4		t_1		t_2		t_3		t_4	
x →	0	1	0	1	0	1	0	1	0	1	0	1	0	1	0	1
a	a	a	a	e	a	e	a	a	0	0	0	0	0	0	0	0
e	–	–	–	–	e	e	a	a	–	–	–	–	0	0	0	1

truth table for a single flip-flop, A, which will handle this problem is given in Table 6-17, and the logic for this flip-flop is shown in equations 6-6, 6-7, and 6-8.

Table 6-17

Present State	Next State								Output							
Time →	t_1		t_2		t_3		t_4		t_1		t_2		t_3		t_4	
x →	0	1	0	1	0	1	0	1	0	1	0	1	0	1	0	1
a 0	0	0	0	1	0	1	0	0	0	0	0	0	0	0	0	0
e 1	–	–	–	–	1	1	0	0	–	–	–	–	0	0	0	1

$$A^{n+1} = \bar{A}(t_2 + t_3)x + At_3 \tag{6-6}$$

$$z^n = (xAt_4)^n \tag{6-7}$$

$$J_A = t_2x + t_3x \qquad K_A = \bar{t_3} \tag{6-8}$$

These two solutions to a single very simple problem emphasize the importance of the definition of the problem. In any decimal computer, there invariably exists somewhere a counter which identifies and distinguishes the four bits of every decimal digit. This counter defines signals t_1, t_2, t_3, and t_4 and therefore permits the designer to use the second of the above solutions rather than the first one. If he defines the problem in such a way as to neglect the bit-time counter he already has, he will derive the first solution and will have more flip-flops than he actually requires. As a matter of fact, referring back to Table 6-11, we see that we can use flip-flops A, B, and C to distinguish the four bit-times as follows:

$$t_1 = \bar{A}\bar{B} \qquad t_3 = \bar{C}$$
$$t_2 = A\bar{B}C \qquad t_4 = BC \tag{6-9}$$

One very interesting aspect of the Huffman-Mealy procedures for circuit synthesis and simplification is that they lend themselves to automatization. An electronic computer might, for example, be used to translate a functional description of a sequential switching circuit into a state table, and then to simplify the table using rules I and II or their equivalents. The use of computers in computer logical design is a very interesting and promising possibility which will receive increased attention in the next few years.

THE INTERCONNECTION OF UNITS SEPARATELY DESIGNED

At the beginning of this chapter we pointed out that the logical designer simplifies his design problem by breaking it down into manageable pieces. Five such pieces have been described (Figure 6-2), and subsequent chapters will discuss each in turn. In practice, we shall see that even these five parts are further subdivided, and the designer is in general faced with the problem of combining the equations describing the various subunits. At the end of Chapter 5 we showed how this combination can be effected in a simple case—the design of an up and down counter and in Chapter 11 we shall return to the problem again. It will, however, be useful to consider a general method for combining such equations. The method requires that a table be drawn up outlining the function of every flip-flop at every bit-time, and noting down carefully all of the conditions which determine the function to be carried out. The use of such a table will be illustrated by a simple example.

Suppose we are to write the logic for flip-flops A, B, C, and D which have the following functions. First, when control flip-flops C_1 and C_2 are both "zero," flip-flops A, B, and C are to carry out the checking function described in the last section (equations 6-3, 6-4, and 6-5) on an input line x. C_1 and C_2 will be "zero" for eight bit-times, or two decimal digits. At the fourth and the eighth of these bit-times, flip-flop D is to be set to the "one" state if the corresponding code was not a permitted decimal digit and to be left in the "zero" state if it was a decimal digit. It must remain in that state for at least four bit-times.

At the eighth bit-time that $C_1 = C_2 = 0$, C_2 is set to "one," and C_1 and C_2 remain in the 01 state for eight more bit-times. During these bit-times, the number appearing on input line x should be transferred into flip-flop A, shifted from A to B, and from B to C.

Table 6-18

Control and Timing Signals			Difference Equations			
C_1	C_2	C_3	A^{n+1}	B^{n+1}	C^{n+1}	D^{n+1}
0	0	0	$\bar{B}C + B\bar{C}\bar{x}$	$\bar{C} + A\bar{B}\bar{x}$	$\bar{A} + B + \bar{C}$	$\bar{A}Bx$
0	0	0	"	"	"	"
0	0	0	"	"	"	"
0	0	0	"	"	"	"
0	0	0	"	"	"	"
0	0	0	"	"	"	"
0	0	0	"	"	"	"
0	0	0	"	"	"	"
0	1	0	x	A	B	D
0	1	0	"	"	"	"
0	1	0	"	"	"	"
0	1	0	"	"	"	–
0	1	0	"	"	"	–
0	1	0	"	"	"	–
0	1	0	"	"	"	–
0	1	0	"	"	"	–
1	1	0	A	B	C	–
1	1	0	"	"	"	–
1	1	0	"	"	"	–
1	1	0	"	"	"	–
1	1	0	"	"	"	–
1	1	0	"	"	"	–
1	1	0	"	"	"	–
1	1	1	0	0	1	–

From the seventeenth to the twenty-fourth bit-times inclusive, C_1 and C_2 are both "one." During this time flip-flops A, B, and C are required to hold whatever they contained the first bit-time that C_1 and C_2 became "one." Thereafter C_1 and C_2 both return to the "zero" state, and the entire cycle is repeated. Because (A, B, C) must be $(0, 0, 1)$ at the first bit-time that $C_1 = C_2 = 0$ (see Table 6-11), it is necessary that they be put in that state during the last bit-time that $C_1 = C_2 = 1$. A third control signal, C_3, is provided to supply a "one" at that bit-time.

The functions described in words above may be set out on two different kinds of truth tables and these are illustrated in Tables 6-18 and 6-19. In Table 6-18, the difference equation for each flip-flop and each bit-time is indicated as a function of the control and timing signals C_1, C_2, and C_3. Note that a convention is adopted here that ditto

Table 6-19

Control and Timing Signals			Input Equations							
C_1	C_2	C_3	J_A	K_A	J_B	K_B	J_C	K_C	J_D	K_D
0	0	0	$\bar{B}$	$BC + \bar{B}\bar{C} + Bx$	$\bar{C} + A\bar{x}$	C	1	$A\bar{B}$	$\bar{A}Bx$	$A + \bar{B} + \bar{x}$
0	0	0	"	"	"	"	"	"	"	"
0	0	0	"	"	"	"	"	"	"	"
0	0	0	"	"	"	"	"	"	"	"
0	0	0	"	"	"	"	"	"	"	"
0	0	0	"	"	"	"	"	"	"	"
0	0	0	"	"	"	"	"	"	"	"
0	0	0	"	"	"	"	"	"	"	"
0	1	0	x	$\bar{x}$	A	$\bar{A}$	B	$\bar{B}$	0	0
0	1	0	"	"	"	"	"	"	"	"
0	1	0	"	"	"	"	"	"	"	"
0	1	0	"	"	"	"	"	"	–	–
0	1	0	"	"	"	"	"	"	–	–
0	1	0	"	"	"	"	"	"	–	–
0	1	0	"	"	"	"	"	"	–	–
0	1	0	"	"	"	"	"	"	–	–
1	1	0	0	0	0	0	0	0	–	–
1	1	0	"	"	"	"	"	"	–	–
1	1	0	"	"	"	"	"	"	–	–
1	1	0	"	"	"	"	"	"	–	–
1	1	0	"	"	"	"	"	"	–	–
1	1	0	"	"	"	"	"	"	–	–
1	1	0	"	"	"	"	"	"	–	–
1	1	1	0	1	0	1	1	0	–	–

marks on a row mean that the entry in that row is the same as the entry directly above it; a dash in a row indicates that we do not care what state the corresponding flip-flop assumes next. Table 6-19 contains the same information as does Table 6-18 except that all the entries are input equations to the corresponding flip-flops. Note that the entry A in column A^{n+1} of Table 6-18 means that flip-flop A must remain in the same state for a series of bit-times and this is reflected in Table 6-19 by the fact that the J and K inputs to flip-flop A are both "zero." The "zero" in the A^{n+1} column of Table 6-18 requires, in Table 6-19, that the K input be "one" and the J input "zero" at that bit-time, so that by the following bit-time flip-flop A will be off.

We may derive the complete input equations from either Table 6-18 or 6-19 by combining the logic shown with the control and timing logic. Doing this for flip-flops A, B, C, and D in Table 6-18 we obtain the following equations.

$$\begin{aligned}
A^{n+1} &= [\bar{C}_1\bar{C}_2(\bar{B}C + B\bar{C}\bar{x}) + \bar{C}_1C_2x + C_1C_2\bar{C}_3A]^n \\
B^{n+1} &= [\bar{C}_1\bar{C}_2(\bar{C} + A\bar{B}\bar{x}) + \bar{C}_1C_2A + C_1C_2\bar{C}_3B]^n \\
C^{n+1} &= [\bar{C}_1\bar{C}_2(\bar{A} + B + \bar{C}) + \bar{C}_1C_2B + C_1C_2\bar{C}_3C + C_1C_2C_3]^n \\
D^{n+1} &= (\bar{C}_1\bar{C}_2\bar{A}Bx + C_2D)^n
\end{aligned} \qquad (6\text{-}10)$$

Note that in the difference equation for flip-flop D we have, in effect, replaced all of the dash marks of Table 6-18 with D's, in the interest of simplifying the difference equation. Other simplifications are, of course, possible.

We can apply this combining technique in an exactly similar way to Table 6-19 with the following results.

$$\begin{aligned}
J_A &= \bar{C}_1\bar{C}_2\bar{B} + \bar{C}_1C_2x \\
K_A &= \bar{C}_1\bar{C}_2(BC + \bar{B}\bar{C} + Bx) + \bar{C}_1C_2\bar{x} + C_1C_2C_3 \\
J_B &= \bar{C}_1\bar{C}_2(\bar{C} + A\bar{x}) + \bar{C}_1C_2A \\
K_B &= \bar{C}_1\bar{C}_2C + \bar{C}_1C_2\bar{A} + C_1C_2C_3 \\
J_C &= \bar{C}_1\bar{C}_2 + \bar{C}_1C_2B + C_1C_2C_3 \\
K_C &= \bar{C}_1\bar{C}_2A\bar{B} + \bar{C}_1C_2\bar{B} \\
J_D &= \bar{C}_1\bar{C}_2\bar{A}Bx \\
K_D &= \bar{C}_1\bar{C}_2(A + \bar{B} + \bar{x})
\end{aligned} \qquad (6\text{-}11)$$

Equations 6-10 and 6-11 can of course be shown to be essentially equivalent.

BIBLIOGRAPHY

D. A. Huffman, "The Synthesis of Sequential Switching Circuits," *J. Franklin Institute*, 161–190, 275–303 (Mar.–Apr. 1954).

G. H. Mealy, "A Method for Synthesizing Sequential Circuits," *Bell System Technical J.*, **34**, no. 5, 1045–1080 (Sept. 1955).

E. F. Moore, "Gedanken-Experiments on Sequential Machines," *Automata Studies*, Princeton University Press, Princeton, 1956.

EXERCISES

1. A sequential circuit having one input (x), one output (z), and containing three flip-flops (A, B, and C) is described by the truth table, Table 6-20. Determine whether the function accomplished by this circuit could be carried out by a circuit containing fewer flip-flops. If so, design the circuit (i.e., write the flip-flop input equations together with the equation defining output z).

Table 6-20

	Time n			Time $(n+1)$			Time n
x	A	B	C	A	B	C	z
0	0	0	0	0	0	0	0
0	0	0	1	0	0	1	1
0	0	1	0	1	0	1	1
0	0	1	1	0	1	1	1
0	1	0	0	0	0	1	1
0	1	0	1	0	1	0	1
0	1	1	0	1	1	0	1
0	1	1	1	1	0	1	1
1	0	0	0	0	0	1	0
1	0	0	1	1	1	1	0
1	0	1	0	0	0	0	0
1	0	1	1	1	0	1	0
1	1	0	0	0	1	0	0
1	1	0	1	0	0	0	0
1	1	1	0	0	1	0	0
1	1	1	1	0	0	0	0

Draw a diagram similar to Figure 6-7 showing the operation of the original circuit and of the simplified one.

2. Each of the following tables, Tables 6-21 through 6-27, describes a sequential switching circuit. Using the Huffman-Mealy simplification procedure, show how the number of states necessary may be reduced for each of the tables. Finally, assign flip-flops to each simplified circuit, and write the input equations for those flip-flops together with the equations defining the outputs.

Table 6-21

Present State	Next State		Output	
	$x = 0$	$x = 1$	$x = 0$	$x = 1$
a	b	c	0	0
b	a	c	1	0
c	e	f	0	1
d	b	d	0	0
e	c	f	1	1
f	b	e	0	0

Table 6-22

Present State	Next State				Output			
$x_1 x_2 \rightarrow$	00	01	10	11	00	01	10	11
a	c	b	a	a	0	0	0	1
b	d	b	b	e	0	1	1	1
c	e	d	e	a	0	0	0	1
d	b	d	b	a	0	1	1	1
e	a	d	c	c	0	0	0	1

Table 6-23

Present State	Next State				Output			
$x_1 x_2 \rightarrow$	00	01	10	11	00	01	10	11
a	a	c	c	f	1	0	0	0
b	b	e	a	f	1	0	0	1
c	a	d	f	e	1	0	0	0
d	d	c	g	b	1	0	0	0
e	d	a	f	e	1	0	0	0
f	b	c	a	f	1	0	0	1
g	d	d	b	e	1	0	0	0

Table 6-24

Present State	Next State				Output			
$x_1\,x_2 \rightarrow$	00	01	10	11	00	01	10	11
a	e	b	d	d	0	0	0	0
b	g	a	h	h	0	0	1	1
c	b	c	d	e	0	0	0	0
d	f	d	c	e	0	0	1	0
e	e	b	f	d	0	0	0	0
f	d	d	g	e	0	0	1	0
g	b	c	f	h	0	0	0	0
h	a	b	f	f	0	0	0	0

Table 6-25

Present State	Next State		Output	
	$x = 0$	$x = 1$	$x = 0$	$x = 1$
a	a	b	0	0
b	c	d	1	1
c	–	b	–	0
d	f	e	1	0
e	c	–	1	–
f	–	d	–	0

Table 6-26

Present State	Next State				Output			
$x_1\,x_2 \rightarrow$	00	01	10	11	00	01	10	11
a	e	a	h	g	1	1	1	0
b	c	d	d	f	0	1	1	1
c	a	b	e	d	1	0	1	1
d	–	c	f	b	–	1	1	0
e	–	–	c	g	–	–	1	1
f	g	a	–	a	0	1	–	0
g	h	e	–	b	1	1	–	0
h	d	e	a	f	1	1	1	1

Table 6-27

Present State	Next State $x_1 x_2 \rightarrow$ 00	01	10	11	Output 00	01	10	11
a	b	c	h	k	0	–	1	0
b	a	e	h	f	0	1	–	–
c	g	e	i	c	1	0	0	–
d	a	i	h	j	–	1	1	0
e	d	e	e	e	–	0	0	1
f	h	c	g	g	0	1	0	–
g	a	i	h	k	0	–	1	0
h	k	c	f	b	1	0	1	–
i	d	i	c	c	1	0	–	1
j	h	c	g	g	0	–	0	1
k	h	e	a	d	–	–	0	1

3. The sequential switching circuit described by Table 6-28 may be simplified in two ways. Simplify it first by combining states d and c. Is any further simplification possible? Next, start again and simplify by combining states d and b. Is any further simplification possible?

Table 6-28

Present State	Next State $x_1 x_2 \rightarrow$ 00	01	10	11	Output z 00	01	10	11
a	a	b	d	c	0	1	0	0
b	d	a	b	c	0	1	0	1
c	c	b	b	c	0	1	0	0
d	d	–	b	c	–	1	0	–

4. A sequential circuit contains three flip-flops, A, B, and C. It has two input lines, x_1 and x_2, and an output line z. The input equations for the three flip-flops are:

$$J_A = \bar{B}x_2 + B\bar{x}_2 \qquad K_A = x_1\bar{x}_2B\bar{C} + \bar{x}_1\bar{x}_2\bar{B}$$

$$J_B = \bar{x}_2 + \bar{A} \qquad K_B = \bar{x}_2\bar{C} + x_2C + \bar{x}_1\bar{A}$$

$$J_C = B\bar{x}_2 + A\bar{x}_1 + A\bar{B}x_1 \qquad K_C = \bar{x}_1 + \bar{x}_2$$

The equation defining the output line is:

$$z = B\bar{x}_2 + Ax_1\bar{x}_2 + \bar{A}x_1x_2 + C\bar{x}_1x_2$$

Certain combinations are known never to occur. These redundancies are:

$$\bar{x}_1BC = \bar{x}_1\bar{B}\bar{C} = x_1\bar{A}B = x_1\bar{B}C = 0$$

Determine whether it is possible to perform this same function with only two flip-flops. If so, write the input equations for the two flip-flops, together with the new equation defining the circuit output z.

5. Redesign the error-detecting circuit described by Table 6-8 so that it gives an error indication at t_1 time instead of t_4 time.

6. Design a sequential switching circuit having one input (x) and one output (z) and having the property that $z = 1$ if and only if the following sequence of bits appears on x: 00101000110**0**. The output z should be "one" for one bit-time only—namely the bit-time corresponding to the last (boldface) 0 in the sequence above. As soon as an output has occurred, the circuit should again seek the desired sequence.

7. A sequential circuit has two inputs (x_1 and x_2) and one output (z). At any bit-time, either x_1 and x_2 are different (01 or 10) or else they are the same (00 or 11). The circuit is originally in state a. The output z is to be "one" when four (or more) consecutive bit-times have elapsed during which x_1 and x_2 were the same, or four or more bit-times have elapsed during which they were different. For example, the following sequence of inputs would have the results shown:

x_1	0 0 1 0 0 1 0 0 1 0 1 0 0 1 1 1 0 ···
x_2	1 1 0 0 1 1 0 0 1 0 1 1 1 0 0 0 0 ···
state	a c f h b e b d g g g g c f h h h
z	0 0 0 0 0 0 0 0 1 1 1 0 0 0 1 1 0

Table 6-29 indicates the operation of the circuit, and it is only partially complete. Fill in the blank spaces in this table, simplify it if possible (using the Huffman-Mealy method), and design a circuit which will mechanize the simplified table; i.e., provide appropriate flip-flops and write their input equations, as well as the equation defining z.

Table 6-29

Present State	Next State				Output z			
$x_1 x_2 \rightarrow$	00	01	10	11	00	01	10	11
a		c				0		
b	d	e			0	0		
c		f				0		
d	g				0			
e				b				0
f			h				0	
g	g	c		g	1	0		1
h	b		h		0		1	

8. In a certain serial computer, decimal digits are represented by a "two-out-of-five" code. That is, five bits represent a decimal digit, and two *and only two* of them are "one." An example of such a code is given in Table 6-30. A number represented in this code appears serially from a flip-flop x. Five timing signals are supplied (t_1, t_2, t_3, t_4, and t_5) with $t_1 = 1$ at the time the least

significant bit of the digit appears in x, and $t_5 = 1$ at the time the most significant bit appears. Derive a Huffman-Mealy table for a sequential circuit whose output at t_5 time will be "one" if the digit being read was an incorrect code (more than or less than two "ones"), and "zero" if the code was correct. Simplify the table whenever possible, and complete the design by providing appropriate flip-flops and writing their input equations. The output is always to be "zero" during t_1 through t_4 times.

Table 6-30

Decimal Digit	Bit-Time				
	t_5	t_4	t_3	t_2	t_1
0	0	0	0	1	1
1	0	0	1	0	1
2	0	1	0	0	1
3	1	0	0	0	1
4	0	0	1	1	0
5	0	1	0	1	0
6	1	0	0	1	0
7	0	1	1	0	0
8	1	0	1	0	0
9	1	1	0	0	0

7 / *Digital computer memories*

INTRODUCTION

In Chapter 6 we defined a digital computer memory unit as a device capable of storing information in such a way that the information can be read, erased, and written automatically by a digital computer. The whole of Chapter 5 was devoted to the study of the logic of one-bit memory elements. Those elements, however, have a very important disadvantage: they are expensive, and therefore cannot be used to store large volumes of data. In this chapter we shall explore the logical properties of cheaper storage devices, suitable for the storage of up to 10^8 binary digits.

The definitions employed here for memory units and for input-output units are entirely arbitrary, as such definitions must be. Since both units handle data employed in computation, it is perhaps to be expected that they have many features in common. In fact, a combination of card-punching equipment, card-reading equipment, printing equipment, and necessary human operators may be looked upon as a memory whose capacity is limited only by the supply of paper available. We distinguish a memory from input-output devices by arbitrarily stating that anything which the computer can automatically read from, erase, and write on is a memory. This leads us to some interesting classifications. A magnetic drum (or portion thereof) on which data is permanently recorded is an input device. A magnetic tape handling device is a memory unit unless it is connected in such a way that it can only be read from or only written into by the computer, in which case it becomes an input or output device respectively.

It is difficult to overstate the importance of the memory unit. A computer memory has two primary functions: the storage of data used in calculations; and the storage of information which determines the sequences of operations (see Chapter 1, p. 1) to be performed on that data. If there were no cheap and convenient large-scale memories, these functions *could* be carried out with the help of input-output equipment. The operations to be performed could be read one by one from an input device, together with the initial data; and intermediate results could be written onto an output device and then transferred back through input equipment as needed. In particular computer applications, these substitutes for a memory may be practical and desirable, and the system and logical designer must be aware of them. However, for many applications they impose severe restrictions on the speed, flexibility, and reliability of the computer system which employs them—restrictions which may be removed if the designer has available a large and cheap memory unit. Obviously, it is more convenient for the computer to write, read, and erase intermediate results itself than to require the manual operations inherent in the use of input-output devices. In addition, the ability of a computer to modify its own sequence of operations is very important in many computer applications; and this ability may be severely limited if the operations are read from an input device.

In discussing computer memories, we shall begin by considering problems common to all kinds of memory units: the problems of providing timing signals, and of locating desired data in a large memory. We shall then describe the principal logical properties of three widely used memories: the magnetic drum, magnetic cores, and magnetic tape.

CLOCK PULSES AND TIMING SIGNALS

In examining one-bit memory elements, we have specified that a clock-pulse signal must be delivered to every flip-flop in order to identify bit-times. Clock pulses must also be supplied to the circuits which read from and write into large-scale memories, so that information stored there is in synchronism with other data in the memory and in other parts of the computer. The source of clock pulses will in general be different for each type of computer memory, and these sources will be discussed later in this chapter when specific memories are discussed.

In addition to the clock pulses, however, it is necessary for the logical designer to have available other timing signals, for use in organizing, sequencing, or identifying data or operations. These timing signals are employed in other parts of the computer besides the memory unit, but their close association with the memory makes it appropriate to discuss them here. They may be obtained as outputs of a counter which changes state every clock-pulse time or every time some event takes place which is to be identified by the timing circuits.

To illustrate how timing signals may be provided, consider the following example. A memory unit is usually subdivided into compartments of from fifteen to fifty bits each, called *words*, and the logical designer often requires a special set of signals which enable him to distinguish various parts of a word. Suppose that words contain only seven bits which appear sequentially from a common source, and that a different timing signal is required for each of the seven bits. We must therefore design a clock-pulse counter, and since it must have seven different states, at least three flip-flops are required. Supposing we decided to count in the ordinary binary number sequence, and gave the three flip-flops the names A, B, and C; we would then derive the truth table shown in Table 7-1, and the logic of equations 7-1.

Table 7-1

	n			$n+1$	
A	B	C	A	B	C
0	0	0	0	0	1
0	0	1	0	1	0
0	1	0	0	1	1
0	1	1	1	0	0
1	0	0	1	0	1
1	0	1	1	1	0
1	1	0	0	0	0

$$\begin{aligned} R_A &= B\overline{C} & S_A &= BC \\ R_B &= AB + BC & S_B &= \overline{B}C \\ R_C &= C & S_C &= \overline{A}\overline{C} + \overline{B}\overline{C} \end{aligned} \qquad (7\text{-}1)$$

The seven bit-times could then be identified as follows:

$$\begin{aligned} t_0 &= \overline{A}\overline{B}\overline{C} & t_4 &= A\overline{B}\overline{C} \\ t_1 &= \overline{A}\overline{B}C & t_5 &= AC \\ t_2 &= \overline{A}B\overline{C} & t_6 &= AB \\ t_3 &= BC \end{aligned} \qquad (7\text{-}2)$$

Note that t_3, t_5, and t_6 require only two diodes every time they are used, because the counter has been designed with the redundancy $ABC = 0$.

If timing signals are used in great numbers throughout the computer, it may be desirable to design the counter so as to introduce redundancies where they will simplify timing signals as much as possible. Suppose, for example, that bits t_6, t_0, and t_1 were used in a great many more input equations than the other four bit-time signals. Then we might rearrange the count sequence for the three-flip-flop counter to make the single redundancy most useful, as illustrated in Table 7-2.

Table 7-2

A	B	C	
1	0	1	$t_0 = AC$
0	1	1	$t_1 = BC$
1	0	0	$t_2 = A\bar{B}\bar{C}$
0	0	0	$t_3 = \bar{A}\bar{B}\bar{C}$
0	0	1	$t_4 = \bar{A}\bar{B}C$
0	1	0	$t_5 = \bar{A}B\bar{C}$
1	1	0	$t_6 = AB$

Alternatively, we might introduce two more flip-flops and design a counter having the count sequence given in Table 7-3.

Table 7-3

A	B	C	D	E	
0	1	0	0	0	$t_0 = B$
0	0	1	0	0	$t_1 = C$
0	0	0	0	0	$t_2 = \bar{A}\bar{B}\bar{C}\bar{D}\bar{E}$
0	0	0	0	1	$t_3 = \bar{D}E$
0	0	0	1	0	$t_4 = D\bar{E}$
0	0	0	1	1	$t_5 = DE$
1	0	0	0	0	$t_6 = A$

Other variations are obviously possible. Each has the effect of adding flip-flops to the counter in order to save on the number of decision elements needed elsewhere.

One important difficulty remains to be overcome. If the timing signals t_0 to t_6 of equations 7-2 are to be of any use, they must, of course, be in phase with the memory. That is, the signal for t_0 ($\bar{A}B\bar{C}$) must actually occur at the time the t_0 bit is read, t_1 when the next bit is read, etc. Of course, once the counter is in phase with the memory, it will always be in phase as long as the clock pulses continue. But when the operation of the computer has been interrupted—e.g., when it is turned off—the designer may have to take steps to force the counter into synchronism with the memory or vice versa. We shall return to this problem later in the chapter, when we discuss magnetic drum memories.

THE SELECTION PROBLEM

There is one problem which arises again and again in connection with large digital memories: the problem of selecting automatically that portion of the memory which is to be examined or into which data is to be written. Considering an ensemble of bits stored somehow in space, we may select one or more of them in one of two distinct ways. We may provide a reading (and writing) station for each bit, so that the problem of selecting the desired bit is reduced to that of selecting one read-write station out of many. Alternatively, we may provide a single reading (and writing) station and cause it to move with respect to the bits in such a way that the bits are read sequentially, so that the problem of selecting the desired bit is reduced to that of determining exactly when that bit passes the read-write station. Let us call these two kinds of selection S-selection (for space) and T-selection (for time) respectively. Each presents certain quite different problems to the logical designer, and these problems will be discussed before we turn to specific memory devices.

S-Selection

Some of the problems of S-selection will be illustrated by use of a specific example. Suppose we are given an array of 256 points, each of which contains a binary signal, "zero" or "one." These binary signals are named P_0, P_1, P_2, $\cdots$, P_{254}, and P_{255}, and we are to provide a means of reading any one of them upon request. Our first step must be to identify the points, which we do by assigning an arbitrary number, the "address," to each. It is convenient and natural to assign the address "zero" to P_0, "one" to P_1, etc., up to 255 for P_{255}. Written as a binary number, the address consists of eight bits. Let us now provide eight flip-flops labeled S_1, S_2, $\cdots$, S_8, and arrange that when we want to examine the state of one of the 256 points, we place its address in the register S_1–S_8. We want to write a Boolean function f

which will have the value P_n when S_1–S_8 contain the address n, and we can draw up a truth table, Table 7-4, and write f as shown there.

Table 7-4

S_1	S_2	S_3	S_4	S_5	S_6	S_7	S_8	f
0	0	0	0	0	0	0	0	P_0
0	0	0	0	0	0	0	1	P_1
0	0	0	0	0	0	1	0	P_2
0	0	0	0	0	0	1	1	P_3
0	0	0	0	0	1	0	0	P_4
. . .	. . .	. . .	. . .	. . .	. . .	. . .	. . .	. . .
1	1	1	1	1	1	1	0	P_{254}
1	1	1	1	1	1	1	1	P_{255}

$$
\begin{aligned}
f = {} & \bar{S}_1 \bar{S}_2 \bar{S}_3 \bar{S}_4 \bar{S}_5 \bar{S}_6 \bar{S}_7 \bar{S}_8 P_0 \\
+ {} & \bar{S}_1 \bar{S}_2 \bar{S}_3 \bar{S}_4 \bar{S}_5 \bar{S}_6 \bar{S}_7 S_8 P_1 \\
+ {} & \bar{S}_1 \bar{S}_2 \bar{S}_3 \bar{S}_4 \bar{S}_5 \bar{S}_6 S_7 \bar{S}_8 P_2 \\
+ {} & \cdots\cdots\cdots\cdots\cdots \\
+ {} & S_1 S_2 S_3 S_4 S_5 S_6 S_7 S_8 P_{255}
\end{aligned}
$$

We might abbreviate the equation for f in Table 7-4 as follows:

$$
f = \sum_{i=0}^{255} (S_1\text{–}S_8)_i \, P_i \tag{7-3}
$$

where $(S_1\text{–}S_8)_i$ designates the particular "and" gate which is true when the S_1–S_8 register contains the address i. We could equally well have written

$$
\begin{aligned}
f = {} & (S_1 + S_2 + S_3 + S_4 + S_5 + S_6 + S_7 + S_8 + P_0) \\
& (S_1 + S_2 + S_3 + S_4 + S_5 + S_6 + S_7 + \bar{S}_8 + P_1) \\
& (S_1 + S_2 + S_3 + S_4 + S_5 + S_6 + \bar{S}_7 + S_8 + P_2) \\
& \cdots\cdots\cdots\cdots\cdots\cdots\cdots\cdots \\
& (\bar{S}_1 + \bar{S}_2 + \bar{S}_3 + \bar{S}_4 + \bar{S}_5 + \bar{S}_6 + \bar{S}_7 + \bar{S}_8 + P_{255}) \\
= {} & \prod_{i=0}^{255} [(S_1 + \cdots + S_8)_{255-i} + P_i]
\end{aligned} \tag{7-4}
$$

which is the equivalent maxterm-type expression. Here $(S_1 + \cdots + S_8)_{255-i}$ designates the "or" gate which is false when the S_1–S_8 register contains the address i.

Equations 7-3 and 7-4 are perfectly good logical expressions for the selection problem, but they have certain unfortunate features from a practical standpoint. Probably the most important disadvantage is the number of diodes required. If we count the diodes necessary to mechanize either equation, we see that each "and" term requires nine diodes, and the entire equation $10 \times 256 = 2560$ diodes. This prohibitively large number of diodes may be reduced by the simple stratagem of building up the address logic in sections, and equation 7-5 indicates how this might be done to equation 7-3. (The same kind of modification can be applied to equation 7-4.)

$$f = \sum_{i=0}^{15} \sum_{j=0}^{15} (S_1\text{–}S_4)_i \, (S_5\text{–}S_8)_j \, P_{16i+j} \tag{7-5}$$

Here, instead of forming 256 addresses using eight diodes each, we break the address in half, and form 16 addresses from S_1–S_4 and 16

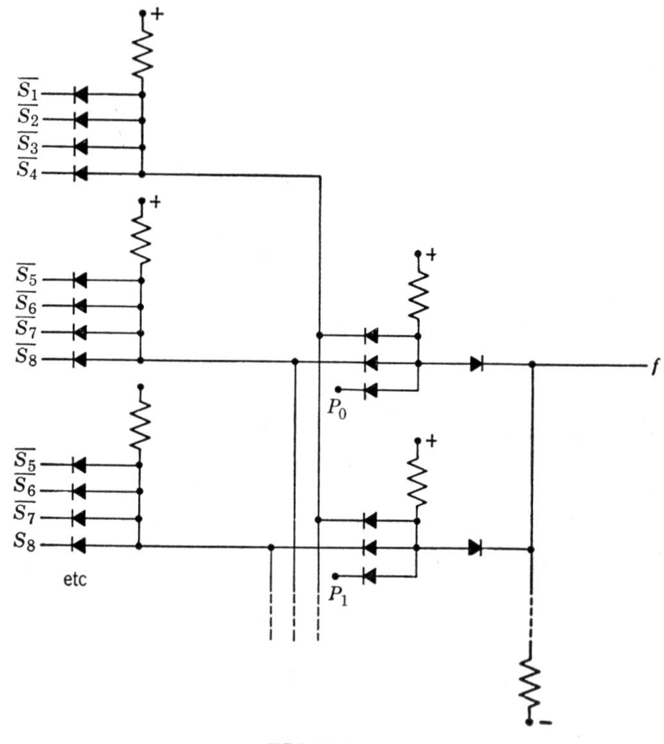

FIGURE 7-1

more from S_5–S_8. Any of the 256 addresses may now be made up by selecting one S_1–S_4 address and one S_5–S_8 address. For example, in Figure 7-1, the combinations which combine with P_0 and P_1 are indicated. We may count the total number of diodes necessary as follows:

$4 \times 16 \quad = 64$ diodes to form the 16 S_1–S_4 addresses

$4 \times 16 \quad = 64$ diodes to form the 16 S_5–S_8 addresses

$3 \times 256 = 768$ diodes to form the 256 "and" gates, each having one P line and two S addresses as inputs

256 diodes for the "or" gate combining all terms

Total $= 64 + 64 + 768 + 256 = 1152$ diodes

We can effect still another minor saving by building up the S_1–S_4 logic in subsections, just as we have already built the S_1–S_8 logic. That is to say, we can write $(S_1\text{–}S_4)_k = (S_1, S_2)_i(S_3, S_4)_j$, and can replace the original $4 \times 16 = 64$ diodes by $2 \times 4 = 8$ diodes to form the S_1, S_2 addresses, another 8 for the S_3, S_4 addresses, and $2 \times 16 = 32$ more to combine those into $(S_1\text{–}S_4)_k$. We therefore need only $32 + 8 + 8 = 48$ instead of the original 64 diodes, and applying this saving to S_5–S_8 as well, we find that a total of only 1120 diodes are needed to form f.

The reading operation requires that 256 different points be joined together in a single function, which we have called f. It is also necessary to provide a method for writing information into any of these 256 points from a single source. Suppose, for example, we have a bit stored in flip-flop W which we want to place in the location specified by an address in S_1–S_8. For each location we must therefore provide a signal, W_i, which will equal W when S_1–S_8 contains the desired address. For example, in equation 7-6 the equations are presented which enable one to write into locations 0, 1, 2, and 255.

$$
\begin{aligned}
W_0 &= \bar{S}_1\bar{S}_2\bar{S}_3\bar{S}_4\bar{S}_5\bar{S}_6\bar{S}_7\bar{S}_8\, W \\
W_1 &= \bar{S}_1\bar{S}_2\bar{S}_3\bar{S}_4\bar{S}_5\bar{S}_6\bar{S}_7S_8\, W \\
W_2 &= \bar{S}_1\bar{S}_2\bar{S}_3\bar{S}_4\bar{S}_5\bar{S}_6S_7\bar{S}_8\, W \qquad (7\text{-}6) \\
&\cdots\cdots\cdots\cdots \\
W_{255} &= S_1S_2S_3S_4S_5S_6S_7S_8\, W
\end{aligned}
$$

The 256 equations may, of course, be written in a general form.

$$W_i = (S_1\text{–}S_8)_iW \qquad (i = 0, 1, 2, \cdots, 255) \qquad (7\text{-}7)$$

Furthermore, although the 256 equations represented by equation 7-7 ostensibly require $9 \times 256 = 2304$ diodes, it should be evident that we can apply here the same diode-saving techniques we applied to equation 7-3. We can therefore write

$$W_{16i+j} = (S_1\text{–}S_4)_i(S_5\text{–}S_8)_jW \tag{7-8}$$

Finally, if both reading and writing logic must be available, which is the usual state of affairs, we can use the address logic employed in

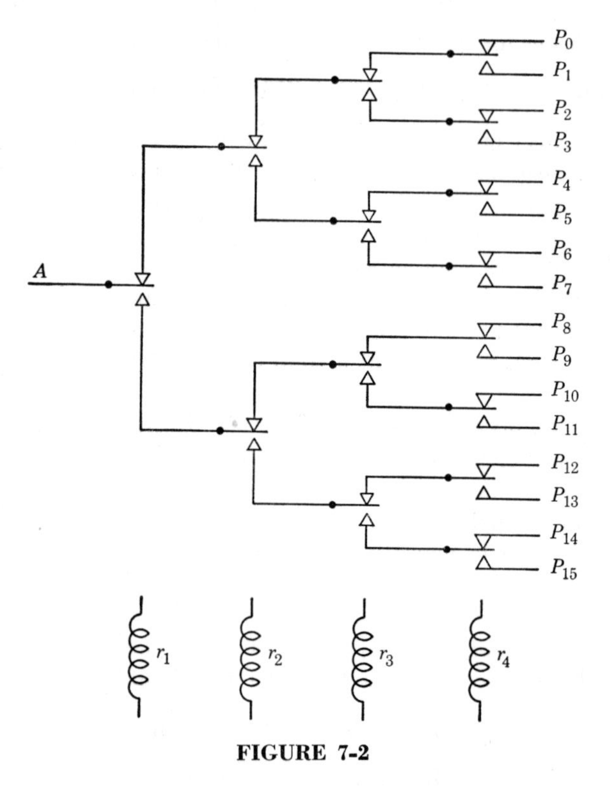

FIGURE 7-2

forming f to form W_i. That is, from each of the 256 $(S_1\text{–}S_4)_i(S_5\text{–}S_8)_j$ combinations we bring two wires: one is combined with W to form W_{16i+j}; the other is combined with P_{16i+j} and then with all the other P-logic to form f.

S-selection for reading and writing may also be accomplished with relays, if enough time is available. The relay-selection scheme shown in Figure 7-2 allows point A to be connected to any one of points P_0 to P_{15}, depending on which of the relays r_1 to r_4 is energized. For example, if current flows in relays r_1 and r_3, the single contact above r_1 moves down and the four contacts above r_3 move down. Under these circumstances, point A is connected to P_{10}. Obviously, we can double the number of points selected by adding another relay having sixteen change-over contacts.

As was indicated above, the principal disadvantage of relay circuits is their operating time. A typical relay having five change-over contacts of the kind illustrated in Figure 7-2 may operate in from 20 to 40 milliseconds, whereas the corresponding diode circuit operates in 10 microseconds or less—more than 1000 times faster. However, in applications where slow operation is permissible, relay selection has certain advantages. Notice, for example, that the relay circuit of Figure 7-2 may be used either for writing or reading: information may flow either from left to right or from right to left. Furthermore, a duplicate and completely separate path allowing the selection of any of points P_0 to P_{15} may be provided simply by increasing the number of contacts on each relay and constructing another circuit in parallel with that of Figure 7-2.

The fact that relays operate slowly does not prevent them from being useful in high-speed computer systems. In some applications, for example, the address of a memory element may be available long before the contents of that location are required. It may then be possible to insert that address in the relay circuitry and to continue with calculations at normal speed during the time required for relay operation.

T-Selection

The central problem of T-selection is that of determining the correct instant to begin reading from (or writing into) a memory in which information is moving with respect to the read or write station. Two different methods for accomplishing this may be distinguished, and each will be discussed in connection with a simple example: suppose we want to choose the proper time to start reading from a line P, which will successively scan each of thirty-two words. The address of the particular word desired will be stored in flip-flops $T_1, T_2, \cdots, T_5$.

The first solution requires that a second address register be provided to identify continuously the word being scanned. We must for this purpose be given another signal which informs this second address register whenever the reading station passes the boundary between words. For example, one of the clock-pulse timing signals of the kind indicated by equations 7-2 can be used. We then introduce this signal (say, t_0) into the logic of a "current-address register," A_1 to A_5. If the words are identified by binary addresses which increase in numerical order from zero to thirty-one, if address zero follows address thirty-one, and if A_1 contains the least significant and A_5 the most significant bit of the address, the logic for flip-flops A_1 to A_5 may be written

$$
\begin{aligned}
J_{A1} &= K_{A1} = t_0 \\
J_{A2} &= K_{A2} = t_0A_1 \\
J_{A3} &= K_{A3} = t_0A_1A_2 \\
J_{A4} &= K_{A4} = t_0A_1A_2A_3 \\
J_{A5} &= K_{A5} = t_0A_1A_2A_3A_4
\end{aligned}
\tag{7-9}
$$

The current-address counter, like the clock-pulse counter, must be synchronized with the memory so that it actually contains the number zero when word zero is being read.

We must now find a Boolean expression which will indicate when the desired address in T_1–T_5 is equal to the actual address, in A_1–A_5. Let us call this expression the coincidence logic, C. The two addresses will be equal, and C will be true, only when A_1 equals T_1, A_2 equals T_2, A_3 equals T_3, A_4 equals T_4, and A_5 equals T_5, all at once. Since two Boolean quantities may be equal only if they are both "zero" or else both "one", we may write

$$
\begin{aligned}
C = (A_1T_1 + \bar{A}_1\bar{T}_1)&(A_2T_2 + \bar{A}_2\bar{T}_2)(A_3T_3 + \bar{A}_3\bar{T}_3) \\
&(A_4T_4 + \bar{A}_4\bar{T}_4)(A_5T_5 + \bar{A}_5\bar{T}_5)
\end{aligned}
\tag{7-10}
$$

Written as a minterm-type expression, this function requires 192 diodes. The maxterm-type expression is much simpler, and is written

$$
\begin{aligned}
C = (A_1 + \bar{T}_1)(\bar{A}_1 + T_1)&(A_2 + \bar{T}_2)(\bar{A}_2 + T_2)(A_3 + \bar{T}_3)(\bar{A}_3 + T_3) \\
&(A_4 + \bar{T}_4)(\bar{A}_4 + T_4)(A_5 + \bar{T}_5)(\bar{A}_5 + T_5)
\end{aligned}
\tag{7-11}
$$

The word-time during which $C = 1$ will be called the *coincidence word-time.*

The second solution to the T-selection problem requires that the complete address of each word be recorded adjacent to the word. For example, let us assume that the address of a word lies in an address channel A, occurring every word-time during the five bit-times which precede t_0. We can then compare these five bits with the five bits of the desired address, in memory elements T_1–T_5. One way of carrying out this serial comparison is indicated in Figure 7-3, which shows clock pulses and flip-flop signals for three particular words in the memory, and which assumes that T_1 to T_5 contain the address eleven. [Note particularly that the address of word $(n + 1)$ in the A-channel appears during the last five bits of word n.] If the A-channel address and the T-address are *not* the same, we turn on auxiliary flip-flop A_1: when A_1

is on at t_0 time, then, we know that we do *not* want to read the subsequent word. The logic for flip-flop A_1 will then be

$$J_{A1} = t_1(A\bar{T}_1 + \bar{A}T_1) + t_2(A\bar{T}_2 + \bar{A}T_2) + t_3(A\bar{T}_3 + \bar{A}T_3) + t_4(A\bar{T}_4 + \bar{A}T_4) + t_5(A\bar{T}_5 + \bar{A}T_5) \qquad (7\text{-}12)$$

$$K_{A1} = t_0$$

Equations 7-12 assume the existence of five clock-pulse timing signals, t_1 to t_5, which are used to "scan" the T_1–T_5 address, one bit at a time. The t_0 signal turns off A_1 in preparation for the next comparison.

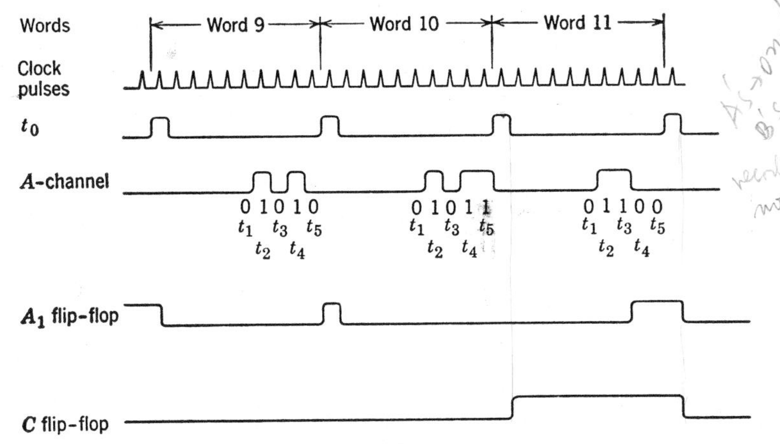

FIGURE 7-3

Inasmuch as the coincidence logic will be used to read or write a word, we must provide a coincidence signal which lasts for an entire word-time—the coincidence word-time. The function C of equations 7-10 and 7-11, which was derived using the first T-selection method, is true as long as the two addresses are equal, i.e., for a whole word-time between one t_0 pulse and another. The A_1 flip-flop of Figure 7-3 and equations 7-12 obviously cannot be used for this purpose, however. The input logic for A_1 could be modified slightly so that A_1 is on from t_0 to t_0 of the coincidence word-time (e.g., simply by adding "$+t_0$" to the equation for J_{A1}), but A_1 would nevertheless still be on during noncoincidence word-times. We must therefore provide another flip-flop and arrange for it to be turned on whenever A_1 is zero at t_0 time, and to be turned off at the following t_0 time. If we call this the C flip-flop,

$$J_C = t_0\bar{A}_1 \qquad K_C = t_0 \qquad (7\text{-}13)$$

The C waveform is shown in Figure 7-3 along with all the other signals.

Combinations of *T*- and *S*-Selection

In any computer having a memory system employing *T*-selection, it is common to find *S*-selection as well. The most important example of such a memory is the magnetic drum, which in general requires *S*-selection of one of the channels along its axis and *T*-selection of one of the words around its circumference in a given channel.

When *S*- and *T*-selection are combined, the address of a storage location must be a combination of the *S*- and *T*-addresses. In a binary

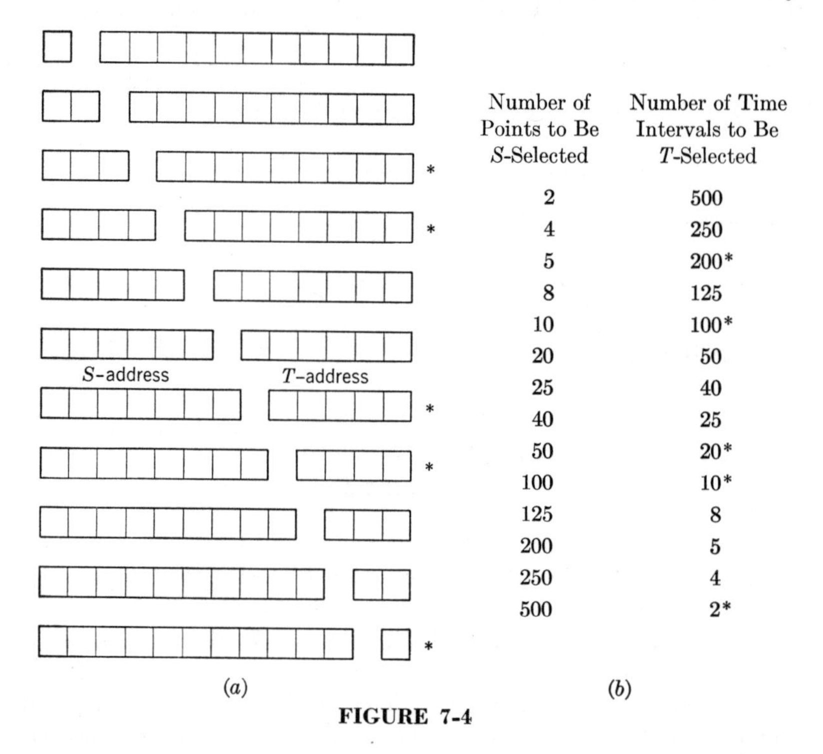

Number of Points to Be *S*-Selected	Number of Time Intervals to Be *T*-Selected
2	500
4	250
5	200*
8	125
10	100*
20	50
25	40
40	25
50	20*
100	10*
125	8
200	5
250	4
500	2*

FIGURE 7-4

computer this is no problem: the capacity of the memory is usually some power of two (say, 2^n) and m of the n bits in the address are used to identify which of the 2^m positions in space are required, whereas the other $(n-m)$ bits identify one of the 2^{n-m} time intervals. For example, a magnetic drum containing $2^9 = 512$ words of storage might have $2^4 = 16$ channels, each containing $2^5 = 32$ words. Ordinarily, the thirty-two words in a channel would be assigned thirty-two consecutive addresses, so that the most significant four bits of an address are the *S*-bits, the least significant five the *T*-bits. The address of location

372, for example, would appear in binary fashion as 101110100. The first four bits, 1011, are sent to flip-flops S_1–S_4, and indicate that the word is in channel eleven. The last five, 10100, are sent to T_1–T_5 and indicate that the twentieth word in that channel is desired. Note that the channel address is independent of the sector address, and vice versa.

In a decimal computer, the assignment of addresses to T- and S-selection is a little more complicated, and depends on what decimal code is to be used for the address. In a decimal computer system, the memory capacity is usually some power of ten. Supposing we have a 1000-word memory which is to be addressed by three decimal digits each coded in the 8-4-2-1 code. The address "372" would then appear as 0011 0111 0010. How many different ways can we apportion T- and S-addresses in such a memory? Noting that there are twelve bits in the decimal address, we might conclude that there are eleven ways, as is indicated in Figure 7-4*a*. On the other hand, examining the factors of 1000 (see Figure 7-4*b*) we might conclude that there are fourteen ways. This apparent contradiction may be resolved by examining carefully each of the proposed subdivisions of Figure 7-4. For example, suppose we would like to use nine bits for the S-address and three for the T-address. Three T-bits might distinguish either five or eight time intervals, and the resulting distribution of addresses is shown in Tables 7-5 and 7-6 respectively.

Table 7-5

S-Address			
Decimal Digits	Binary Bit	T-Address	Total Address
0 0	0	0 to 4	0 to 4
0 0	0	5 to 7 }	5 to 9
0 0	1	0 to 1 }	
0 1	0	0 to 4	10 to 14
0 1	0	5 to 7 }	15 to 19
0 1	1	0 to 1 }	
. . .	. . .	. . .	. . .
9 9	0	0 to 4	990 to 994
9 9	0	5 to 7 }	995 to 999
9 9	1	0 to 1 }	

200 channels of five words each

Examining these tables, we find two rather unexpected difficulties. In the first place, the count sequence for T-selection depends on which part of the S-selected memory we are scanning. For example, the count sequence in alternate points in space in Table 7-5 can be seen to be

0	0	0		1	0	1
0	0	1		1	1	0
0	1	0	and	1	1	1
0	1	1		0	0	0
1	0	0		0	0	1

The second difficulty is that the particular point in space desired cannot be determined by examining the S-address alone; the T-address must also be examined. For example, the S-address 03-0 in Table 7-6

Table 7-6

S-Address		*T*-Address	Total Address
Decimal Digits	Binary Bit		
0 0	0	0 to 7	0 to 7
0 0	1	0 to 1	8 to 15
0 1	0	0 to 5	
0 1	0	6 to 7	16 to 23
0 1	1	0 to 1	
0 2	0	0 to 3	
0 2	0	4 to 7	24 to 31
0 2	1	0 to 1	
0 3	0	0 to 1	
0 3	0	2 to 7	32 to 39
0 3	1	0 to 1	
0 4	0	0 to 7	40 to 47
0 4	1	0 to 1	48 to 55
0 5	0	0 to 5	
. . .	. . .	. . .	. . .
9 8	0	4 to 7	984 to 991
9 8	1	0 to 1	
9 9	0	0 to 1	
9 9	0	2 to 7	992 to 999
9 9	1	0 to 1	

125 channels of eight words each

may refer to an address in the channel which contains addresses 24 to 31, or to the channel containing addresses 32 to 39.

Although both of these difficulties can be surmounted with a little effort, they can be eliminated altogether by using one of the address breakdowns marked by an asterisk in Figure 7-4. For example, if five bits are used for the *T*-address and seven for the *S*-address, each *S*-location will contain twenty words, and the address of the twenty will always be between zero (00000) and nineteen (11001). The *S*- and *T*-addresses are completely independent, just as they were for the binary addressing system. (See Table 7-7.)

Table 7-7

S-Address			
Decimal Digits	Binary Bits	*T*-Address	Total Address
0	000	0 to 19	0 to 19
0	001	0 to 19	20 to 39
0	010	0 to 19	40 to 59
0	011	0 to 19	60 to 79
0	100	0 to 19	80 to 99
1	000	0 to 19	100 to 119
. . .	. . .	. . .	. . .
9	011	0 to 19	960 to 979
9	100	0 to 19	980 to 999

LOGICAL PROPERTIES OF TYPICAL DIGITAL MEMORIES

The descriptions which follow were drawn up with the intention of presenting examples of the chief properties, advantages, and limitations of commonly used digital computer memories. However, the characteristics discussed are primarily those of interest to the logical designer, who must be given a clear picture of just how the memory fits in with his other circuit components.

Magnetic Drums

General description. A magnetic drum is a metal cylinder which revolves about its axis and on the surface of which information is recorded. It is probably today's most widely used memory device, for it is reliable and it provides a large storage capacity at very low cost.

Reading and writing. As was indicated in the previous section, a combination of *S*- and *T*-selection is usually used with a magnetic drum. The drum is divided into channels along the drum axis, and each channel is divided into sectors or words. Figure 7-5*a* indicates the arrangement of a hypothetical, sixteen-word serial memory on the surface of a drum, and Figure 7-5*b* shows schematically where the words are located and how they are addressed. Note that each word contains six bits and that there is a "space bit" between adjacent words around the drum circumference.

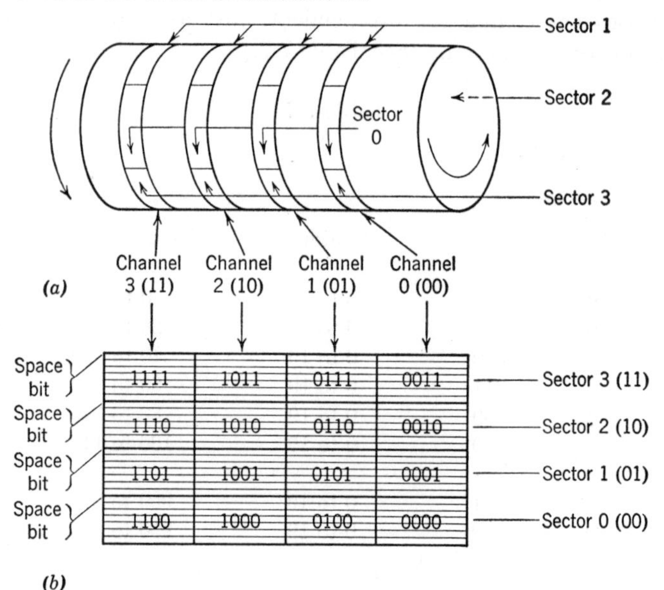

FIGURE 7-5

Reading from a drum is accomplished by placing a small coil called the *read head* near the drum surface where it can intercept magnetic flux, and connecting the read head to an amplifier. A similar coil similarly located, called the *write head,* will erase old data and simultaneously write new information in the channel beneath the head, if current is forced through the coil. Depending on what limitations are imposed by the circuit designer, the logical designer may have a good deal of flexibility in placing read and write heads. In Figure 7-6*a*, for example, read and write heads are separated by 45° of arc. Notice that this configuration of heads has the rather undesirable feature that the sector address (*T*-address) for a sector being *read* is different from the address for the same sector when new data is *written* into it. This difficulty may be circumvented as shown in Figure 7-6*b*, where now the read and write coils are mounted on a single head. It is, of course, also possible to mount a number of read and write heads around a single channel, as shown in Figure 7-6*c*.

The read and write amplifiers may themselves hold one bit of information each and thus introduce an extra delay between the drum and the amplifier output. If the circuit designers supply such a flip-flop amplifier, the logical designer must be careful to provide correct timing

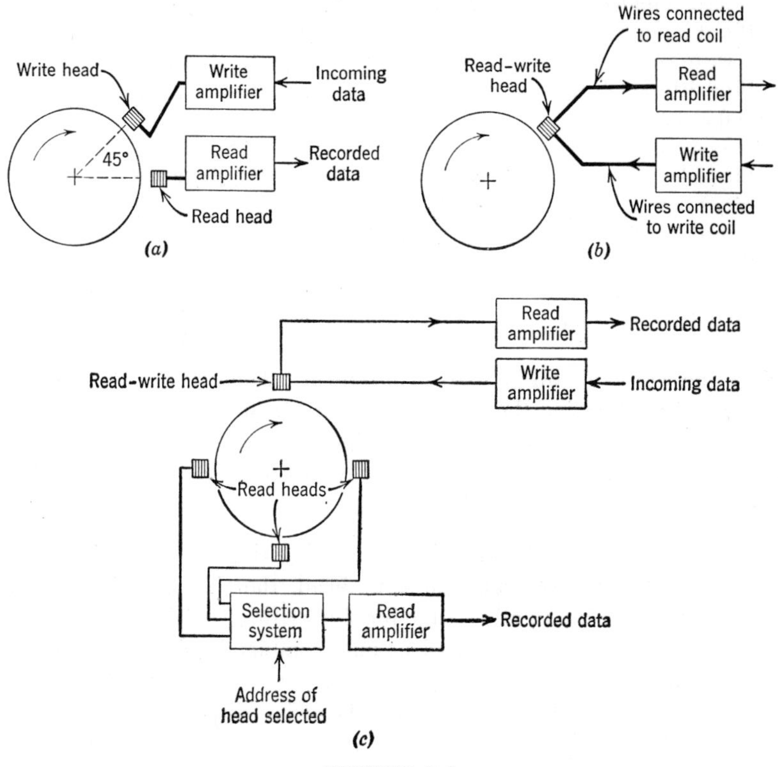

FIGURE 7-6

for read and write operations. Table 7-8 illustrates the problems which arise when flip-flop amplifiers are used with a serial memory and it is necessary to read and write during adjacent word-times. Note first that the flip-flop amplifiers require a T-selection (coincidence) signal from bit-time 1 to bit-time 0 for reading, and from bit-time 6 to bit-time 5 for writing. A single signal from bit 0 to bit 6 is all that is necessary if no-delay amplifiers are used. Note also that bit 0′ of the second word must be inserted in the write amplifiers before bit 6 of the first word can be read.

This overlapping of words is one reason for providing a space bit between words. If bits 0 and 0′ of Table 7-8 are not data bits, then they need be neither read nor written, and the most significant bit read

Table 7-8

No-delay amplifiers															
Bit under head	0	1	2	3	4	5	6	0′	1′	2′	3′	4′	5′	6′	0″
Bit at output of read amplifier	0	1	2	3	4	5	6								
Bit at input to write amplifier								0′	1′	2′	3′	4′	5′	6′	
Bit written on drum								0′	1′	2′	3′	4′	5′	6′	
Flip-flop amplifiers (one-bit delay)															
Bit under head	0	1	2	3	4	5	6	0′	1′	2′	3′	4′	5′	6′	0″
Bit at output of read amplifier		0	1	2	3	4	5	6							
Bit at input to write amplifier							0′	1′	2′	3′	4′	5′	6′		
Bit written on drum								0′	1′	2′	3′	4′	5′	6′	

from one word (6) can be used in whatever calculations are being carried out to determine the least significant bit of the next (1′). Even if no-delay reading and writing amplifiers are used, however, it is in general desirable to place space bits between words for three other reasons:

1. If it is necessary to write in a channel during one word-time, and read from it the next using the same head, the circuit designer may insist on a space bit to allow time for the writing transients to die out in the head—a very small transient may completely obscure the read-out signal.
2. The magnetic flux from a write head is usually not confined to the particular cell immediately under that head, so that the process of writing a bit on the drum may alter the next bit. When a whole word is to be inserted into the memory, this overlapping does no harm until the last bit of the word is written, but if no space bit were present the first bit of the next word might then be altered.
3. The presence of an extra bit-time between words is often a great convenience to the logical designer, who may need time between words to complete logical or arithmetic operations.

Clock pulses and timing signals. A special channel on the drum, containing permanently recorded pulses, is employed as the clock-pulse source for a magnetic drum and supplies pulses to the read and write amplifiers as well as to the computer flip-flops. Furthermore, there may be one or more additional special channels containing such timing pulses as the logical designer finds useful or necessary. For example, somewhere on the drum there must be recorded information which will insure that the clock-pulse counter (equations 7-1) is synchronized with the bits of a word, and that the word counter (equations 7-9) is synchronized with the words in the memory. In a magnetic drum computer both synchronizations may be accomplished by providing a channel which contains a single, permanently recorded bit. This bit can be placed so that it is read at some particular bit-time (e.g., t_0), of a particular sector on the drum (e.g., the sector whose address is zero). If this index bit is read from an amplifier labeled X, it can be used to modify equations 7-1 so as to set the bit-counter to 001.

$$\begin{aligned} R_A &= B\bar{C} + X & S_A &= BC\bar{X} \\ R_B &= AB + BC + X & S_B &= \bar{B}C\bar{X} \\ R_C &= C\bar{X} & S_C &= \bar{A}\bar{C} + \bar{B}\bar{C} + X \end{aligned} \tag{7-14}$$

It can also be used to modify equations 7-9 so as to set the current-address counter to zero.

$$\begin{aligned} J_{A1} &= t_0\bar{X} & K_{A1} &= t_0 + X \\ J_{A2} &= t_0A_1\bar{X} & K_{A2} &= t_0A_1 + X \\ J_{A3} &= t_0A_1A_2\bar{X} & K_{A3} &= t_0A_1A_2 + X \\ J_{A4} &= t_0A_1A_2A_3\bar{X} & K_{A4} &= t_0A_1A_2A_3 + X \\ J_{A5} &= t_0A_1A_2A_3A_4\bar{X} & K_{A5} &= t_0A_1A_2A_3A_4 + X \end{aligned} \tag{7-15}$$

Note that $t_0 = \bar{A}\bar{B}\bar{C}$.

Access time. When some form of T-selection is used in a digital computer memory, the time which elapses between the instant a word is requested from (or ready for transmittal to) the memory and the instant the required address appears under a read or write station is called the *access time.* In a magnetic drum memory, the access time of words in a given channel on the drum depends on how the heads are located around that channel. Referring again to Figure 7-6*b*, for example, we see that the maximum access time for either writing or reading is one drum revolution; but in Figure 7-6*c* the maximum access time for reading a word has been reduced to one-quarter of a drum revolution by the addition of three read heads around the circumference. It would be possible to reduce the access time to one *word-time* by placing as many heads around a channel as there are words in that channel.

The addition of heads, of extra reading and writing circuits, and of the logic necessary to select the head nearest to the word desired is unfortunately somewhat expensive; and the system designer and logical designer must in general decide on some compromise between cost and access time.

Serial and parallel operation. In a *parallel* computer, all the bits of a word are operated upon and must be available simultaneously; in a *serial* machine they are operated on sequentially, one at a time. A magnetic drum, on which bits and words are scanned in sequence by a read-write head, is inherently a good serial memory device. However, it may also be used as a parallel memory by recording all n bits of a word in n separate channels on the drum and by reading them simultaneously. Note that it may then be necessary to make every other bit in each channel a space bit because of the overlap which occurs in writing on the drum. Because of this, and because the speed inherent in parallel arithmetic operations may be lost as a result of the drum access time, drum memories are usually employed to store serial information.

Memory capacity. The number of bits which may be stored on a magnetic drum depends, of course, on the size of the drum and on the density with which bits may be recorded on it. One drum has been built which is 2 inches in diameter and 2 inches long, and which has a storage capacity of some 20,000 binary digits. At the opposite extreme is a drum 4 feet in diameter with a storage capacity of 20,000,000 bits. A capacity of 100,000 bits is common, and may be obtained from a drum 6 inches in diameter and 6 inches long if a storage density of 50 bits per inch around the circumference and 20 bits per inch along the axis is assumed.

Some of the factors affecting bit density are especially important to the logical designer. Around the circumference of the drum, densities of from 50 to 200 bits per inch are in common use, and densities up to 500 bits per inch have been achieved. Along the axis of the drum, it is common to employ a density of twenty channels to the inch. The choice of circumferential and axial bit densities depends not only upon the ability of the read-write system to resolve recorded signals, but also upon the dimensions of the read-write heads and on the head-spacing requirements of the logical designer. If, for example, the designer needs one channel containing a read head one word away from a write head, if a word happens to contain 20 bits, and if the heads are 0.2 inch thick, then the maximum possible drum bit density will be $20/0.2 = 100$ bits per inch. The practical density achieved may be less than that to permit the heads to be mounted independently (so that their distance from the drum and from one another may be adjusted when the drum is first put together), with a shielding material between them (to prevent magnetic flux leakage directly from the write head to the read head).

Circulating registers. A read head and a write head on the same channel n bits apart may obviously be used to introduce n bits of delay into a computing system. Supposing, for example, that 150 bits are stored on the surface of the drum between the write head and the read head of Figure 7-6a, the write and read amplifier will have exactly the same function as 150 flip-flops connected as a shifting register—with the contents of each one shifting to the next in line every clock-pulse time. If we now connect the read amplifier to the write amplifier, we have a *circulating register* which will store 150 bits for as long as we leave them connected.

Such registers are much cheaper than equivalent flip-flop shifting registers, and are used wherever short access to a word or to a few words is required. The arithmetic unit of a digital computer often requires several such registers, for example.

Complete logic. The complete logic for a magnetic drum memory is given in Chapter 11 as part of the design of a drum computer.

Magnetic Cores

General description. A typical magnetic memory core is a doughnut-shaped object between 0.1 and 0.4 inch in diameter made of a special magnetic material having a high remanent flux density. The core may be magnetized in either a clockwise or a counterclockwise direction, and it is therefore capable of storing one binary digit. An array of these cores, together with the necessary circuits for reading and writing information, is called a *core matrix* and is fast becoming the most widely used memory after the magnetic drum. It has the advantage that access time to information is very small—words are S-selected from the matrix—and the disadvantage of relatively high cost.

Reading and writing. As might be expected in view of the discussion earlier in this chapter, the problem of S-selecting one word out of hundreds or thousands is so difficult that it dominates the design of any core memory. The generally accepted solution to this problem is to make S-selection an integral part of the reading and writing circuitry of the core memory. Reading and writing are both accomplished by driving electric current through wires which thread the core. When the current exceeds a certain minimum value, say I_m, the core is reset to "zero." Figure 7-7 illustrates how this minimum value of operating current may be used to select one core out of sixteen. Two wires must be threaded through each core, one being common to all cores in that row, the other to all cores in that column. We may reset any one of the sixteen cores to the "zero" state by driving a current $I_m/2$ through each of the two wires which thread that core; nine of the sixteen will then be completely undisturbed, six more will be partly disturbed by current $I_m/2$, which is not enough to make them change state, and the last core will receive the sum of the currents from the two wires, and

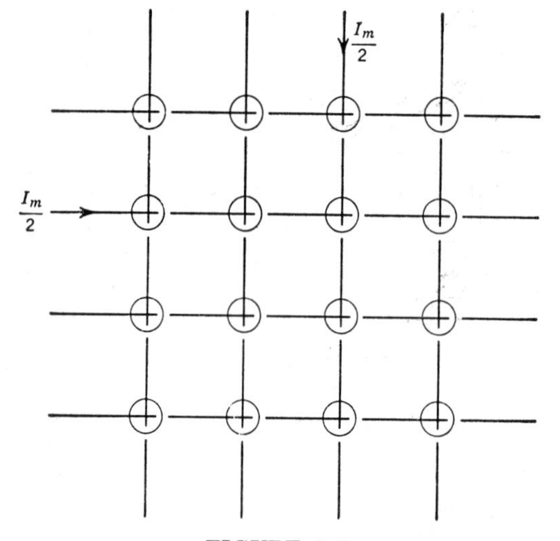

FIGURE 7-7

will be reset to "zero." To set a core to the "one" state, the current may be reversed in the two selecting wires, or additional wires may be provided which link the cores in the other direction.

Reading is accomplished by providing another wire (not shown in Figure 7-7) which links *all* the cores. When the bit in a particular core is to be read, a "zero" is written in that core, and the change of state, which will occur only if the core had contained a "one," causes an output pulse on the read wire. Of course, after reading, the core contains a "zero," whatever it had contained before. If a "one" was read, that "one" must be inserted back into the core from which it came, and this means that the core address must be retained until that insertion, generally called *regeneration,* is complete. Regeneration is often accomplished automatically by the read-write circuits, so that the logical designer need do nothing more than allow time for it to occur. Alternatively, it may be necessary for the logical designer to arrange that every word read from the memory is immediately written back into the position from which it came.

The operation of writing into a magnetic core memory may be

accomplished by carrying out the read-regenerate cycle, but regenerating the new word to be written rather than the old word just read.

Clock pulses and timing signals. The magnetic core memory neither supplies nor requires a synchronizing clock-pulse signal, and so the clock pulses required by the other computer circuits may be obtained from an oscillator or from a synchronized memory device (e.g., a drum) if one is present.

The operations of reading, regenerating, and writing in a core memory generally require a rather complicated series of timing signals. However, such signals are usually generated and supplied by the circuit designer rather than the logical designer for two reasons: (1) because the timing signals must often have special shapes; (2) because the timing signals must often last a nonintegral multiple of a clock-pulse interval. The sequence of timing signals may, for example, take place entirely during one clock-pulse interval, so that if a word is requested from (or ready to be written into) the memory at one bit time, it will be available (or stored away) by the next clock-pulse time. Alternatively, it may be desirable to run the arithmetic and control circuits at a clock-pulse rate so fast that the memory cannot operate reliably in one clock-pulse interval. The memory reference cycle may then extend over several clock-pulse times.

Access time. There is, of course, no T-selection involved in the use of a core memory. The only delay necessary between the time an address is available and the time the corresponding word is read from or written into that address is the delay occasioned by the operation of the S-selection circuits and the regeneration cycle. Such a delay is typically between one and five clock-pulse intervals, depending on how fast the circuit designer can make the memory cycle operate.

Serial and parallel operation. Because a core memory is inherently able to make any bit available in a few clock-pulse times, at most, it is usually used in a parallel computer where quick access to information is important. In such a parallel memory, a "plane" of cores of the kind shown in Figure 7-7 contains one bit from each word stored. For example, Figure 7-7 would represent one plane in a sixteen-word memory; and if each word in this memory had to contain twenty bits, there would have to be nineteen other identical planes. The S-selection circuits would then always stimulate the two corresponding selection wires in twenty planes every time a word must be read or written, and twenty memory elements must be available to store the word read out or the word to be written in.

It is, however, also possible to employ a magnetic core memory to read out and write in words one bit at a time so that they may be handled serially by the computer. In such a memory it is very desirable that the computer clock-pulse interval and the interval between bits in a word read from or written into the memory be the same. If they are not the same, e.g., if a bit comes from the memory every two clock-pulse times, the arithmetic logic must be arranged accordingly. Even if they are the same, there may be a delay of several bit-times between the time an address is presented to the memory and the time the least significant bit of the selected word is available. The logical designer must allow for this delay.

Memory capacity. The size of a core memory is limited by circuit design considerations. A large number of cores involves a large driving current, a cumbersome amount of S-selection equipment, and a decrease in the ratio of wanted to unwanted signal in the read-out operation. An experimental memory has been built having 65,536 37-digit words—2.5 million bits! Common core memories range in size from 20,000 to 200,000 bits.

Complete logic. Let us take a look at a typical parallel magnetic core memory, from the point of view of the logical designer. Suppose

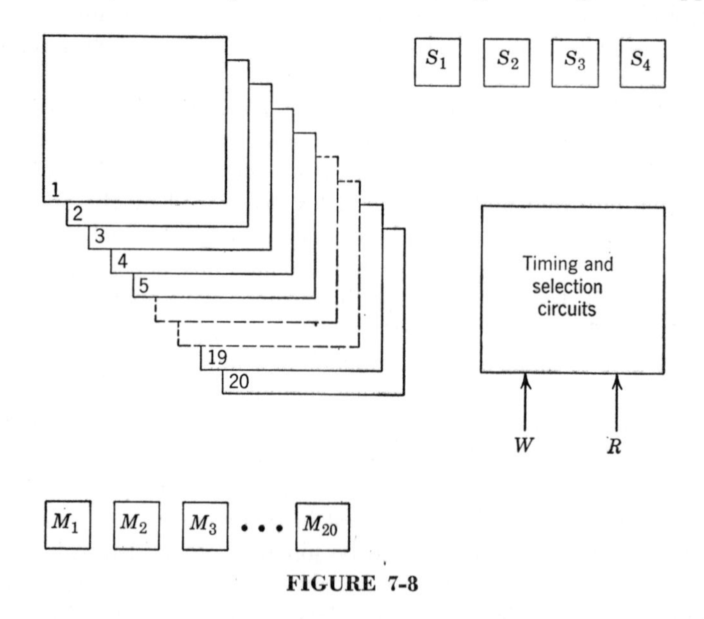

FIGURE 7-8

we want a sixteen-word memory where each word is to contain twenty bits. We must then provide twenty planes of the size illustrated in Figure 7-7. Suppose also that the circuit designer specifies the following procedure for reading and writing (see Figure 7-8). For reading:

a. The address of the word to be read must be in the S-selection register (flip-flops S_1–S_4) before the reading cycle can begin. Let us assume that the address is inserted at bit-time n.

b. If the R input to the timing circuits is stimulated at bit-time $(n + 1)$, the word whose address is in the S-register will be shifted into the twenty-flip-flop M-register by bit-time $(n + 3)$.

c. The M-register must be in the zero state in order for a new word to be read in. It must therefore be reset to zero no later than $(n + 1)$ bit time.

d. Regeneration of the information just read is not complete until bit time $(n + 5)$. Therefore the M-register must not be reset again, nor can the S-address be changed before that bit-time.

For writing:

a. The word to be written and the address into which it is to be sent must be written into the M- and S-registers, respectively, before the writing operations can start—say, at bit-time n.

b. At bit-time $(n + 1)$ the W input to the timing circuits may be stimulated.

c. The writing cycle is not complete until bit-time $(n + 3)$. Neither the S-register nor the M-register may be altered until then.

This combination of restrictions and rules establishes the framework within which the logical designer must work. Perhaps the most important feature of this framework is the sequence of delays involved in reading from and writing into the memory. A sequence of four bit-times is required for reading, two for writing. A four-bit cycle may therefore become the basic timing cycle for the computer associated with this memory, and the designer will try to arrange that basic computer operations take four bit-times. A faster computing time is expensive, and means that the computer will spend much of its time idle, waiting for the memory cycle to be complete. A slower computing time wastes some of the speed available from the core memory. It is thus a matter of "matching" computer to memory. (These remarks assume that the reading operation occurs more frequently than the writing one, and this is in general true.)

A basic part of the computer will thus be a four-bit-time counter, which will then be used to schedule all other computer operations, and which can, of course, be used in the equations defining the operation of the S- and M-registers. These equations are perfectly straightforward to write, and will not be presented here.

Magnetic Tape

General description. A magnetic tape is a flexible plastic or metal strip from 0.001 to 0.010 inch thick, from 0.25 to 2 inches wide, and as long as can conveniently be handled—2500 feet of tape can be stored on a reel with ease. The tape, like the drum described earlier, is prepared in such a way that information may be recorded on its surface. A magnetic tape provides a cheap means of obtaining a very large storage capacity, but has the disadvantage of long access time. It is ideally suited for applications which require the sequential scanning of a very large file of data, e.g., in the insurance business where thousands of policies must be recorded and manipulated every year.

Magnetic tape is usually stored on a reel, and is transferred from one reel to another in the course of reading or writing data. Figure 7-9 illustrates a typical reel system. The tape is moved past the read-

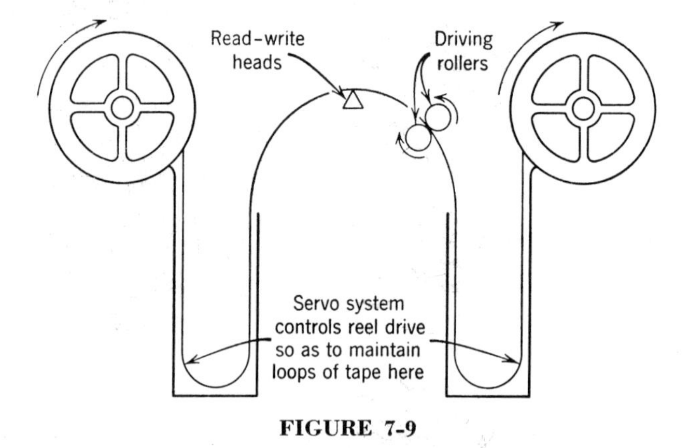

FIGURE 7-9

write heads by means of the two driving rollers, which are pressed against the tape from each side. The tape loops and the servomechanisms which maintain them are provided so that the tape may be moved a few inches at a time very quickly without waiting for movement of the high-inertia reels.

Any tape reel equipment has the disadvantage that it is necessary either to read the tape backwards or to rewind it completely after it has been run from one reel onto the other. This troublesome feature may be eliminated by providing an endless tape, dumped into and drawn from a box as shown in Figure 7-10. Note that the tape loop is driven by a pair of rollers, just as the reel tape was. If the loop is moved a few inches at a time and then stopped, it has many of the characteristics of a tape reel. If the tape is run continuously, it has many of the characteristics of a magnetic drum.

S-selection arises in connection with a magnetic tape memory whenever several tapes are connected to a computer and one or more must be operated on at a time. *T*-selection arises when a tape loop is run continuously, or when an automatic search is carried out for some particular word on a long tape. However, as will be seen in a moment, tapes are often moved on a start-stop basis rather than continuously, so that the *T*-selection problem does not usually arise.

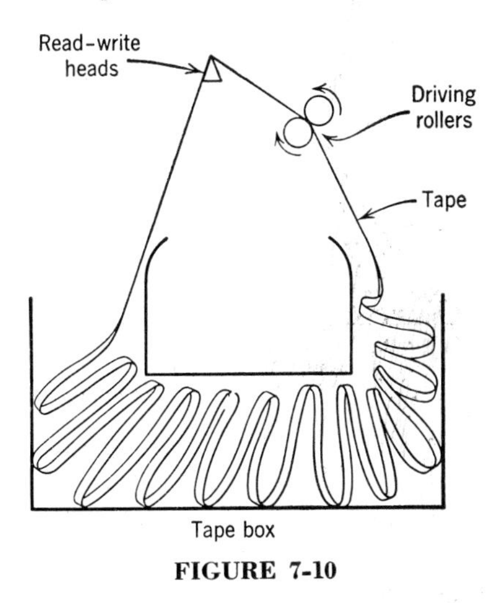

FIGURE 7-10

Reading and writing. The reading and writing heads and circuitry for magnetic tapes are much like those used for a magnetic drum. However, certain new difficulties arise when a tape must be started, stopped, and rewound. Suppose the tape has stopped, and we want to read the next word or the next few words from the tape. First, a signal must be provided to operate a solenoid which in turn operates the driving rollers. Second, a delay occurs while the rollers engage and while the tape accelerates to operating speed. The end of this wait time is often signaled by a magnetically recorded or mechanically sensed marker (e.g., a hole punched in the tape and read photoelectrically) which indicates that the desired information is about to appear under the read heads. Third, the information is read from the tape. Fourth, an end-of-information signal must be read or derived, and used to disengage the rollers and stop the tape.

The writing operation may be described in similar terms. The driving roller is energized, a suitable delay is introduced, the information is then written, and an end-of-information signal recorded. The "suitable delay" is very important and should be described in more detail. If writing were begun as soon as the driving roller was energized, it is obvious that until the tape started moving, all the bits would be recorded on one spot—or rather, the *only* bit recorded would be the one written just as the tape started to move. It is necessary, therefore, to introduce a delay *at least* as long as the time required to get the tape moving at full speed. The delay must be even longer than that, however, to allow for variations in tape acceleration and deceleration times. Figure 7-11 indicates the difficulty which arises. Suppose that

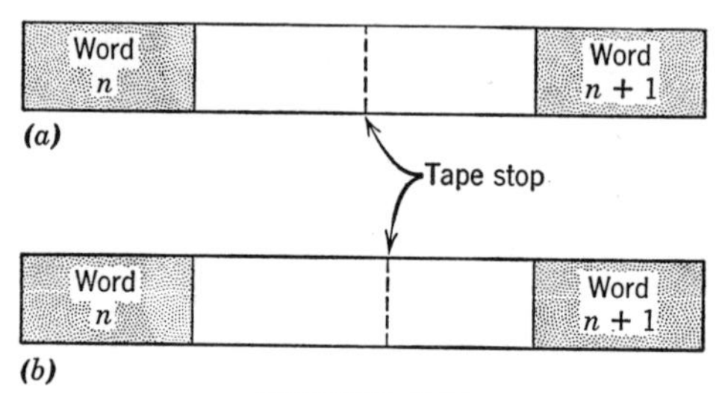

FIGURE 7-11

when two particular words are written on the tape, the driving rollers happen to operate a little faster than usual. Then the distance between the end of word n in Figure 7-11a and the place where the tape stops (dotted line) is particularly short. If no extra delay is introduced before writing word $(n + 1)$, the space between tape stop and the beginning of the next word may also be particularly short. Next, suppose that these two words are read some time later, when the driving rollers happen to be a little sluggish. Referring to Figure 7-11b, we see that the tape will then not stop in the same place it did when the words were written; and if its acceleration is somewhat slow, it will not have attained full speed by the time the first part of word $(n + 1)$ crosses the read head, and the first bits may not be read correctly.

This detail, which is typical of the kind of problems encountered in magnetic tape memory design, is eliminated if the tape contains a series of permanent beginning-of-word markers spaced far enough apart to allow for the slowest driving-roller action.

When a tape reel is to be used, some means must be provided for detecting the tape ends before they come off the reels. This is often done mechanically, e.g., by closing a contact through a hole punched near the end of the tape. When the tape is read from or written on in the forward direction, this end-tape signal must be checked during every read or write operation. If the tape is to be rewound all at once, the beginning-of-tape marker determines when to stop rewinding; and if the tape is to be read from and written on backwards (which in general introduces a great many new problems) the beginning-of-tape marker obviously has the same function as an end-tape marker has.

Clock pulses and timing signals. If a tape loop is run continuously, so that it looks to the logical designer like a very short drum of very large diameter, it may contain a special clock-pulse channel which supplies clock pulses continuously to the entire computer. If, on the other hand, the tape is started and stopped frequently, the computer itself must have a source of clock pulses different from those recorded on the tape.

The latter situation is by far the more usual one. A magnetic tape, or a number of magnetic tapes, may be employed as auxiliary memories to a magnetic core or drum memory which has its own set of clock pulses. However, in order to read data from the tape there must be some form of tape clock pulse, usually recorded on a special channel on the tape. These clock pulses occur at a much slower rate than the main memory clock pulses (10,000 per second compared with 200,000 per second is typical) and cannot easily be synchronized with them. The transfer of information between two storage media which operate at different information rates is very complicated, and will be discussed in the next chapter.

Access time. Supposing a tape reel is stopped with no tape on the take-up reel, so that the first word on the reel is next to be read. If the computer asks for that word, its access time depends only on the space between the read head and the word, and on the speed with which the driving roller moves the tape. This access time is commonly from 5 to 10 milliseconds.

At the other extreme, suppose the computer asks for the very *last* word on the tape. The time necessary for the driving rollers to make contact with and accelerate the tape is then insignificant compared with the time necessary to run through the entire tape. A reel might contain 2500 feet of tape, and if the tape moves at 150 inches per second, the access time to the last word will be 200 seconds.

Serial and parallel operation. A magnetic tape generally contains several channels, recorded side by side, each with its own read-write head. One or two of these channels, as we have seen, are reserved for clock pulses and word markers. The remaining channels are used for data. If there is but one more, data can only be recorded serially. If there are twenty or more, data may be recorded in parallel. If there are fewer than twenty, some series-parallel arrangement of data is indicated. For example, with ten data channels it will take two tape clock-pulse times to read a complete twenty-bit binary word; with four data channels in a decimal machine, each group of four bits might represent a decimal digit, and a word ten digits long would require ten tape clock-pulse times.

Memory capacity. The capacity of a magnetic tape memory depends on tape length and width, on the storage density of data, and on

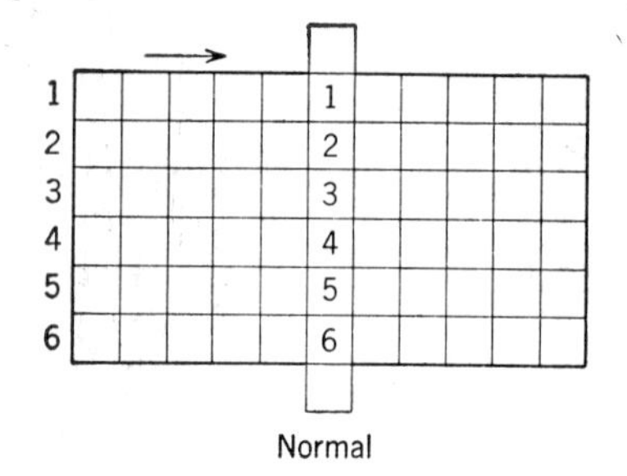

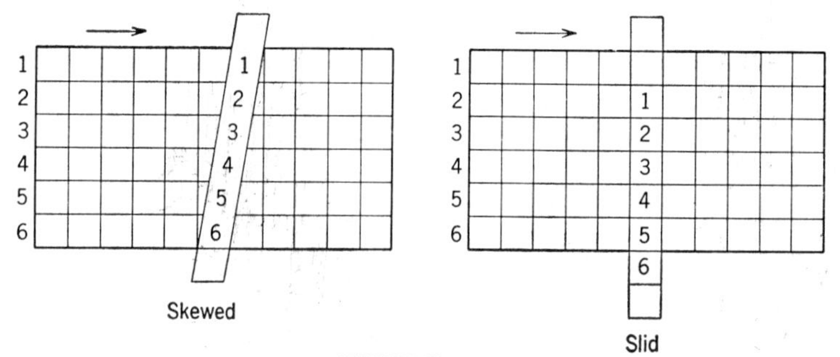

FIGURE 7-12

the way data is arranged on the tape. Increases in length give added capacity at the expense of access time. Increased width or increased storage density along the length gives added capacity but makes necessary an extremely accurate alignment of tape and heads. If the tape twists or skews with respect to the heads, the head at one side of the tape may be reading from one line while the head on the other side reads from the next line (see Figure 7-12). Increased storage density across a narrower tape tends to reduce the effect of a slight skew. However, it makes tape alignment more of a problem in another respect: if the tape slides to one side, for example, head n may be reading channel $(n + 1)$.

The arrangement of data on the tape has a very great influence on storage capacity in that it determines the relative proportions of data and word spacing (see Figure 7-11). We have seen that the space allowed for the tape to stop and start must be long enough to allow for the greatest expected deceleration and acceleration times of the driving rollers. A tape driven at 150 inches per second moves more than $\frac{1}{8}$ inch in a millisecond, and because millisecond operation and precision are very difficult to obtain from a mechanical device, it may be necessary to provide a 2-inch space between pieces of data on the tape.

This being so, it is obviously desirable to read or write as much data as possible every time the tape is started, so that the ratio of data space to start-stop space will be as large as possible. So far, all our discussions have been based on storage of a single computer word—say twenty bits. Supposing there were five data channels across the tape, and 100 bits to the inch stored along the tape, each word would occupy 0.04 inch. If 2-inch separation between words is dictated by the operating speed of the driving rollers, then only $(0.04/2.04) \times 100$ or about 2% of the tape could contain data. We are therefore forced to record several computer words at a time in order to make really efficient use of the tape. For example, if fifty words at a time are recorded on the tape, a "block" of words will occupy 2 inches, and 50% of the tape will contain data.

Used in this way, a 2500-foot reel of tape will contain about $(0.5)(2500)(12)(100)(5) = 7.5 \times 10^6$ bits. Ten tape units of the kind indicated in Figure 7-9 can be provided at reasonable cost and make almost 10^8 bits directly and automatically accessible to the computer proper.

Complete logic. To the logical designer, a tape reel has the appearance of a black box having a number of inputs and outputs, as is indicated in Figure 7-13. Let us suppose that two of these tapes are to be connected to a computer through appropriate selection devices and amplifiers. It is the logical designer's job to write equations defining the input lines as functions of other computer variables, and to use the output lines in the input logic to other flip-flops.

Let us suppose this tape system is to be connected to a computer in such a way that the computer may request one of three things: that the next block of data be read from the tape into the computer's magnetic core memory; or that a block of data in the core memory be written on the tape starting with the next tape block marker (and writing over the block already there); or that the tape be reversed and rewound to the beginning.

Before describing these three operations in detail, it is desirable to point out how complicated the timing problem is. There will be two sets of clock pulses involved in the computer circuits: the computer clock pulses, which occur (say) every 5 microseconds; and the tape clock pulses, which are derived from the tape when data is being read, and must be supplied to the tape when data is written. If the tape

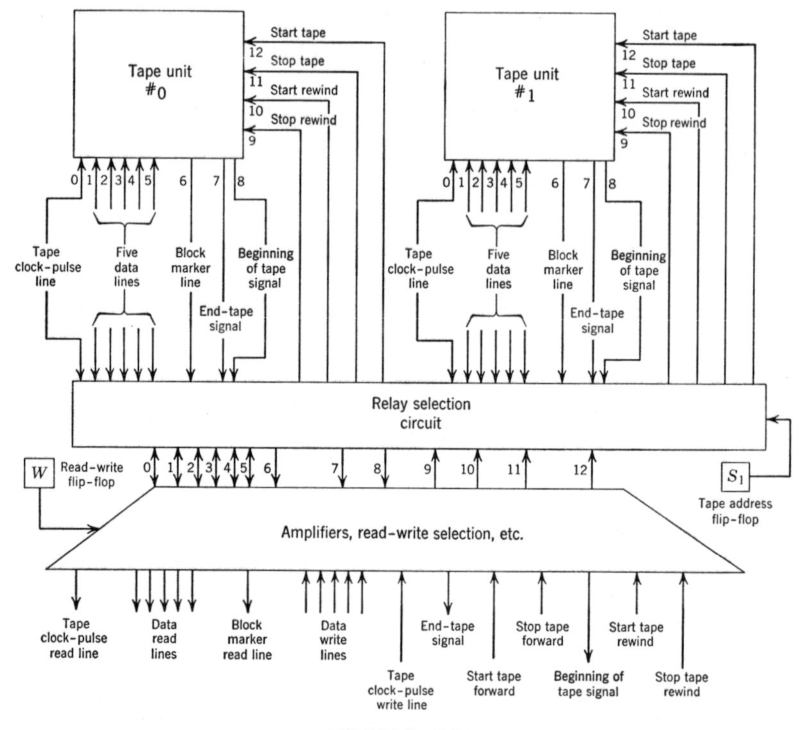

FIGURE 7-13

travels at 100 inches per second and has a bit density of 100 bits to the inch, five data bits will be read from the tape, in parallel, approximately every 100 microseconds. Although the 5-microsecond basic clock-pulse interval may be very carefully stabilized, the 100-microsecond clock interval derived from the tape is somewhat imprecise and may vary from 90 to 110 microseconds. It is, of course, not in synchronism with the computer clock pulse during reading, though the tape *writing* clock pulse may be derived from and exactly synchronized with the computer clock pulses. The logical designer must arrange to compensate for the differences in information flow rate implied by the difference in clock-pulse intervals, and may even have to use flip-flops synchronized with tape clock pulses. The problem of effecting transfers between devices whose operating times do not "match" will be discussed in the next chapter.

In order that the computer can carry out a "read tape" operation, the logical designer must arrange for the following sequence of functions to take place at the request of the computer:

a. The S-address must be in flip-flop S_1, so that the proper tape is connected to the amplifiers. If this address is different from the last tape address used, the logical designer must allow time for the relay selection network to set up before going on to step c below.

b. The W flip-flop must be in the 0 or read state.

c. The "start tape forward" line must be stimulated.

d. The "block marker read" line must be monitored. When a "one" is read, the "data read" lines are connected to appropriate flip-flops, and every time a signal appears on the "tape clock-pulse" line, the five "data" lines are read and their contents placed in the appropriate place in the core memory.

e. When the last five bits of the block have been read (which may be signaled in a number of ways, e.g., by the appearance of another signal from the "block marker read" line), the "stop tape forward" line must be stimulated.

f. If the "end-tape signal" occurs any time during the read operation, the computer must be notified.

For the "write tape" operation, the following sequence of functions is prescribed.

a. The S-address must be in S_1.

b. The W flip-flop must be in the 1 or write state.

c. The "start tape forward" line must be stimulated.

d. The "block marker read" line must be monitored. When a "one" is read, the writing operation starts and the first five bits of the first word are written on the tape via the "data write" lines. At the same time, the first tape clock pulse is written. The logical designer's independent source of tape clock pulses (derived, e.g., by counting twenty computer clock pulses, if the computer clock pulses occur every 5 microseconds and the tape clock pulses should occur every 100) then is used to tell when the second, third, etc., group of five bits should be written, and each group in turn is connected to the write lines.

e. When the last five bits of the block have been written on the tape (and the logical designer must determine this, perhaps by counting bits as he writes them) the end-block marker is written, and the "stop tape forward" line is stimulated.

f. If the "end-tape signal" occurs any time during the write operation, the computer must be notified.

For the "rewind tape" operations, the following sequence of functions is prescribed.

a. The S-address must be in S_1.
b. The "start tape rewind" line must be stimulated.
c. The "beginning of tape" line must be monitored, and when a signal appears, the "stop tape rewind" line must be stimulated. The start and stop rewind lines may operate a motor which drives the feed reel in the backward (rewind) direction.

Because these tape-using operations take a long time as measured by computer clock-pulse times, the logical designer may take steps to permit the computer to go on with other operations after a tape-using operation has been initiated. In order that computer and tapes work independently, their control logic must be largely independent. Furthermore, some care must be taken that the two activities do not interfere with one another: for example, when a relatively slow tape operation is complete, it must not interrupt the computer; and when one tape operation is under way and the computer requires another (for instance, when the tape is rewinding and the computer asks that the first block be read), the computer must then be forced to wait until the current tape command is complete.

This discussion of magnetic tape memories has necessarily been abbreviated. It is not possible to describe all the problems and all solutions in such a short space. Furthermore, it must be realized that many of the features described here may not appear in any particular tape system, and that many others may appear which are not described here.

Other Memories

In addition to the drum, cores, and tape, many other digital memories have been and are being used. Mercury delay lines and magnetostrictive delay lines have characteristics much like the magnetic drum. The electrostatic and diode-capacitor memories resemble a magnetic core memory, and the magnetic wire resembles a magnetic tape. However, these and other memories are not now widely used in the design of new computers and have for that reason been omitted. The three memories described here will probably, in their turn, be displaced by more efficient storage media of one kind or another as time goes on, but for the moment their advantages are dominant.

BIBLIOGRAPHY

Surveys of Digital Memories

I. L. Auerbach, "Digital Memory Systems," *Electrical Manufacturing*, 100–107 (Oct. 1953); 136–143 (Nov. 1953).

G. L. Hollander, "Bibliography on Data Storage and Recording," *Communications and Electronics*, 49–58 (Mar. 1954).

F. Fowler, Jr., "The Computer's Memory," *Control Engineering*, **3**, no. 5, 93–101 (May 1956).

Magnetic Drums

A. S. Hoagland, "Magnetic Drum Recording of Digital Data," *Communications and Electronics*, 381–385 (Sept. 1954).

F. C. Williams, T. Kilburn, and G. E. Thomas, "Universal High Speed Digital Computers: A Magnetic Store," *Proc. of the I.E.E.*, **99**, Part II, 94 (1952).

H. W. Fuller, P. A. Husman, and R. C. Kelner, "Techniques for Increasing Storage Density of Magnetic Drum Systems," *Proc. Eastern Joint Computer Conference*, 16–20, Dec. 1954.

M. K. Taylor, "A Small High-Speed Magnetic Drum," *Computers and Automation*, 18–19 (Jan. 1955).

I. W. Merry and B. G. Maudsley, "The Magnetic Drum Store of the Computer Pegasus," *Proc. of the I.E.E.*, **103**, Part B, Suppl. 2, 197–202 (Apr. 1956).

A. J. Strassman and R. E. King, "Counters Select Magnetic Drum Sectors," *Electronics*, **38**, no. 4, 161–163 (Apr. 1957).

Magnetic Cores

J. W. Forrester, "Digital Information Storage in Three Dimensions Using Magnetic Cores," *J. Applied Physics*, **22**, no. 1, 44–48 (Jan. 1951).

R. Thorensen and W. R. Arsenault, "A New Nondestructive Read for Magnetic Cores," *Proc. Western Joint Computer Conference*, 111–115, 1955.

R. C. Minnick and R. L. Ashenhurst, "Multiple Coincidence Magnetic Storage Systems," *J. Applied Physics*, **26**, no. 5, 575–579 (May 1955).

G. E. Valenty, "A Medium-Speed Magnetic Core Memory," *Proc. Western Joint Computer Conference*, 57–67, 1957.

R. L. Best, "Memory Units in the Lincoln TX-2," *Proc. Western Joint Computer Conference*, 160–166, 1957.

E. L. Younker, "A Transistor-Driven Magnetic Core Memory," *I.R.E. Trans. on Electronic Computers*, **EC-6**, no. 1, 14–20 (Mar. 1957).

E. Foss and R. S. Partridge, "A 32,000-Word Magnetic-Core Memory," *I.B.M. J. of Research and Development*, **1**, no. 2, 102–109 (Apr. 1957).

Magnetic Tapes

"Super-Speed Tape Puller for Computer Memory," *Electronics*, 222 (Jan. 1952).

M. V. Wilkes and D. W. Willis, "A Magnetic-Tape Auxiliary Storage System for the EDSAC," *Proc. of the I.E.E.*, **103**, Part B, Suppl. 2, 337–345 (Apr. 1956).

S. Baybick and R. E. Montijo, Jr., "An RCA High Performance Tape-Transport System," *Proc. Western Joint Computer Conference*, 52–56, 1957.

Cathode-Ray Tube Storage

F. C. Williams and T. Kilburn, "A Storage System for Use With Binary Digital Computing Machines," *Proc. of the I.E.E.*, **96,** Part II, 183 (1949).

D. B. G. Edwards, "The Design and Operation of a Parallel-Type Cathode-Ray Tube Storage System," *Proc. of the I.E.E.*, **103,** Part B, Suppl. 2, 319–326 (Apr. 1956).

Ultrasonic Delay Lines

M. V. Wilkes and W. Renwick, "An Ultrasonic Memory Unit for the EDSAC," *Electronic Engineering*, **20,** 208 (1948).

Ferroelectric Memories

C. F. Pulvari, "Memory Matrix Using Ferro-Electric Condensers as Bistable Elements," *J. Association for Computing Machinery*, **2,** no. 3, 169–185 (July 1955).

Cryotrons

D. A. Buck, "The Cryotron—A Superconductive Computer Component," *Proc. I.R.E.*, **44,** no. 4, 482–493 (Apr. 1956).

Magnetic Disks

D. Royse, "The IBM 650 RAMAC System Disk Storage Operation," *Proc. Western Joint Computer Conference*, 43–48, 1957.

EXERCISES

1. Write the simplest input equations for the five-flip-flop timing counter of page 177. Use T flip-flops. Assume that a synchronizing signal, X, occurs every t_0 time, and use it to ensure that the counter remains in synchronism with the computer. (Cf. equation 7-14.)

2. The relay selection "tree" of Figure 7-2 employs four relays. The first has one change-over contact, the second two, the third four, and the fourth eight. Redesign the "tree" so that the same selection function is performed, but no relay has more than five contacts.

3. Four flip-flops S_1–S_4 contain the 8-4-2-1 code for a decimal digit addressing one of ten points $P_0, P_1, P_2, \cdots, P_9$. S_1 is the most and S_4 the least significant bit of this code. Write the equation which mechanizes the S-selection of these ten points. Use redundancies to simplify wherever possible, and write out the equation both as a minterm-type expression (similar to equation 7-3) and as a maxterm-type expression (similar to equation 7-4).

4. A particular channel on a magnetic drum contains eight words. Four read heads are equally spaced around this channel and are connected to four no-delay read amplifiers R_1, R_2, R_3, and R_4. Thus a word read from R_1 at one word-time may be read from R_2 two word-times later, from R_3 four word-times later, from R_4 six word-times later, and again from R_1 eight word-times later.

A flip-flop counter W_1-W_2-W_3 contains the address of the word currently being read from the R_1 amplifier. The words have consecutive addresses, and while $W_1 = W_2 = W_3 = 0$, word zero is read from R_1, word two from R_4, word four from R_3, and word six from R_2.

Every word contains sixteen bits. The word counter W_1-W_2-W_3 has one added to it every time the space bit is in the read amplifiers—i.e., every t_{15} time. The least significant bit of each word appears in the read amplifiers at the *next* bit-time, t_0.

An address register A_1-A_2-A_3 contains the address from which it is desired to read. A new address will always appear in A_1-A_2-A_3 at bit-time t_{15}. For example, supposing that word six is required at the time that word one is being read from R_1, the contents of A_1–A_3 and W_1–W_3 will change as shown below:

···	W_1	W_2	W_3	A_1	A_2	A_3	
t_{13}	0	0	1	×	×	×	old desired address
t_{14}	0	0	1	×	×	×	
t_{15}	0	0	1	1	1	0	new desired address
t_0	0	1	0				
t_1	0	1	0				
···	0	1	0				

(*a*) Make no change in the operation of the W and A flip-flops, and write logic for a new flip-flop B from which the desired word is to be read at the earliest possible time after its address appears in A_1–A_3. Note that "the earliest possible time" implies that any of the four heads may have to be connected to B. B is a delay-type flip-flop.

(*b*) Now forget about solution (*a*) and assume that the input logic for B will be

$$D_B = \bar{A}_1\bar{A}_2A_3R_1 + \bar{A}_1A_2A_3R_4 + A_1\bar{A}_2A_3R_3 + A_1A_2A_3R_2$$

Write whatever logic is necessary to modify flip-flops A_1, A_2, and A_3 at t_{15} time in such a way that the above logic for B will read the desired word at the earliest possible time.

5. Write the complete logic for the sixteen-word magnetic-core memory described in the text. Assume that a twenty-flip-flop A-register is to be used as a destination for data read from the memory and as a source of data to be written in the memory. Assume also that two signals, R and W, determine what action is to be taken, as follows:

R	W	Action
0	0	M-register does not change
0	1	Write contents of A-register into memory location whose address is stored in flip-flops S_5–S_8
1	0	Read the contents of the memory location whose address is stored in flip-flops S_5–S_8, and store the result in the A-register
1	1	(Never occurs)

(Note that the address for reading and writing must be shifted from S_5–S_8 to S_1–S_4 at the appropriate time.) R and W may only change state at multiples of four bit-times, so it is possible to design a continuously operating control counter which identifies four bit-times and can be used to sequence the reading and writing operations.

8 / *Input-output equipment*

INTRODUCTION

In Chapter 6 we discussed input and output equipment briefly, and showed how it fits in with the other important functional parts of a digital computer. We defined an input device as a piece of equipment which makes it possible for a computer to accept information from outside itself; and an output device as a mechanism of some kind which translates and in some sense displays information sent from the computer. In this chapter we shall discuss such devices in more detail, and dispose of some of the problems which arise in employing them.

Because memories and input-output devices are very similar in nature (in the last chapter we saw, for example, that a magnetic tape might be used as a memory unit, an input device, or an output device) it is not surprising that problems arising in the use of one are very similar to those found in the use of the other. In Chapter 7 we discussed the problem of selecting automatically one of many sources of data, and showed how S-selection and T-selection are used in typical memory devices. In some computer systems, the logical designer must provide similar equipment for selecting one of many input or output devices and connecting it to the computer.

In the next two sections of this chapter we shall consider two problems—the use of asynchronous devices and the design of buffers—which, like the problems of S- and T-selection, are common to many input-output devices and which sometimes arise in connection with the design of memory devices as well. We shall then go on to describe how particular kinds of input and output equipment look to the logical designer.

ASYNCHRONOUS DEVICES

For the most part, the logical designer works with decision elements and memory elements whose outputs are defined at every clock-pulse time. In this section we discuss briefly the characteristics of relays, switches, push buttons, and similar devices, which provide signals which are not so well-defined. Such components are often used when essentially slow, mechanical pieces of equipment must communicate with a computer, or when the logical designer must provide for intervention by an operator who wants to stop, start, or otherwise exercise direct control over the computer.

Two principal characteristics make these components difficult to use. The first and obvious one is that their output signals may change at any time with respect to the principal computer timing signals. The contacts of a push button, for example, may close between any clock pulses, or during any clock pulse, at a time which is perfectly random as far as the general operations of a computer are concerned. The second awkward characteristic common to all these components is a tendency to "bounce"—to provide a more or less random series of "zeros" and "ones" before stabilizing. This characteristic is illustrated in Figure 8-1, where a signal from a pair of relay contacts is compared with a sequence of computer clock pulses. Such a signal may be very difficult for the logical designer to use, and he may require instead a signal which goes on and stays on from the time the contacts first close, and which goes off and stays off from the time the contacts first open after having been on. This "desired signal" is indicated on the third line of Figure 8-1.

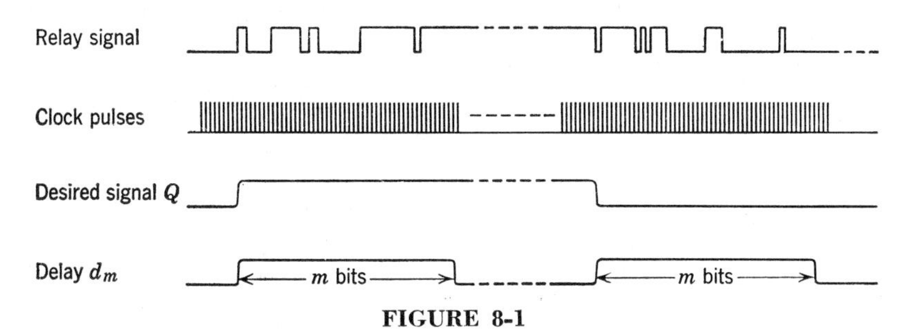

FIGURE 8-1

The derivation of this "smoothed" signal may be accomplished by introducing a delay longer than the longest possible mechanical transient expected. Suppose, for example, that experiments disclose the time between the first closure (or opening) of a pair of contacts and their final bounce is always less than m clock-pulse times. If the circuit designer can provide a circuit component d_m (Figure 8-1, bottom) which may be set by turning on its J input, but which then automatically remains in the "one" state for m or more bit-times before turning itself off automatically, the logical designer can provide the desired signal, Q, very easily. The delay is simply used to prevent the "noisy" bouncing contacts from affecting Q, and the logic is as follows:

$$J_Q = r\bar{d}_m \qquad K_Q = \bar{r}\bar{d}_m$$
$$J_{dm} = r\bar{Q} + \bar{r}Q \tag{8-1}$$

As might be expected, there are other generally cheaper methods of eliminating the effect of contact bounce. Other special circuits provided by the circuit designer have roughly the same effect: they all introduce a delay which masks the random variations in contact signals.

BUFFERS

A *buffer* is a device used to compensate for a difference in rate of flow of information or time of occurrence of events when transmitting information from one device to another. The transfer may be between an input device and a memory, between two memories, between an input device and an output device, or between a memory and an output device. A buffer must in general be capable of accepting information at one frequency and of transferring it out at a completely different frequency. The two frequencies may be multiples of one another, but more often are completely asynchronous with respect to one another. The input frequency may be lower than the output frequency (as for a transfer between a punched card reading device and a magnetic drum memory); or it may be higher than the output frequency (as for the transfer between a magnetic core memory and a paper tape punch); or the two frequencies may be nominally the same (as for a transfer between two magnetic drums which are not mechanically coupled to one another).

In addition to compensating for a difference in information flow rates, a buffer may have any one or all of a large number of additional functions to carry out. Among these other functions are:

a. *The rearrangement of information.* A buffer may be required to accept information in serial form and to transfer it out in parallel or vice versa. It may also be required to reverse the digits within a particular word while carrying on a straight serial transfer. For example, data is most conveniently punched or typed most significant digit first by a human operator, and it will therefore appear that way on a punched paper tape or punched card. However, arithmetic operations must be carried out on numbers least significant digit first, and it may be the function of the buffer to carry out this reversing action.

b. *The prevention of interference with a computer.* Input and output devices generally produce or accept data at a much lower frequency than that at which the computer operates. Therefore, between the time the computer requests information from an input device and the time the input device has delivered that information to the buffer, the computer may be able to carry out several hundred or several thousand computations. Similarly, between the time the computer transfers a block of data to an output buffer at the computer frequency and the time the last part of this information is delivered to the output device, the computer may again be able to carry out several thousand computations. It is therefore often the function of the buffer to control the entire information transfer independently of the computer proper, thus allowing the computer to go on with other work except during the time it communicates with the buffer at computer frequency.

c. *The insertion or deletion of information.* A buffer may be required to insert data into or delete data from a block of information passing through it. For example, a buffer which handles the transfer of information from a magnetic core memory to a magnetic tape output device may be required to insert an identifying number or address into each block so transferred. Another buffer which handles the transfer of information from a card reader into a computer memory may be required to delete certain combinations of holes which were originally punched on the card to edit the card for printing, or to correct a mistake in punching.

Study of the functions described above will indicate that the most important single characteristic of a buffer is its memory capacity. The capacity required depends on the number of digits or words transferred at one time, on the relative information rates of the devices which feed data to and take data from the buffer, and on the access time (if any) of whatever memory device takes part in the data transfer. If only a few digits are to be transferred at a time, the required memory may be made up of a few memory elements. On the other hand, the required capacity may be so large that separate memory elements are uneconomical and part of the main computer memory may be used in the buffer. Note that if part of the main memory is to be used and the buffer is required not to interfere with other operations taking place in the computer, the buffer memory must be independent of the operation of the rest of the main memory. This implies that any S- or T-selection required for the buffer must be separate from similar selection equipment operating in the rest of the memory.

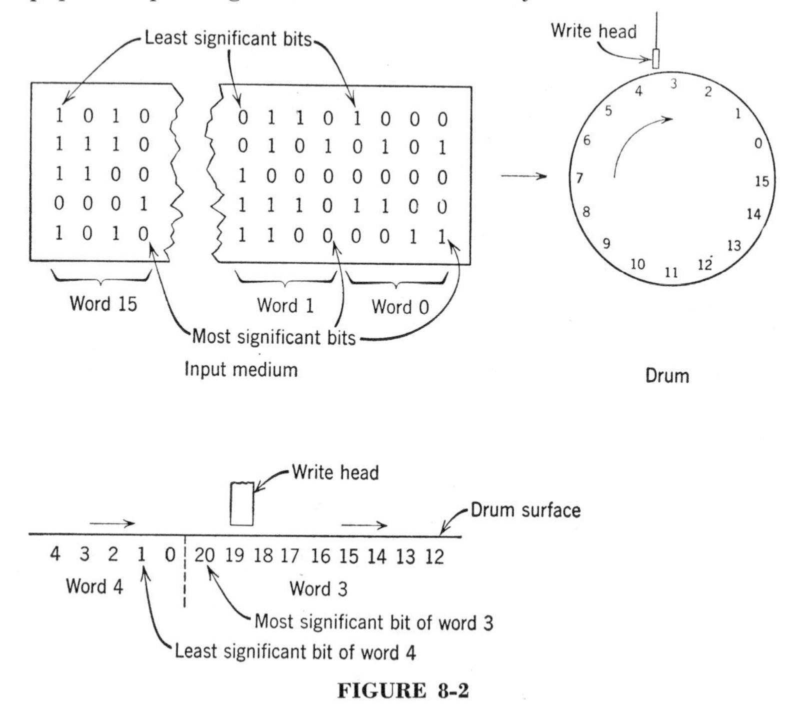

FIGURE 8-2

As an example of the problems which arise in buffer design, let us construct a buffer which will handle the following problem. An input to a magnetic drum computer transfers information in blocks of sixteen words at a time. Each word consists of twenty binary digits and the input device supplies them five digits at a time, most significant digits first. It thus takes four input intervals to read a single word from the input device and sixty-four such intervals to read the complete "block" of sixteen words. The major part of buffer storage into which the sixteen words are to be written is a channel on the magnetic drum. It takes 8 milliseconds for the drum to make one rotation and words are, of course, to be stored on the drum least significant bit first. The digits are delivered by the input device at a nominal rate of 200 lines—i.e., 200 groups of five bits each—per second, so that one arrives approximately every 5 milliseconds. However, the drum and input device are not synchronized, and there is no way for the logical designer to predict which drum word-time a group of five bits will be read from the input device.

The general arrangement of bits and words on the input device and on the drum is indicated in Figure 8-2. Note particularly that the most significant bits of each word appear first from the input device, but that the least significant bit of each word must be written first on the drum. In the enlarged view of the drum surface shown in Figure 8-2, the arrangement of bits in a word is indicated. The twenty bits of each word are identified there by numbers, and the position of the drum indicates that the nineteenth bit of the third word is under the write head. A space bit (labeled zero) separates words, and we shall assume that the computer supplies twenty-one timing signals, t_0, t_1, t_2, $\cdots$, t_{20}, one for each bit-time. We shall also assume that there is associated with the main computer memory a T-register (flip-flops T_1, T_2, T_3, and T_4) which at any time gives the address of the word under the write head. The T-register changes state every t_{20} time, and we are permitted to use the output signals from these flip-flops, but may not change the T-register input logic or the logic of any of the rest of the computer.

The general logical structure of the buffer may now be described, and is illustrated in Figure 8-3. The transfer of bits from input device to drum is accomplished in three steps. First, each set of five bits is read in parallel into flip-flops B_1–B_5 from the input lines L_1–L_5. Second, groups of five bits are shifted serially, most significant bit (B_5) first, into a group of twenty D flip-flops Q_1–Q_{20}. Third, when an entire word has been collected in the Q-register, and when the space into which that word is to be inserted appears under the drum write head, the contents of the Q-register are shifted out serially to the write amplifier A_w. If we shift bits into the Q-register from left to right, and shift them out from right to left, we automatically take care of the fact that words must be read in most significant bit first and read out least significant bit first.

We can now describe each of these three steps in some detail, and write the resulting equations which describe the operation of the buffer. The input lines L_1–L_5 are assumed to be asynchronous with respect to the drum memory, and there must therefore be an input indexing signal of some kind which tells us when five bits are ready to be read into the buffer. This index line is called L_6, and we assume that it is a one-bit-time signal occurring at some random t_0 time. We provide a control flip-flop C_0 which is set when the five bits are transferred into B_1–B_5, and which continues in the "one" state until they have all been transferred into the Q-register. This transfer always takes place during the first five bit-times which follow their arrival in the B flip-flops unless the Q-register contains a complete word at that time. If the Q-register is full and is awaiting the arrival of the proper word

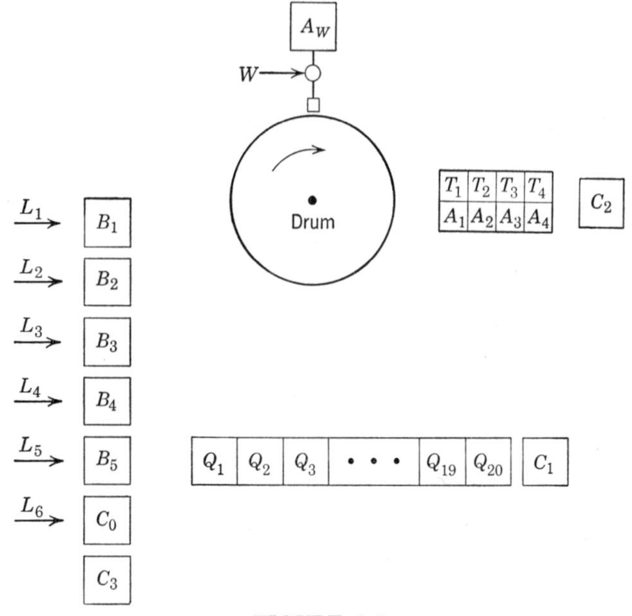

FIGURE 8-3

under the drum write head, the five bits remain in B_1–B_5 (and C_0 remains set) until the word in Q has been shifted onto the drum. Although bits are written into the B flip-flops in parallel, they are read out serially by shifting B_1 into B_2, B_2 into B_3, B_3 into B_4, B_4 into B_5, and B_5 into Q_1 during the five successive bit-times t_1–t_5.

In shifting groups of five bits into the Q-register, it is important to keep track of how many bits have been inserted so that a transfer to the drum can be effected whenever a complete word has been accumulated. One way to do this would be to provide three flip-flops which count up one every time we transfer five bits from B_1–B_5. The state of these flip-flops would then indicate whether zero, five, ten, fifteen, or twenty bits are in the Q-register, and when twenty bits are present we can begin to look for the appropriate word on the drum. We can do the same thing in a simpler way by making a counter out of the

Q-register itself and an additional flip-flop C_1. We arrange that flip-flop Q_1 contains a "one" and flip-flops Q_2–Q_{20} and C_1 contain "zeros" when no bits are in the Q-register. Every time we shift a bit into the Q-register from B_5, we shift Q_1 into Q_2, Q_2 into Q_3, $\cdots$, Q_{19} into Q_{20}, and Q_{20} into C_1. If this is done, C_1 remains "zero" until the last bit of a word is shifted into Q_1, at which time the "one" originally in Q_1 will have shifted through all the Q flip-flops into C_1.

When C_1 is "one," then, we are ready to shift the word in Q onto the drum. We must provide four flip-flops A_1–A_4 to hold the address of the next word to be inserted on the drum, and must compare this address with the address in the T flip-flops. When the two addresses are equal and C_1 is "one," we set a writing control flip-flop C_2, and shift the contents of the Q-register out to write amplifier A_w.

The complete logic may now be written as follows:

$$\begin{aligned} J_{B1} &= L_1 L_6 \\ K_{B1} &= C_0 \overline{C}_1 \overline{t}_0 \end{aligned} \qquad (8\text{-}2)$$

$$\begin{aligned} J_{Bi} &= L_i L_6 + B_{(i-1)} C_0 \overline{C}_1 \overline{t}_0 \\ K_{Bi} &= \overline{B}_{(i-1)} C_0 \overline{C}_1 \overline{t}_0 \end{aligned} \qquad i = 2, 3, 4, 5 \qquad (8\text{-}3)$$

$$\begin{aligned} J_{C0} &= L_6 C_3 \\ K_{C0} &= t_5 \overline{C}_1 \end{aligned} \qquad (8\text{-}4)$$

$$D_{Q1} = B_5 C_0 \overline{C}_1 + Q_1 \overline{C}_2 (\overline{C}_0 + C_1) + Q_2 C_2 + t_{20} C_2 \qquad (8\text{-}5)$$

$$D_{Qi} = Q_{i-1} C_0 \overline{C}_1 + Q_i \overline{C}_2 (\overline{C}_0 + C_1) + Q_{i+1} C_2 \qquad i = 2, 3, 4, \cdots, 19 \qquad (8\text{-}6)$$

$$D_{Q20} = Q_{19} C_0 \overline{C}_1 + Q_{20} \overline{C}_2 (\overline{C}_0 + C_1) \qquad (8\text{-}7)$$

$$\begin{aligned} J_{C1} &= Q_{20} \\ K_{C1} &= t_{20} C_2 \end{aligned} \qquad (8\text{-}8)$$

$$\begin{aligned} J_{C2} &= C_1 t_0 (A_1 + \overline{T}_1)(\overline{A}_1 + T_1)(A_2 + \overline{T}_2)(\overline{A}_2 + T_2) \\ &\quad (A_3 + \overline{T}_3)(\overline{A}_3 + T_3)(A_4 + \overline{T}_4)(\overline{A}_4 + T_4) \\ K_{C2} &= t_{20} \end{aligned} \qquad (8\text{-}9)$$

If A_1 is the most significant and A_4 the least significant bit of the A-address,

$$J_{A1} = K_{A1} = A_2 A_3 A_4 t_{20} C_2 \qquad (8\text{-}10)$$

$$J_{A2} = K_{A2} = A_3 A_4 t_{20} C_2 \qquad (8\text{-}11)$$

$$J_{A3} = K_{A3} = A_4 t_{20} C_2 \tag{8-12}$$

$$J_{A4} = K_{A4} = t_{20} C_2 \tag{8-13}$$

$$J_{Aw} = \overline{K}_{Aw} = Q_1 \tag{8-14}$$

$$W = C_2 \tag{8-15}$$

(When $W = 1$, the write amplifier A_w is connected to the write head). Finally, we must supply a flip-flop C_3 which will be set when the buffering operation is to begin, and reset when the last of the sixteen words have been transferred to the drum.

$$\begin{aligned} J_{C3} &= \text{(start input)} \\ K_{C3} &= A_1 A_2 A_3 A_4 C_2 t_{20} \end{aligned} \tag{8-16}$$

Several comments on this buffer will conclude this section.

1. The $\bar{t}_0$ which appears in equations 8-2 and 8-3 is necessary to prevent the bits in B from shifting at t_0 time when a word has just been transferred from the Q-register to the drum.

2. The terms containing $\overline{C}_2(\overline{C}_0 + C_1)$ in equations 8-5, 8-6, and 8-7 are necessary so that the contents of the Q-register do not change when no shift is taking place.

3. In designing any buffer, one must be very careful that the asynchronous devices do not supply (or require) information before the buffer is able to receive (or supply) it. The input device for this buffer supplies five bits approximately every 5 milliseconds. When the Q-register is not full, these bits are shifted out immediately, and the B flip-flops are then ready to receive the next five bits. Suppose, however, that the last five bits of a word appear in the B flip-flops at t_0 time of the word that $T = A$. They will immediately be shifted into the Q-register and C_1 will become "one," but the buffer will have to wait an entire drum revolution (8 milliseconds) before Q may be shifted into the appropriate place on the drum. During that 8 milliseconds the most significant five bits of the *next* word will be read from the input device into B_1–B_5, but the Q-register will always empty and these new bits will be inserted in it before the second set of five bits appears on L_1–L_5. The buffer thus has adequate storage capacity for the worst combination of input circumstances.

4. Other forms of storage might be used in place of the twenty D flip-flops used to form the Q-register. For example, a special circulating register on the drum might provide a cheaper means of accumulating digits.

5. Note that the computer proper starts the input operation by setting C_3 and starting the device which supplies data to lines L_1–L_6; but that during the approximately $5 \times 64 = 320$ milliseconds (or about forty drum revolutions) the input data is being read, the computer is free to do other things.

INPUT DEVICES

In this section and in the section which follows on output devices, we shall discuss typical input-output equipment from the point of view of the logical designer, just as we discussed typical digital memories from this point of view in the last chapter. As before, we shall be interested only in the properties which the logical designer must understand if he is to incorporate one of these devices into a computer he is designing. It will, of course, not be possible to discuss all of the widely different kinds of input and output devices which have been and can be used for digital computers. Rather, our intention is to describe with a certain amount of detail the most important and the most widely used of these devices. The descriptions will be of a basic and very general nature, principally because input and output devices are so little standardized. For example, many card punches and card readers can be purchased with certain buffer memory capacity supplied as an integral part of the equipment. The logical designer must then adapt or modify this buffer to suit his own purposes.

The data which is supplied to a computer originates in one of two forms: digital or analog. For most business and many scientific applications, the input data is generated in digital form. The premium to be paid on an insurance policy, the hourly wage rate of an employee, the face value of a bank check, and the sales price of an item sold in a department store are examples of data which must be read into a business computer and which originates in digital form out of some transaction or other operation. Examples of digital data fed into a scientific or engineering computer include the tensile strength of steel, the numerical value of π, all prime numbers under 1000, and the coefficients of the Taylor series for e^x. Data of this kind may be entered into a computer by way of punched cards, punched paper tape, a keyboard, or a print reader, and we shall discuss card and tape readers in this chapter.

Much of the data fed to computers, however, originates in analog rather than digital form. That is, the original material is a measurement made on some physical characteristic such as length, speed, time, pressure, temperature, electric current or voltage, viscosity, weight, chemical composition, etc. Data in this form may arise out of laboratory experiments, guided missile tests, petroleum refinery operations, the inspection of manufactured articles, and a host of other sources. In every instance, some instrument is used to make the measurement. It is of course possible for an operator to read the instrument, translate the reading into digital form, and enter the digits into a computer by way of one of the devices mentioned in the last paragraph. It is, however, often desirable to carry out the analog-to-digital conversion automatically and to feed the converted data directly to the computer. The problems of analog-to-digital conversion will be briefly discussed later in this section.

Paper-Tape-Reading Devices

The punched paper tape commonly used as an input medium for digital computers ordinarily comes in one of three sizes having either five, six, or seven bit positions per row. In addition, such a tape has a small guide hole at each bit position, used to locate the rows along the length of the tape. Figure 8-4 illustrates a five-hole tape. A tape-

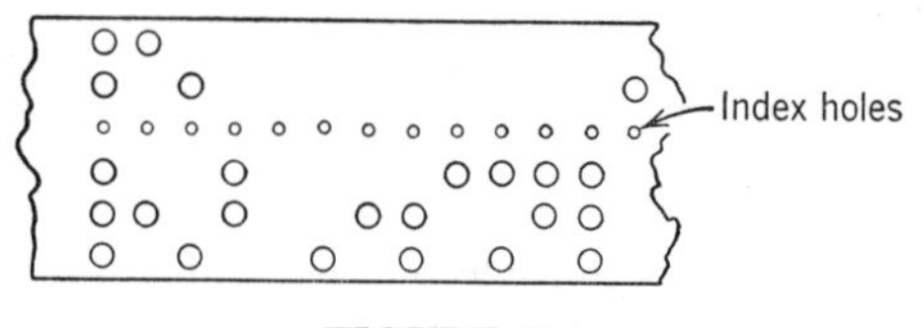

FIGURE 8-4

reading device comprises a reading station able to sense all five (or six or seven) hole positions across the tape, a tape-driving mechanism capable of moving the tape past the sensing device, and some sort of index mechanism which delivers a signal whenever the tape is in such a location that the line of hole positions is opposite the sensing device.

Sensing of the holes is usually done either mechanically, by running five small reading fingers or brushes along the tape, or photoelectrically, by shining light through the tape and detecting the presence or absence of a hole with photoelectric cells. The latter method is generally preferred because it causes the minimum possible amount of wear on the tape. The tape-driving mechanism may be either a sprocket which engages the small guide hole in the tape, or some kind of friction-driving device similar to the driving roller described in the last chapter in connection with the magnetic tape memory. The tape may be started and stopped for every row of holes on the tape, in which case a tape speed of some 20 rows per second is common. Alternatively, the tape may be driven continuously, so that a great many lines may be read between the time it starts and the time it stops. A tape speed of some 400 lines or characters per second is possible with continuous operation.

The indexing mechanism performs an important function in the tape reader and in many kinds of input equipment. It is logically similar to a clock pulse derived from a magnetic drum or magnetic tape memory, and its purpose is the same: to distinguish the time when data may be read from the times in between when the device is changing over from one signal position (line of holes) to another. On a paper tape, the indexing device is generally associated with the small guide holes found on the tape. These holes, which obviously correspond to a clock-pulse channel, may be sensed mechanically or photoelectrically, just as the signal holes themselves are sensed.

The important features of a tape-reading device are shown in Figure 8-5. The logical designer must in general provide something like the

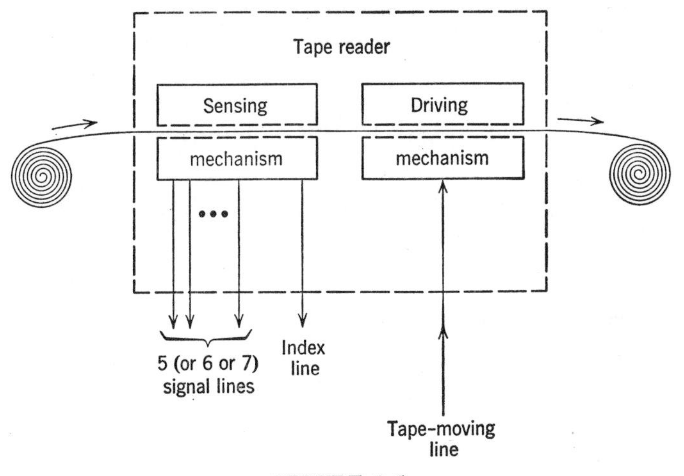

FIGURE 8-5

following simple sequence of operations in order to connect a tape reader to a computer or computing system.

1. As soon as a piece of data is called for from the tape, the index line must be sensed to determine whether the tape has stopped moving since the last time there was a read operation. If the index indicates it is not yet possible to read, the logical designer must arrange for the computer to wait until the next line of holes is opposite the sensing mechanism.

2. When the index indicates it is permissible to read from the tape, the signal lines are sensed, and the input bits are sent on to the computer.

3. At the same time, a stimulating signal is provided to initiate the mechanical operation of moving the tape from one row of holes to the next. Thus, if the computer does not call for another read operation too soon, the tape may reach the next hole position in time for the next tape-reading operation to be performed instantaneously with no wait for a signal from the index line.

A punched tape may be prepared by a computer using the output punch described in the next section. It may also be prepared manually using a device similar to a typewriter, which ordinarily produces a printed record at the same time as the tape is punched. If decimal numbers or a combination of decimal numbers and alphabetical characters are punched on the tape, each tape line corresponds to one such character. There may thus be as many as 32 characters on a five-hole tape and as many as 128 on a seven-hole tape. (Often one of the seven holes on the latter tape is used as a checking bit, thus reducing the number of characters represented to 64. We will discuss such checking codes in Chapter 10.) If a binary input is required from the paper tape, the binary numbers must be split up into groups of five to seven bits each and each group must be represented by the corresponding decimal number or alphabetical character. The keyboard operator then types this sequence of characters and the proper binary numbers are automatically obtained.

Punched Card Readers

The punched card is the most widely used input and output medium for digital computers. The Sperry-Rand Corporation and the International Business Machines Corporation (IBM) are the principal manufacturers of punched card equipment in the United States, and both use a card $3\frac{1}{4}$ inches wide by $7\frac{3}{8}$ inches long, 0.0067 inch thick. Each such card is divided into "rows" and "columns," and in each column there are as many hole positions as there are rows. An IBM card, for example, contains 12 rows and 80 columns, and therefore has $80 \times 12 = 960$ hole positions.

Like paper tape, the cards may be prepared either by a computer output punching device or by a manual punching operation. Data may be arranged on a card in many different ways. If the card contains information in binary form, each row or column may contain a complete number or part of a number. If the data is in decimal form, some binary-decimal code must be employed. The code most commonly used for IBM cards requires one column (twelve bits) for each decimal digit or alphabetical character and a card may therefore contain eighty such characters. This code is shown in Table 8-1, where the "ones" represent holes. Note that only a fraction of the 4096 possible twelve-bit codes are employed, so that the potential data-handling capacity of such a card is largely unused.

Table 8-1

Symbol	12	11	0	1	2	3	4	5	6	7	8	9
0			1									
1				1								
2					1							
3						1						
4							1					
5								1				
6									1			
7										1		
8											1	
9												1
A		1		1								
B		1			1							
C		1				1						
D		1					1					
E		1						1				
F		1							1			
G		1								1		
H		1									1	
I		1										1
J	1			1								
K	1				1							
L	1					1						
M	1						1					
N	1							1				
O	1								1			
P	1									1		
Q	1										1	
R	1											1
S			1		1							
T			1			1						
U			1				1					
V			1					1				
W			1						1			
X			1							1		
Y			1								1	
Z			1									1

Many different types of punched card readers are available from the manufacturers. They differ from one another primarily in the speed at which they operate and the manner in which they read the card. Speeds of from 100 to 1000 cards per minute are readily available. Note that 1000 cards per minute corresponds to 16,000 bits (hole posi-

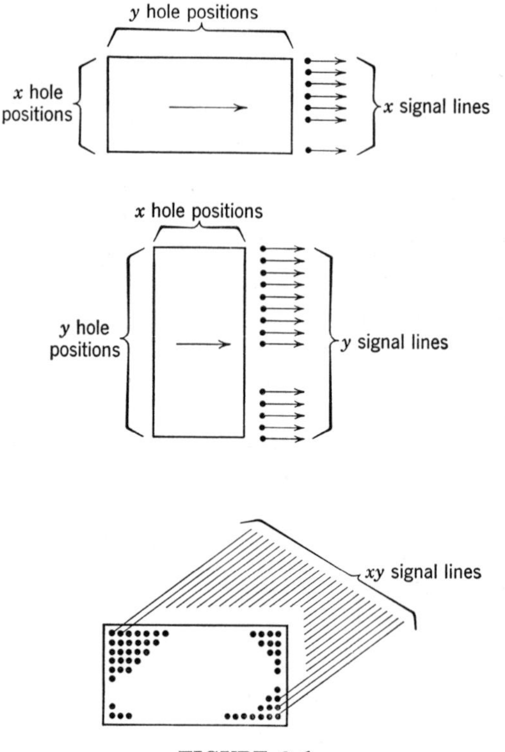

FIGURE 8-6

tions) per second, compared with 2800 bits per second for a fast, seven-hole paper tape reader.

The manner in which the card is read by the card reader is of great importance to the logical designer. Some card input devices read the card endways, so that one column (twelve bit positions for the IBM card) is read at a time. Other card readers take the card sideways, reading one row (eighty bit positions for the IBM card) at a time. Still other card readers read the entire card at once. These three reading modes are illustrated schematically in Figure 8-6. Most card input devices handle an entire card at a time, so that once the reading operation is begun all hole positions on that card are read. It is not possible to read a few rows or columns at one time and the remainder at some later time.

We can now describe the general logical properties of a card reader. As indicated in Figure 8-7, it consists of a card-feeding mechanism capable of separating one card from a stack and feeding it past the reading station, a sensing mechanism able to read holes in a card (mechanically or photoelectrically), and an indexing or synchronizing mechanism which delivers a signal whenever a card or a particular row or column on a card is in a position to be read. The feeding mechanism requires an input signal labeled F which starts the reading operation.

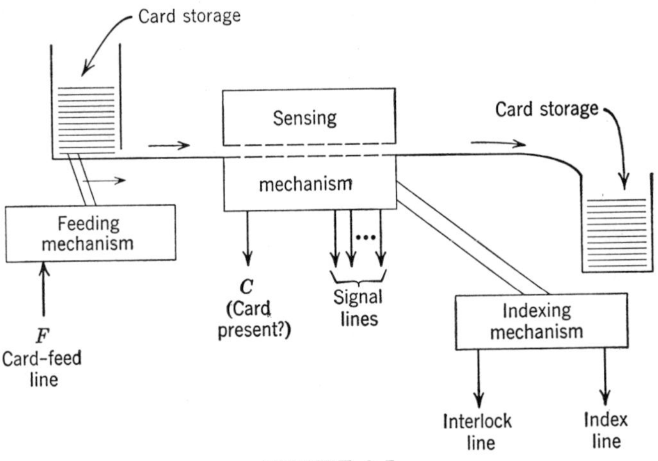

FIGURE 8-7

The sensing mechanism provides as many output lines as there are holes to be read simultaneously (see Figure 8-6) and also provides a signal C which indicates whether a card has, in fact, been fed from the stack. The indexing mechanism provides the equivalent of a clock pulse, determining when the output lines from the sensing device contain signals corresponding to their respective hole positions. A card does not, of course, have index holes like a paper tape has, and the indexing mechanism is usually mechanically linked to the edge of the card. In addition to the index signal, the indexing mechanism provides an interlock signal which determines when the card reader cycle is complete and a new card may be read.

The logical designer must provide equipment able to carry out the following operations if he is to connect a punched card reader into a computing system:

1. When the computer requires a card full of data, equipment must be provided to examine the interlock signal to see whether or not the

card reader has finished its last reading cycle. If the cycle is not complete, the new card-reading operation may not begin and the computer must go on with some other operation, or must wait until the card reader is ready.

2. When the interlock signal indicates that it is proper to read from the card reader, the card-feed line F may be stimulated (see Figure 8-8). This will cause the next card to be sent to the sensing mechanism.

3. The index and card lines must now be observed. First, a signal should appear on the card line C, indicating that a card is in the sensing mechanism. Next, there will appear on the index line as many equally spaced signals as there are rows or columns to be read sequentially.

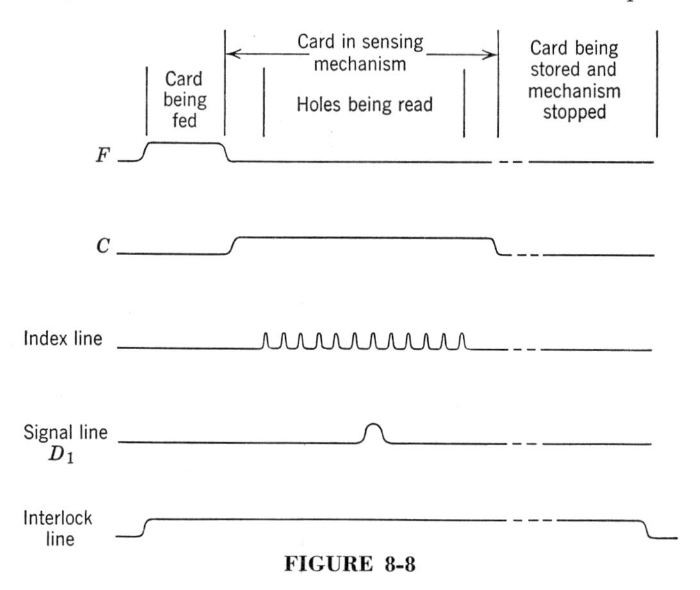

FIGURE 8-8

For example, Figure 8-8 shows the twelve index signals which would appear if an IBM card were read sideways. The signal labeled D_1 in Figure 8-8 is a typical data line—one of eighty—which happens to have a hole punched in the seventh hole position.

If no signal appears on line C, the logical designer must provide logic which ignores the index signals, and gives an alarm indicating that no cards are present or that the card reader is not operating correctly.

4. As the data is read from the card, it must enter a buffer whose primary purpose is to compensate for the difference between card reader and computer information rates. The buffer must also rearrange the card data according to whatever rules relate computer words with card data; and translate the information from card code to com-

puter code. Note that this translating operation may be very complicated. If the card code of Table 8-1 must be translated into a six-bit computer code, for example, the buffer is required to carry out eighty simultaneous translations, where in general it cannot complete the translation of any column until all twelve holes in that column have been read.

Analog-to-Digital Converters

As was mentioned at the beginning of this section, computer input data sometimes arises in analog form and must be converted into digital form and read into a computer automatically. A typical configuration

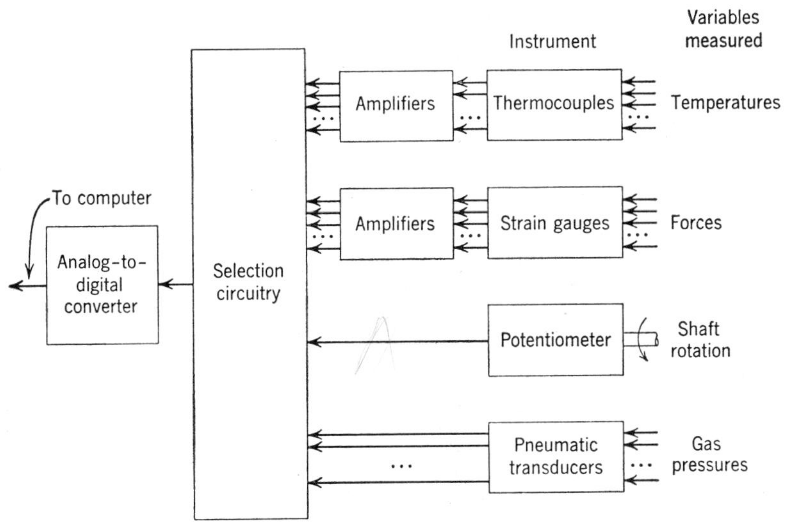

FIGURE 8-9

of such inputs is shown in Figure 8-9. There, each primary measurement is converted into an electrical signal, amplified if necessary, and fed to a selection circuit which connects one of the signals to an analog-to-digital converter. Note that the selection problem here is somewhat different from the selection problems we have discussed before, because an analog signal must be connected to the converter without attenuating it or changing it in any way. It is obviously more difficult to do this than to select a simple on-off signal. Normally, the connection is made through some kind of relay circuit (see Figure 7-2, e.g.), but if high speeds are required, special electronic selection circuits may be designed.

A great deal of work has been done on the design of analog-to-digital converters. One converter of a kind commonly used is shown in Figure 8-10, and its operation may be described as follows. A binary number in the counter register C_1–C_7 is converted to an analog signal by means of a digital-to-analog converter (p. 238). This signal is compared with the voltage being converted into digital form by means of a comparator, whose output is "zero" if the converter voltage is larger than the input voltage, and "one" if it is smaller. The output of the comparator is thus a binary signal which may be used to cause the counter to count up or down. If the counter contains a number smaller than that represented by the input voltage, for example, the comparator output is

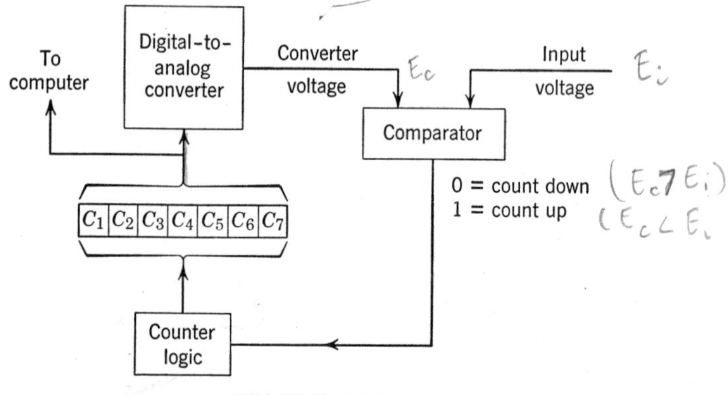

FIGURE 8-10

"one" and the counter begins to count up · · · to increase by unity every bit-time. As the number in C_1–C_7 increases, the converter voltage increases until it is slightly greater than the input voltage. The comparator output then becomes "zero," and causes the counter to count down until the converter voltage is less than the input voltage. The counter is thus forced to oscillate about a value equivalent to the input voltage. The magnitude of the oscillation will depend on the speed of the counter compared with the speed of the converter and comparator. The conversion is of course not complete until this oscillation begins, and the converted number cannot be read from the C-register until the counter has had time to count from its initial value to a number equivalent to the input voltage.

It is possible and sometimes useful to convert a shaft position or linear position directly into digital form instead of translating it into an analog voltage, as was shown in Figure 8-9. In Figure 8-11, for example, the rotational position of a shaft is converted into a digital signal by means of a code wheel fastened to the end of the shaft. The code wheel in the figure has printed on it a series of concentric bands, each one of which represents one bit position. If the 360° shaft rotation is to be measured to within one-eighth of a revolution, three concentric bands are required as shown. Each band is transparent wherever the corresponding bit position should be "one," and is opaque where it should be "zero." A light source is placed on one side of the wheel, and three photocells are mounted on the other side, each one behind one of the three bands, and all three located along a radius of the wheel.

Whatever the position of the code wheel, each of the three photocells will either be stimulated by light passing through the transparent

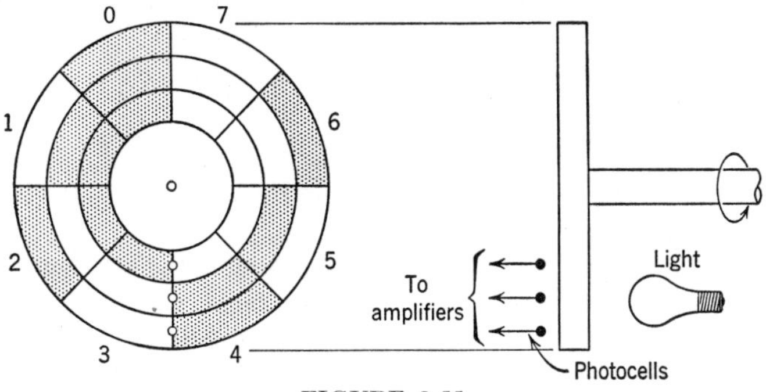

FIGURE 8-11

part of a band or will not be stimulated if an opaque part of the band is in the way. These three on-off signals thus correspond to a particular position of the wheel, and may be read directly into a computer. Note that any desired accuracy may be achieved by increasing the size of the wheel and by providing more concentric bands and more photocells.

An unexpected problem arises, however, when a *binary* code is used in a converter of this kind. Notice that the code wheel of Figure 8-11 has rotated so that sector 3 is just opposite the photoelectric cells, and suppose the shaft is moving so that sector 4 is about to appear. In sector 3, the cells are reading the binary number 011; in sector 4 they will be reading 100. At the boundary between the two sectors, all photocells should therefore reverse themselves. If, however, the wheel stops on or very near the boundary, the reading from each photocell depends on the sensitivity of the various cells, on the precision of the markings on the code wheel, and on the accuracy with which the cells and wheel are aligned. In fact, each of the cells may read either "zero" or "one," and the observed code may thus be anything between zero and seven.

One common way of overcoming this difficulty is shown in Figure 8-12, where two photocells are employed for each band except the one corresponding to the least significant digit. The pairs of cells are located symmetrically about a radius through the single cell, and the arc between each pair of cells is equal to the width in degrees of an opaque section in the next band out. The three binary bits, B_0, B_1, and B_2, corresponding to the position of the wheel, may be derived from the photocell signals as follows:

$$
\begin{aligned}
B_0 &= P_0 \\
B_1 &= \bar{B}_0 P_{10} + B_0 P_{11} = \bar{P}_0 P_{10} + P_0 P_{11} \\
B_2 &= \bar{B}_1 P_{20} + B_1 P_{21} = P_{20}\overline{(\bar{P}_0 P_{10} + P_0 P_{11})} \\
&\qquad + P_{21}(\bar{P}_0 P_{10} + P_0 P_{11}) \\
&= \bar{P}_0 \bar{P}_{10} P_{20} + P_0 \bar{P}_{11} P_{20} + \bar{P}_0 P_{10} P_{21} + P_0 P_{11} P_{21}
\end{aligned}
\tag{8-17}
$$

These equations state that the lagging photocell in the nth band should be used if a "one" is read in the $(n - 1)$st band, and the leading cell should be used if a "zero" is read. This procedure eliminates uncertainties because of a peculiar characteristic of the binary number system: if a "one" appears in any bit position, the next most significant bit position will not change in the next lower numbers; if a "zero" appears, the next most significant bit position will not change in the next *higher* numbers.

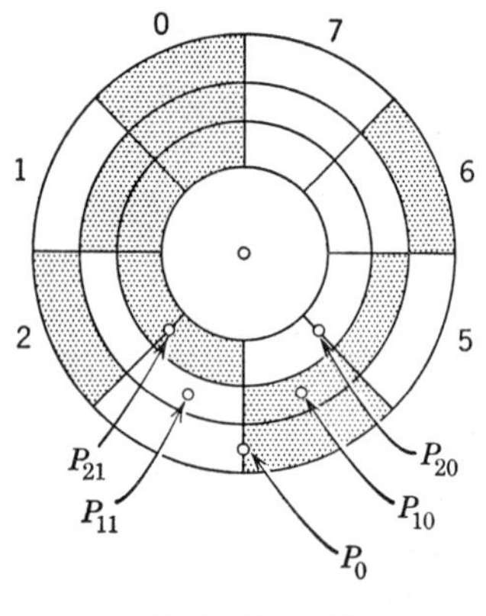

FIGURE 8-12

The ambiguity which arises in the use of an ordinary binary code may be eliminated in another way which does not require the addition of double read stations in each band. Since the difficulty is caused by a change of more than one bit position from one number to the next, it may be removed by using a code in which only *one* bit position changes from one number to the next. Such a code is called a Gray code and four examples of binary Gray codes are given in Table 8-2. The derivation of such codes is easily accomplished with reference to a Veitch diagram, for a change of one bit in a code corresponds to a move from one square to an "adjacent" one. The four codes of Table 8-2 are plotted in Figure 8-13. The code wheel of

Table 8-2

	Code 1	Code 2	Code 3	Code 4
0	0 0	0 0 0	0 0 0 0	0 0 0 0
1	0 1	0 0 1	0 0 0 1	0 0 1 0
2	1 1	0 1 1	0 0 1 1	0 1 1 0
3	1 0	0 1 0	0 0 1 0	0 1 1 1
4		1 1 0	0 1 1 0	0 0 1 1
5		1 1 1	0 1 1 1	0 0 0 1
6		1 0 1	0 1 0 1	0 1 0 1
7		1 0 0	0 1 0 0	0 1 0 0
8			1 1 0 0	1 1 0 0
9			1 1 0 1	1 1 1 0
10			1 1 1 1	1 1 1 1
11			1 1 1 0	1 1 0 1
12			1 0 1 0	1 0 0 1
13			1 0 1 1	1 0 1 1
14			1 0 0 1	1 0 1 0
15			1 0 0 0	1 0 0 0

Figure 8-14 employs the second of these four codes and it will be seen that, at the border line between any two sectors (two numbers), there is uncertainty in the signal from only one photocell, and this uncertainty can cause no ambiguity.

Gray codes are, unfortunately, difficult to compute with, and it is therefore necessary to have available a method for translating a Gray code number into an ordinary binary number. If the Gray code number is available in parallel form, it is an easy matter to write a Boolean

2	1
3	0

4	5	2	3
7	6	1	0

8	11	4	7
9	10	5	6
14	13	2	1
15	12	3	0

8	9	2	7
11	10	3	6
12	13	4	5
15	14	1	0

FIGURE 8-13

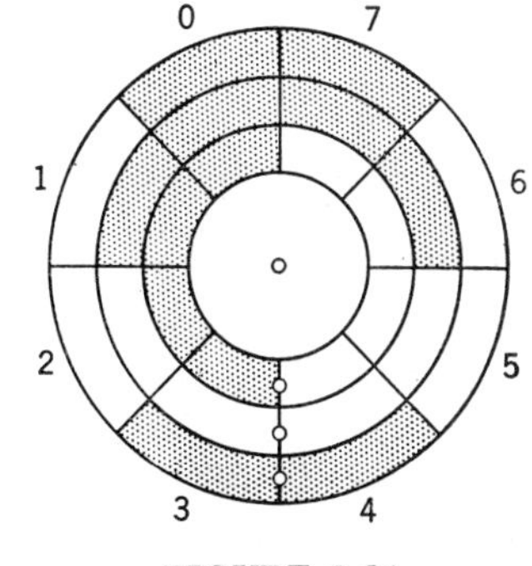

FIGURE 8-14

equation defining each binary digit as a function of the Gray code digits. If the translation is to be done serially, however, the problem is more complicated. Certain Gray codes, called *reflected binary codes,* have properties which make their serial translation fairly simple. The first three codes of Table 8-2 are reflected codes, and are given that name because all digit positions except the most significant one in each code are symmetrical with respect to the dotted line. It can be shown (Appendix 2) that if R_j is the jth bit in the reflected code for a number and B_j the corresponding bit in the binary code for the same number,

$$\begin{aligned} B_n &= R_n \\ B_j &= B_{j+1}\bar{R}_j + \bar{B}_{j+1}R_j \qquad j \neq n \end{aligned} \tag{8-18}$$

where the nth digit is the most significant one in both codes.

If a reflected Gray code number appears *most* significant bit first on a line R, we can interpret equation 8-18 as a difference equation defining the operation of a flip-flop B which carries out a serial Gray-to-binary conversion. That is,

$$B^{n+1} = (B\bar{R} + \bar{B}R)^n \tag{8-19}$$

where B must be in the "zero" state before the first digit appears on R. If B is a J-K flip-flop, then

$$J_B = K_B = R \tag{8-20}$$

and the binary translation of R appears in B.

OUTPUT DEVICES

Just as the data fed into a computer may originate in either analog or digital form, so the results obtained by a computer may be most useful in either analog or digital form, depending on the application. In most business applications, digital results are a necessity. In many scientific and engineering calculations also, it is essential that the precision obtained through the use of a digital computer be retained in the output, so the results must be made available in decimal form before they can be interpreted. Digital output equipment includes tape and card punches, printers, cathode-ray tube displays, gas tubes, and many other special devices. We shall discuss punches and printers in this section.

However, in much scientific and engineering work, the operator is best able to interpret results if they are given to him in analog form.

He may be interested only in the general form of a curve, or he may want to compare several time-varying functions in order to establish their phase relationships. He prefers an answer in the form of a curve plotted on graph paper, and if the computer provides him with a digital result, his first action is to plot it himself. It is therefore often very useful to provide a digital computer with an analog output device like a curve plotter or a trace on a cathode-ray tube.

In addition, in some computer applications the results of calculations are used to control the flow of liquids, to move the control surfaces of aircraft, to position the bed of a machine tool, or to take some other action which requires the adjustment of some physical variable. For control systems involving digital computers it is, in general, essential that analog outputs be made available which correspond exactly to digital results. The last section in this chapter will discuss one kind of digital-to-analog converter.

Paper-Tape Punches

A schematic diagram of a paper tape punch is shown in Figure 8-15. Such a punch punches one line (five, six, or seven holes plus the index

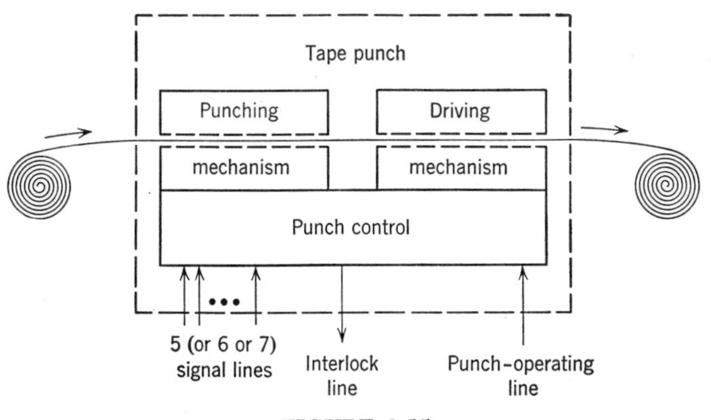

FIGURE 8-15

hole) at a time, the code punched depending on which of the input signal lines are stimulated. The operation of the punch may be described as follows. If the punch is ready to begin a punch cycle, there will be a signal on the interlock line. When such a signal exists, the logical designer provides circuits which connect the signal lines to a register whose contents must be punched and stimulate the punch-operating line. The signal on the interlock line then becomes "zero," and the punching cycle begins. The punch control first activates those punch plungers whose signal lines are "one." The plungers cut through the tape and are withdrawn, and then the punch control stimulates the driving mechanism so that the tape moves on to the next tape line which is to be punched. When the tape stops moving (or just before), the interlock line becomes "one" again, and the logical designer provides circuits which may change the information on the signal lines and start a new punching operation.

Paper tape punches are fairly slow devices, operating at speeds between 10 and 200 lines (characters) per second.

Card Punches

Just as the paper tape punch is very similar in structure to the paper tape reader (cf. Figures 8-5 and 8-15), so the card punch is very similar to the card reader. The operation of the punch of Figure 8-16 may be

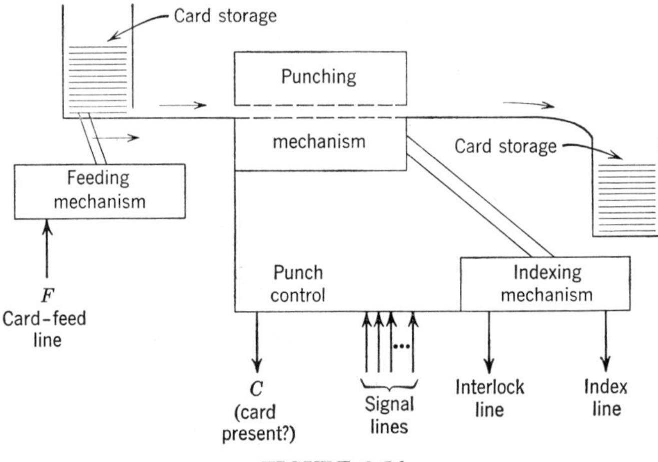

FIGURE 8-16

described as follows. The punching cycle is begun whenever there is a signal on the interlock line by stimulating line F, which operates the card-feeding mechanism. The interlock line then becomes "zero," and a card is fed to the punching mechanism. The signal lines must be connected to an appropriate buffer register in which the first row or column to be punched is stored. When the card reaches the punch, the card line C becomes "one," and as soon as the proper position on the card is reached the punch control energizes the appropriate punch plungers in accordance with the data on the signal lines. When the plungers have been extracted from the card, a signal appears on the index line, and the information to be punched in the next row or column must be entered into the buffer register. This sequence of events is then repeated until the last row or column has been punched, whereupon the card line will again become "zero." When the cycle is complete, the interlock line will again become "one," indicating that the punch is ready for another output cycle.

The card punch buffer is similar in complexity to the card-reading buffer, and must translate from computer to card code as well as absorb the difference in information rates. Punches ordinarily operate at a speed of some 100 cards per minute.

Printers

From the logical designer's point of view there are basically two kinds of printers: one prints one character at a time, the other one line of type at a time. The one-character-at-a-time printer has one set of type dies and a means of moving the paper and the type dies with respect to one another. For example, an electric typewriter may be adapted to be an output printer. A printer of this kind has logical properties very much like the tape punch of Figure 8-15. There must be from five to seven signal input lines which determine what character is next to be printed. There must be a stimulating line to start the printing operation, and there must be an index line to notify the logical designer when he may introduce another character onto the signal lines. A one-character printer is, however, complicated by one other factor not present in a punch: when the end of a line has been reached, some action must be taken to make sure that the next character is printed at the left-hand side of the page and on the line below the previous line. In an electric typewriter this means that a carriage-return and line-space operation must be provided. Commonly the carriage-return–line-space operation is either carried out automatically when the carriage reaches a certain position; or supplied by the logical designer, who must therefore provide circuits which count characters and determine when the carriage has gone far enough; or supplied as part of the incoming information—i.e., certain combinations of binary digits on the input lines have the meaning "carriage return and line space" and the computer user must insert these at proper places in his output data.

The line-at-a-time printer has a set of character or print dies for each column to be printed. If, for example, any of 64 different symbols is to be printed simultaneously in each of 120 columns, there must be $64 \times 120 = 7680$ such dies, and means must be provided for selecting 120 of these every time a line must be printed. A buffer similar in complexity to the card punch buffer is required.

Digital-to-Analog Converters

Most digital-to-analog converters work by converting numbers into actions digit by digit and then using some simple analog system of adding the actions. In Figure 8-17, a four-digit binary number is shown connected to a converter which consists of four current-generators and a current-adder. The current-generator corresponding to the

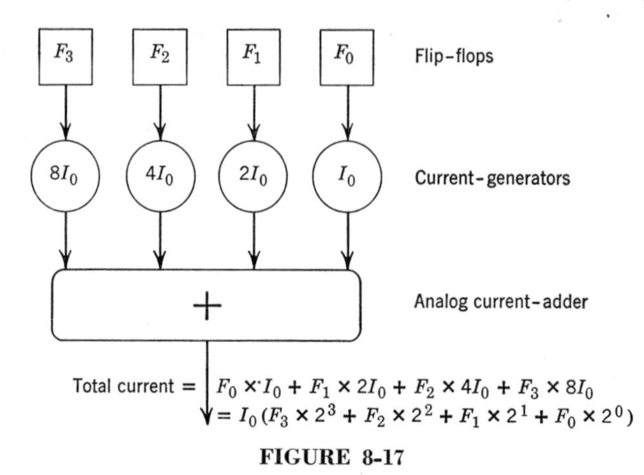

FIGURE 8-17

least significant bit is assumed to produce a current I_0 when flip-flop F_0 contains a "one." The second current-generator produces a current $2I_0$, the third generator a current $4I_0$, and the fourth a current $8I_0$, each when the corresponding flip-flop contains a "one." The currents then need only be added together to give a total current equal to I_0 times the binary number in the flip-flop register. This current may then be used for whatever purpose is desired.

BIBLIOGRAPHY

Surveys of Digital Input-Output Equipment

J. M. Carroll, "Trends in Computer Input-Output Devices," *Electronics*, 142–149 (Sept. 1956).

M. Rubinoff and R. H. Beter, "Input and Output Equipment," *Control Engineering*, **3**, no. 11, 115–123 (1956).

J. Gibbons, "How Input-Output Affects Performance," *Control Engineering*, **4**, no. 7, 97–106 (1957).

E. M. Grabbe, *Automation in Business and Industry*, John Wiley & Sons, New York, 251–302, 1957.

Punched Paper Tape

B. G. Welby, "Intermittent-Feed Computer Tape Reader," *Electronics*, 115–117 (Feb. 1953).

W. P. Byrnes, "Teletype High Speed Tape Equipment and Systems," *Proc. Eastern Joint Computer Conference*, 35–38, 1954.

Punched Cards

Engineering Research Associates, Inc., *High Speed Computing Devices*, McGraw-Hill Book Company, New York, 146–181, 1950.

Analog-Digital Input-Output Systems

F. Gray, U. S. Patent No. 2632058, Mar. 17, 1953.

H. E. Burke, "A Survey of Analog-to-Digital Converters," *Proc. I.R.E.* **41,** no. 10, 1455–1462 (Oct. 1953).

L. P. Retzinger, Jr., "An Input-Output System for a Digital Control Computer," *Proc. Wescon Computer Sessions*, 67–76, 1954.

T. C. Fletcher and N. C. Walker, "Analog Measurement and Conversion to Digits," *I.S.A. Journal*, 341–345 (Sept. 1955).

I. Flores, "Reflected Number Systems," *I.R.E. Trans. on Electronic Computers*, **EC-5,** no. 2, 79–81 (June 1956).

S. K. Feingold, "The Logic of V-Brush Analog-to-Digital Converters," *I.S.A. Journal*, 66–68 (Feb. 1957).

G. G. Bower, "Analog-to-Digital Converters—What Ones are Available and How They are Used," *Control Engineering*, **4,** no. 4, 107–118 (Apr. 1957).

High-Speed Printers

E. Masterson and A. Pressman, "A Self-Checking High-Speed Printer," *Proc. Eastern Joint Computer Conference*, 22–30, 1954.

E. M. DiGuilio, "Burroughs G-101 High Speed Printer," *I.R.E. Convention Record*, **4,** Part 4, 94–100 (1956).

Other Input-Output Systems

G. W. King, G. W. Brown, and L. N. Ridenour, "Photographic Techniques for Information Storage," *Proc. I.R.E.*, **41,** no. 10, 1421–1428 (Oct. 1953).

G. L. Hollander, "Design Fundamentals of Photographic Data Storage," *Proc. Wescon Computer Sessions*, 44–49, 1954.

H. M. Smith, "The Typotron, A Novel Character Display Storage Tube," *I.R.E. Convention Record*, **3,** Part 4, 129–134 (1955).

D. H. Shepard and C. C. Heasly, "Photoelectric (Print) Reader Feeds Business Machines," *Electronics*, 134–138 (May 1955).

EXERCISES

1. Redesign the paper-tape-to-drum buffer described in this chapter so that it employs a drum circulating register rather than the twenty-flip-flop shifting register.

2. The twelve-bit punched-card code of Table 8-1 is to be translated into the six-bit code shown in Table 8-3. Assume that the original code resides in

Table 8-3

0	0	0	0	0	0	0	I	1	0	1	0	0	0
1	0	0	0	0	0	1	J	1	0	1	0	0	1
2	0	0	0	0	1	0	K	1	0	1	0	1	0
3	0	0	0	0	1	1	L	1	0	1	0	1	1
4	0	0	0	1	0	0	M	1	0	1	1	0	0
5	0	0	0	1	0	1	N	1	0	1	1	0	1
6	0	0	0	1	1	0	O	1	0	1	1	1	0
7	0	0	0	1	1	1	P	1	0	1	1	1	1
8	0	0	1	0	0	0	Q	1	1	0	0	0	0
9	0	0	1	0	0	1	R	1	1	0	0	0	1
A	1	0	0	0	0	0	S	1	1	0	0	1	0
B	1	0	0	0	0	1	T	1	1	0	0	1	1
C	1	0	0	0	1	0	U	1	1	0	1	0	0
D	1	0	0	0	1	1	V	1	1	0	1	0	1
E	1	0	0	1	0	0	W	1	1	0	1	1	0
F	1	0	0	1	0	1	X	1	1	0	1	1	1
G	1	0	0	1	1	0	Y	1	1	1	0	0	0
H	1	0	0	1	1	1	Z	1	1	1	0	0	1

twelve *R-S* flip-flops $C_1, C_2, C_3, \cdots, C_{12}$, and that in one bit-time the translation is to appear in the last six of these flip-flops. For example, the translation of the code for "M" would appear as follows:

	C_1	C_2	C_3	C_4	C_5	C_6	C_7	C_8	C_9	C_{10}	C_{11}	C_{12}
time n	1	0	0	0	0	0	1	0	0	0	0	0
time $(n+1)$							1	0	1	1	0	0

Write the input logic for flip-flops C_7–C_{12}. Simplify as much as possible, taking into consideration the redundancies in the punched-card code.

3. Using the Huffman-Moore model, design a sequential switching circuit whose input, x, is the signal from the comparator of Figure 8-10, and whose only output, z, is "zero" when the converter is searching for equality between counter and analog voltage, and "one" when the converter has completed the search and the counter has begun to oscillate about the converted number. (Assume that this oscillation is one bit-time's duration.) The output z may thus be used to signal when the contents of the counter may be read into the computer.

4. A 480-bit circulating register contains eighty six-bit digits in the code of Exercise 2. These eighty digits are to be punched on a card, coded as shown

in Table 8-1. The card moves past the punch dies sideways, so that the "twelve" row of each of the eighty characters is to be punched during the first punch cycle, the "eleven" row during the second, the "zero" row during the third, etc., until the "nine" row is punched during the twelfth punch cycle. A punch cycle is very long compared with the time the circulating register circulates, so that during each punch cycle all eighty characters can be read, and one bit of the card code for these eighty characters can be derived. For example, if the three characters "ME2," somewhere in the circulating register, were read serially from amplifier A_R, the translation would take place as shown in Table 8-4.

Table 8-4

Read from A_R Each Time the Register Circulates···	2	E	M
	0 0 0 0 1 0	1 0 0 1 0 0	1 0 1 1 0 0 ···
"12" row	0 0 0 0 0 0	0 0 0 0 0 0	1 0 0 0 0 0
"11" row	0	1	0
"0" row	0	0	0
"1" row	0	0	0
"2" row	1	0	0
"3" row	0	0	0
"4" row	0	0	1
"5" row	0	1	0
"6" row	0	0	0
"7" row	0	0	0
"8" row	0	0	0
"9" row	0	0	0

Design a sequential circuit having a single input line A_R and a single output line z. The output z is to be "zero" during the first five bits of any character, and is to be the "twelve" row bit for that character during the sixth bit-time. The output of this circuit for the three characters "ME2" is given in the line labeled "12 row" in Table 8.4. (A complete translator would obviously contain, in effect, twelve of these sequential circuits, one of which would be used during each card punch cycle.)

9 / *The arithmetic unit*

We stated in Chapter 6 that the arithmetic unit is the name given to that part of a digital computer which performs operations on numbers or words to change them or to combine them according to given rules. The most important of these operations are usually the common arithmetic ones: the operations of adding, subtracting, multiplying, and dividing numbers. However, it may, in special cases, be desirable to carry out more complicated mathematical operations such as finding square roots, or integrating, or finding the sine or log of an argument. In addition, there are other operations which may be useful but cannot properly be called arithmetic. For example, it is sometimes convenient to be able simply to compare two numbers to discover which is the greater, or it may be useful to perform a digit-by-digit multiplication so that the result of "multiplying" the two numbers 001110 and 123456 is 003450. Because arithmetic operations are the most important and the most complicated operations which normally need be carried out, the main body of this chapter will be devoted to a description of the various methods used in the logical design of circuits which carry out the common and simple arithmetic operations. At the end of the chapter we shall also discuss the logical operation of comparison.

NUMBER REPRESENTATION

Number Codes

We have already seen (Chapter 2) the necessity for representing numbers by means of a sequence of "zeros" and "ones." However, given the desirability of representing numbers in this fashion, there are still an incredibly large number of different ways we can encode the numbers we have to represent. If we wish, for example, to represent each of the numbers from 0 to 999, we can simply choose 1000 different combinations of ten or more binary digits each, and assign them arbitrarily and at random to the numbers 0 through 999. The difficulty with such a procedure, of course, is that it might prove very difficult to perform even the simplest arithmetic operation. Because of this, some form of weighted number system is most often used in digital computer design and we shall here emphasize such number codes.

The general form for a weighted number code is given in equation 9-1.

$$X = x_n w_n + x_{n-1} w_{n-1} + \cdots + x_i w_i + \cdots + x_1 w_1 + x_0 w_0 + x_{-1} w_{-1} + \cdots + x_{-m} w_{-m} \quad (9\text{-}1)$$

where x_i = coefficient of ith position

w_i = weight associated with ith position

and the number is normally written

$$X = x_n x_{n-1} \cdots x_2 x_1 x_0 \cdot x_{-1} x_{-2} \cdots x_{-m} \quad (9\text{-}2)$$

In most common number systems, $w_i = r^i$, and r is known as the base of the number system. (Cf. Chapter 2, p. 16.) Each coefficient x_i may then have any value from zero to $(r - 1)$. The number system is called binary if $r = 2$, ternary if $r = 3$, octal if $r = 8$, decimal if $r = 10$, and hexadecimal if $r = 16$.

Strictly speaking, because of the circuit designer's restriction that the coefficients x_i can only have the value "zero" or "one", the binary number system is the only one that may be used which has the property that all weights are powers of a single base r. However, as was indicated in Chapter 2, it is possible to employ other bases by using a binary code for each digit x_i. The most common example is the use of a decimal base where each decimal digit x_i is represented by four or more binary digits. This has the effect of making w_i a product of the decimal weight and whatever weight is given to the binary digits. In the 8-4-2-1 decimal code mentioned in Chapter 2, where every decimal digit is represented by its ordinary binary equivalent,

$$w_{4i+j} = 10^i 2^j \qquad i = 0, \pm 1, \pm 2, \pm 3, \pm 4, \pm 5, \cdots; j = 0, 1, 2, 3$$

That is to say,

$$\begin{aligned} X = \cdots &+ x_8 \times 100 + x_7 \times 80 + x_6 \times 40 + x_5 \times 20 + x_4 \times 10 \\ &+ x_3 \times 8 + x_2 \times 4 + x_1 \times 2 + x_0 \times 1 \\ &+ x_{-1} \times 0.8 + x_{-2} \times 0.4 + x_{-3} \times 0.2 + x_{-4} \times 0.1 \\ &+ x_{-5} \times 0.08 + x_{-6} \times 0.04 + x_{-7} \times 0.02 + x_{-8} \times 0.01 + \cdots \end{aligned}$$

For example, the number 27.93 is represented by the sequence of binary digits 00100111.10010011. In Table 9-1, five examples of weighted decimal codes and two of nonweighted codes are given. The variety

Table 9-1

Weighted Decimal Codes

	8 4 2 1	6 3 1 1	5 2 1 1	3 3 2 1	5 1 1 1 1
0	0 0 0 0	0 0 0 0	0 0 0 0	0 0 0 0	0 0 0 0 0
1	0 0 0 1	0 0 0 1	0 0 0 1	0 0 0 1	0 0 0 0 1
2	0 0 1 0	0 0 1 1	0 1 0 0	0 0 1 0	0 0 0 1 1
3	0 0 1 1	0 1 0 0	0 1 1 0	0 0 1 1	0 0 1 1 1
4	0 1 0 0	0 1 0 1	0 1 1 1	0 1 0 1	0 1 1 1 1
5	0 1 0 1	0 1 1 1	1 0 0 0	1 0 1 0	1 0 0 0 0
6	0 1 1 0	1 0 0 0	1 0 0 1	1 1 0 0	1 1 0 0 0
7	0 1 1 1	1 0 0 1	1 0 1 1	1 1 0 1	1 1 1 0 0
8	1 0 0 0	1 0 1 1	1 1 1 0	1 1 1 0	1 1 1 1 0
9	1 0 0 1	1 1 0 0	1 1 1 1	1 1 1 1	1 1 1 1 1

Nonweighted Decimal Codes

	"Excess Three"	"Two-Out-Of-Five"
0	0 0 1 1	0 0 0 1 1
1	0 1 0 0	0 0 1 0 1
2	0 1 0 1	0 0 1 1 0
3	0 1 1 0	0 1 0 0 1
4	0 1 1 1	0 1 0 1 0
5	1 0 0 0	0 1 1 0 0
6	1 0 0 1	1 0 0 0 1
7	1 0 1 0	1 0 0 1 0
8	1 0 1 1	1 0 1 0 0
9	1 1 0 0	1 1 0 0 0

of decimal codes possible is so wide that no attempt will be made here to evaluate them.

In many machines it is necessary not only to be able to represent the ten decimal digits, but also to distinguish the twenty-six letters of the alphabet. A "digit" which distinguishes between one of these thirty-six combinations is commonly called an alphanumeric digit and must obviously be composed of at least six binary digits.

An important consideration in choosing the code used to represent numbers in a computer is the question of the precision desired from computations, i.e., the number of distinct states which must be distinguishable. The desired precision determines the number of digits in a word, which we have seen (Chapter 7) is the unit in which information may be stored in a computer memory, and is also the unit of data upon which arithmetic and other operations are carried out. The required precision depends on the job which the computer is to do and must be determined by the system designer. Once it has been determined, the number of digits in a word is fixed. However, the range of the numbers represented by the machine is still undetermined. If a precision of forty binary digits is required in a binary machine, it is possible to consider the numbers as being in range 2^{40} to 2^{80}, 1 to 2^{40}, or 2^{-40} to 1, among others. The numerical result obtained from performing the simple arithmetic operations of addition, subtraction, multiplication, or division on two numbers in one of these ranges is the same no matter which range is chosen. To this extent, the design of the arithmetic unit is independent of the placement of a binary or decimal point. However, when the result of one of these arithmetic operations is to be used in further computation with other numbers, care must be taken in the design of the arithmetic unit to see that relative binary or decimal points are taken into consideration. If, for example, we multiply 54 by 72 we get the same numerical result as we get in multiplying .54 by .72. It is only when we add 65 or .65 to the product that the location of the decimal point is important, as is illustrated below:

$$\begin{array}{r} 72. \\ \times 54. \\ \hline 288 \\ 360 \\ \hline 3888. \\ +65. \\ \hline 3953. \end{array} \qquad \begin{array}{r} .72 \\ \times .54 \\ \hline 288 \\ 360 \\ \hline .3888 \\ +.65 \\ \hline 1.0388 \end{array}$$

One other form of computer code should be mentioned briefly: the "floating-point" code. It is possible to encode a number N by assigning two numbers, a and b, to it such that $N = a \times r^b$, where r is the base and is constant. In a decimal machine where $r = 10$ it would thus be possible to represent 171,286,392 by 1.71286392×10^8 and 0.00843972681 by $8.43972681 \times 10^{-3}$. The design of arithmetic units for numbers encoded in this way involves some complications which will not be treated here.

Machine Representation of Numbers

The computer designer has a choice not only of the code used to represent numbers in the machine, but also of whether the coded numbers are operated on in serial or parallel form (p. 194) or in some combination thereof. If an arithmetic operation is performed serially, the necessary equipment may be relatively simple, for the logical equations are dependent on only a few bits of each word at one time. The unit of computation time is then one word-time, or the time it takes for a word to shift serially through the arithmetic unit. If a computation is performed in parallel, the equipment necessary is more complicated, for the most significant digits of a number may be a function of all of the less significant ones. As a result, some memory element input equations will be very complicated and may be functions of a great many variables. In addition, there must be an equation for each digit in the answer, where in a serial unit one equation determines all the various digits of the answer, one by one, over the word-time. The parallel arithmetic unit, though it is more complicated and therefore more expensive than its serial counterpart, is also faster. A complicated operation may be carried out in one bit-time or a few bit-times in a parallel machine, as compared to a word-time for a serial unit. If a word contains forty bits, this may mean an increase in speed by a factor of ten to forty.

In the remainder of this chapter we shall derive the logic for circuits which perform serial and parallel arithmetic operations. In order to simplify the descriptions which follow, we shall, wherever possible, adopt the standard notation of Figures 9-1 and 9-2 as a means of representing serial and parallel numbers respectively. A serial number denoted by X will ordinarily be assumed to be available in some sort of circulating or shifting register, such as that shown in Figure 9-1. The word is read one bit at a time, starting with the least significant bit, from a read amplifier known as X_r. This same word or some other word which replaces it may be inserted in this same register by writing it into write amplifier X_w. Between amplifiers X_w and X_r is a register of some kind which retains and shifts $(n - 2)$ bits, if the total number of bits in the register is n. The time at which the least significant bit, of the desired word is in X_r is denoted t_1. For example, if the circulating register contains a word of four binary digits, together with a fifth

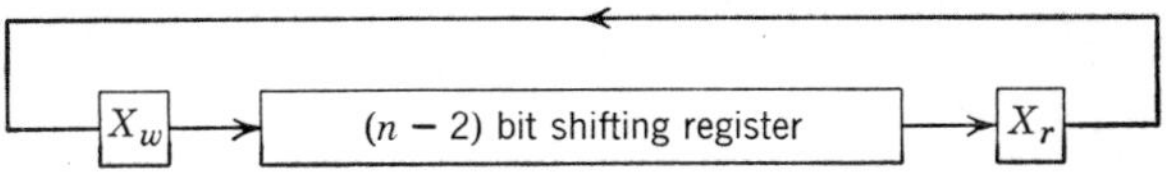

X_w → $(n - 2)$ bit shifting register → X_r

	X_w				X_r
t_0	0	0	0	1	s
t_1	s	0	0	0	1
t_2	1	s	0	0	0
t_3	0	1	s	0	0
t_4	0	0	1	s	0
t_0	0	0	0	1	s
t_1	s	0	0	0	1

(s = space bit)

FIGURE 9-1

or space digit inserted between the least and the most significant bits of the word, the content of this circulating register during seven consecutive bit-times is shown in Figure 9-1. The number which circulates in this register is taken to be the number 0001, and it will be noted that X_r contains the digits of this number sequentially during the four bit-times starting with t_1. The space bit is denoted by s. If a word containing more than four bits were to be represented serially, the only change necessary would be a lengthening of the circulating register and an increase in the number of bit-times necessary for one complete circulation of a word.

If a number is stored in parallel, all of its bits must be available in memory elements at one time. In Figure 9-2, we see an eight-bit

X_8	X_7	X_6	X_5	X_4	X_3	X_2	X_1		
0	1	1	0	0	0	0	1	= 97	(binary code)
1	0	0	1	0	1	1	1	= 97	(8-4-2-1 decimal code)

FIGURE 9-2

number stored in eight memory elements labeled $X_1, X_2, X_3, \cdots, X_8$. Flip-flop X_1 holds the least significant bit of the word and flip-flop X_8 the most significant. Note that these eight flip-flops may be considered to hold an ordinary binary number, or else two decimal digits of four bits each. The number 97 encoded in these two ways is illustrated in Figure 9-2.

CIRCUITS FOR ARITHMETIC OPERATIONS

It should be evident from what has been said above that the designer has a great deal of freedom in planning an arithmetic unit. The equipment necessary to carry out operations will be different for every different code employed and for every different method of representing numbers. There are perhaps four major categories of computers—serial binary, serial decimal, parallel binary, and parallel decimal—but there are obviously a host of others besides. Even when a particular code has been chosen and a particular method for representing numbers has been adopted, there are many different schemes the designer can use to mechanize even the simplest operations. In designing a circuit to perform any given function, then, the designer must first choose a scheme; i.e., he must list a set of rules specifying a sequence of operations which perform the function. As an obvious example of how two quite different rules may be used to accomplish the same result on numbers expressed in identical codes and stored in a machine in the same way, consider addition. The addition of two numbers may be carried out either by use of an addition table which determines what sum digit is obtained when two corresponding digits of the two numbers are combined; or else by simultaneously adding unity to the addend while subtracting unity from the augend until the number which was the augend becomes zero, at which time the number which was the addend is the sum.

There is not space in this chapter to discuss all the possible rules which may be employed in mechanizing the various arithmetic operations. It will not even be possible to carry through the detailed design of all of the very simplest arithmetic units. Rather, the aim here will be to familiarize the reader with the principal and most widely used arithmetic rules, and to show him how the techniques which have been discussed in previous chapters can be applied to arithmetic unit design.

Counters

Counting is the operation of adding unity to or subtracting unity from the least significant digit position of a number. Counters are most often used in computers as indices or timers (see Chapter 7, p. 176), whose purpose it is to define the state of the computer and therefore to help control its operation. However, as indicated above, a counter may also form the basis for an arithmetic unit. Counting, up or down, may be looked upon as the basic arithmetic operation in terms of which all other operations may be carried out. Designers do not usually employ counters in this way, however, because of the very long time required to add, subtract, multiply, or divide by counting.

We shall first design a circuit which will add unity every word-time to the number held in a circulating register such as that shown in Figure 9-1. If the word consists of four bits and a space bit and the number initially in the register is 0001, the contents of the register during three successive word-times are shown in Table 9-2. If we look

Table 9-2

	X_w	1 2 3	X_r
t_0	0	0 0 1	s
t_1	s	0 0 0	1
t_2	0	s 0 0	0
t_3	1	0 s 0	0
t_4	0	1 0 s	0
t_0	0	0 1 0	s
t_1	s	0 0 1	0
t_2	1	s 0 0	1
t_3	1	1 s 0	0
t_4	0	1 1 s	0
t_0	0	0 1 1	s
t_1	s	0 0 1	1
t_2	0	s 0 0	1
t_3	0	0 s 0	0
t_4	1	0 0 s	0
t_0	0	1 0 0	s
t_1	s	0 1 0	0

etc.

(s = space bit)

at the register every t_0 time, we see that the number in it steadily increases by "one". We shall assume that at any given bit-time four of the five bits are inaccessible to the logical designer. The only one he can read is that in read amplifier X_r, for the others are in the circulating register and are being shifted, one position to the right, every bit-time.

Let us begin the design of this simple counter by examining with some care the operations we actually perform when we continually add "one" to a binary number. The three additions of Table 9-2 are recorded below.

$$\begin{array}{rl} & \overbrace{\;\;\;\;\;\;}^{\bar{t}_1}\; t_1 \\ X_r & 0\,0\,0\,1 \\ & 1 \\ \hline & 0\,0\,1\,0 \end{array} \qquad \begin{array}{rl} & \overbrace{\;\;\;\;\;\;}^{\bar{t}_1}\; t_1 \\ X_r & 0\,0\,1\,0 \\ & 1 \\ \hline & 0\,0\,1\,1 \end{array} \qquad \begin{array}{rl} & \overbrace{\;\;\;\;\;\;}^{\bar{t}_1}\; t_1 \\ X_r & 0\,0\,1\,1 \\ & 1 \\ \hline & 0\,1\,0\,0 \end{array}$$

Note that the least significant bit-time is labeled t_1, and the other bit-times are labeled $\bar{t}_1$. The rules which we follow out in counting may be expressed as follows. At t_1 time we look at X_r, and if it is a "zero" we write "one" in X_w; if it is a "one" we write "zero." If X_r was "one" at t_1 time we generated something which we called a "carry" which was employed at bit-time t_2 and compared with the second digit from X_r to determine the next bit to be written in the write amplifier. The rules already stated which determine X_w and the carry when $t_1 = 1$ are shown in tabular form in Table 9-3. Also shown are the

Table 9-3

	n		$n+1$	
X_r	t_1	C	C	X_w
0	0	0	0	0
1	0	0	0	1
0	1	0	0	1
1	1	0	1	0
0	0	1	0	1
1	0	1	1	0

rules which apply during subsequent bit-times, when $t_1 = 0$ and the carry generated during the previous bit-time may be either "zero" or "one." The table as written contains only six entries although we know there are eight possible combinations of the three variables X_r, t_1, and C. The two missing combinations are those for which both t_1 and C are simultaneously equal to "one," and they are omitted from the table because the carry is never "one" at t_1 time.

In order to construct this counter, we shall supply a flip-flop C which will hold the carry from one bit-time to the next. In order that the state of this flip-flop at t_1 time need not concern us, let us add two entries to Table 9-3 which will ensure that the counter works as desired no matter what C is at t_1 time. These two entries are shown below.

n			$n+1$	
X_r	t_1	C	C	X_w
0	1	1	0	1
1	1	1	1	0

Notice that X_w and C at bit-time $(n + 1)$ have the same values when t_1 and C are both "one" as they do when t_1 is "one" and C is "zero."

We are now in a position to write the application equations for write amplifier X_w and for flip-flop C. From the amended truth table we can plot the appropriate difference equations, and these are given in Figure 9-3. The input logic can be written from these two Veitch diagrams as follows:

$$J_{Xw} = \overline{K}_{Xw} = X_r\bar{t}_1\overline{C} + \overline{X}_rC + \overline{X}_rt_1 \quad \text{(9-3)}$$

$$J_C = X_rt_1 \qquad K_C = \overline{X}_r \quad \text{(9-4)}$$

The reader should verify that this logic actually accomplishes the counting operation illustrated in Table 9-2.

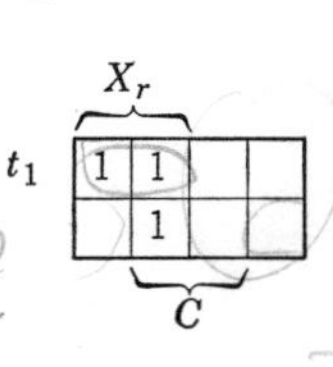

X_w^{n+1}

C^{n+1}

FIGURE 9-3

Before leaving this section on counter design, let us carry out the design of a parallel decimal counter employing the ordinary 8-4-2-1 decimal code. Let us assume that our counter will contain two decimal digits stored in flip-flops X_1–X_8, where X_1 is the least significant bit of the less significant decimal digit. In Table 9-4 the truth table which

Table 9-4

n				$n+1$			
X_4	X_3	X_2	X_1	X_4	X_3	X_2	X_1
0	0	0	0	0	0	0	1
0	0	0	1	0	0	1	0
0	0	1	0	0	0	1	1
0	0	1	1	0	1	0	0
0	1	0	0	0	1	0	1
0	1	0	1	0	1	1	0
0	1	1	0	0	1	1	1
0	1	1	1	1	0	0	0
1	0	0	0	1	0	0	1
1	0	0	1	0	0	0	0

governs the counting operation of the less significant four binary digits is shown. Note that the register changes from one number in the count sequence to the next higher one every bit-time, where the serial counter only changed once every word-time. The difference equations for flip-flops X_1–X_4 as derived from Table 9-4 (and taking into con-

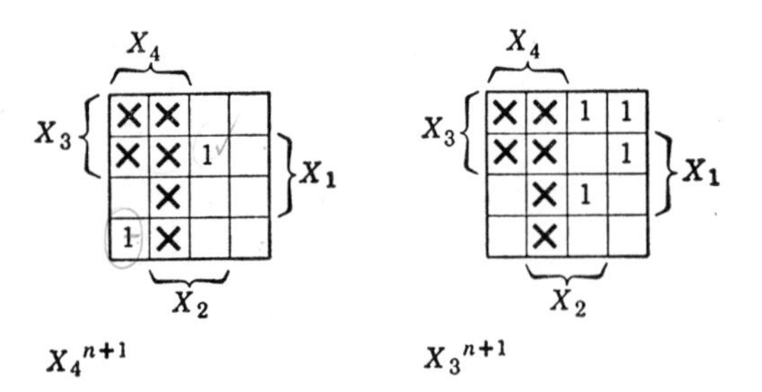

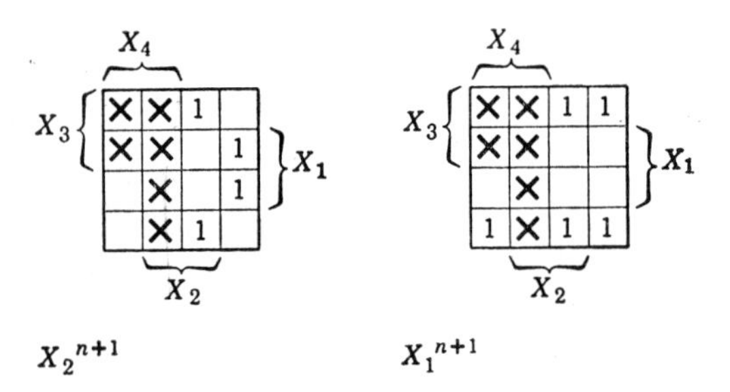

FIGURE 9-4

sideration the redundancies caused by the fact that six four-bit combinations never occur) are shown in Figure 9-4, and the resulting input equations for these four flip-flops are:

$$J_{X1} = K_{X1} = 1 \tag{9-5}$$

$$J_{X2} = X_1\overline{X}_4 \qquad K_{X2} = X_1 \tag{9-6}$$

$$J_{X3} = K_{X3} = X_1X_2 \tag{9-7}$$

$$J_{X4} = X_1X_2X_3 \qquad K_{X4} = X_1 \tag{9-8}$$

It is obvious that we want flip-flops X_5–X_8 to count in the same sequence as do flip-flops X_1–X_4. However, instead of counting every bit-time as the least significant part of the counter does, we want the second decimal digit to change only when the least significant one changes from nine to zero. In other words, where flip-flop X_1 changes every bit-time, we want flip-flop X_5 to change only when both X_1 and X_4 are "one," indicating that flip-flops X_1–X_4 contain a nine. We thus see that the logic for flip-flops X_5–X_8 corresponds exactly to the logic for the other four flip-flops, except that it must be multiplied by a logical combination which recognizes a nine. The input equations can be written directly:

$$J_{X5} = K_{X5} = X_1X_4 \tag{9-9}$$

$$J_{X6} = X_1X_4X_5\overline{X}_8 \qquad K_{X6} = X_1X_4X_5 \tag{9-10}$$

$$J_{X7} = K_{X7} = X_1X_4X_5X_6 \tag{9-11}$$

$$J_{X8} = X_1X_4X_5X_6X_7 \qquad K_{X8} = X_1X_4X_5 \tag{9-12}$$

Addition

If numbers are represented by a code in which each digit represents the coefficient multiplying a weight associated with that digit position, the first step in carrying out an addition is to arrange the numbers being added so that digits having the same weight may be compared. The second step is to compare these coefficients and, by following a set of rules, to derive therefrom a new coefficient for that digit position in the sum, as well as data determining how other coefficients in the sum should be affected. The set of rules will in general be different for every different method of assigning the weights w_i in equation 9-1.

The first step, that of lining up corresponding digits of the numbers being added, is usually no problem in the design of an adder. If the numbers are represented in serial fashion, the corresponding digits must be made to appear simultaneously for comparison. If numbers are represented in parallel, the first step is automatically accomplished. The second step, that of comparing corresponding digits and of applying a rule which will determine the digit of the sum, is more complicated. In this section we shall illustrate different methods which can be used to attack and to solve this problem and shall design several different adders.

A two-input serial binary adder. The first and simplest adder to be designed forms the sum of two numbers expressed in the binary code, which appear serially, least significant bit first, from two flip-flops or amplifiers. A typical configuration is illustrated in Figure 9-5. A

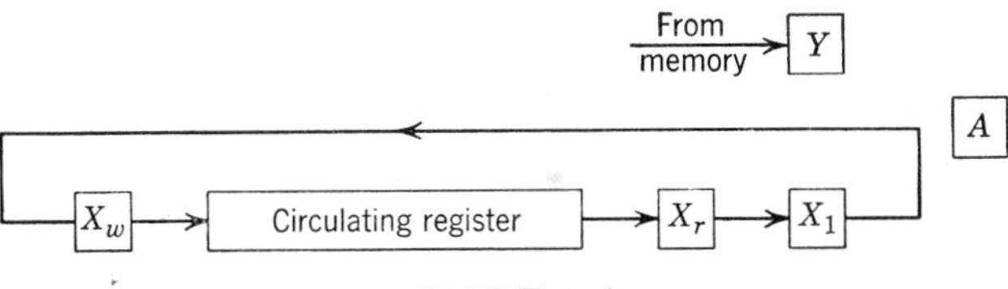

FIGURE 9-5

number, X, is available in a shifting register or a circulating register, where the least significant bit of the number appears in read amplifier X_r at t_0 time. Normally, this number is to be circulated upon itself, so that X_r shifts into X_1 and X_1 is to be transferred directly to X_w. However, when an addition is to be performed, a controlling flip-flop A goes to the "one" state for one word-time. At t_1 time of that word-time the least significant bit of a new number appears in flip-flop Y and during subsequent bit-times the other bits of this binary number appear in Y. A circuit must be designed to form the sum of the number appearing in flip-flop Y and the number in the circulating register, and to transfer that sum back into the circulating register, replacing the original number X.

The least significant bit of X is in read amplifier X_r at t_0 time and therefore must be in flip-flop X_1 at t_1 time. In order to satisfy the first requirement, that of comparing bits which have the same weight, we must therefore perform the addition by comparing the digit in flip-flop Y with that in flip-flop X_1 at every bit-time. If we examine memory elements X_1 and Y at some bit-time n, we are examining two coefficients which must be interpreted as follows:

$$\begin{aligned} &\cdots + (X_1)^{n+1}2^{n+1} + (X_1)^n 2^n + \cdots \\ &\cdots + (Y)^{n+1}2^{n+1} + (Y)^n 2^n + \cdots \end{aligned} \tag{9-13}$$

where $(X_1)^{n+1}$, $(Y)^{n+1}$, $(X_1)^n$, and $(Y)^n$ represent the values ("zero" or "one") of flip-flops X_1 and Y at bit-times $(n + 1)$ and (n) respectively, and the equations are to be interpreted as arithmetic, not Boolean, relations. Ignoring for the moment what has gone at previous bit-times, we see that if X_1 and Y are both "zero" at time n the sum digit which represents the coefficient of 2^n will be "zero." If either X_1 or Y is "one" and the other is "zero," their sum is "one" and the term (1×2^n) must appear in the sum. If both X_1 and Y are "one," their sum is 2 and we must add to the sum $(2 \times 2^n) = 2^{n+1}$. In this case the sum coefficient of 2^n is "zero" and we have generated a "carry," which is unity in the $(n + 1)$st binary position. In fact, in every binary position in the word except the least significant one, the sum digit is determined not only by the contents of flip-flops X_1 and Y but also by the carry which may have been generated by the combination of previous digits in the two numbers being added.

The various combinations of X_1, Y, and the carry C, together with the resulting sum bit and the new carry, are shown in Table 9-5. (Notice the similarity of this table to that for the counter whose logic is shown in Table 9-3.) If we now provide a carry flip-flop C, just as we did for the counter, we can immediately write the input equations for that flip-flop and for the write amplifier X_w. These equations are:

$$J_C = X_1 Y \qquad K_C = \overline{X}_1 \overline{Y} \tag{9-14}$$

$$J_{Xw} = \overline{K}_{Xw} = \overline{X}_1 \overline{Y} C + \overline{X}_1 Y \overline{C} + X_1 \overline{Y} \overline{C} + X_1 Y C \tag{9-15}$$

The design of the serial binary adder may now be completed by modifying slightly the basic equations 9-14 and 9-15. In the first place, we

Table 9-5

	n		$n+1$	
X_1	Y	C	C	X_w
0	0	0	0	0
1	0	0	0	1
0	1	0	0	1
1	1	0	1	0
0	0	1	0	1
1	0	1	1	0
0	1	1	1	0
1	1	1	1	1

must recognize and remember that the adder will not work correctly unless the least significant carry bit is "zero"; that is to say, flip-flop C must contain a "zero" when the addition begins, at t_1 time. We can accomplish this by adding t_0 to the K input of the C flip-flop, if we make sure that the J input to that flip-flop is "zero" at that time. The modified input equations are

$$J_C = X_1 Y \bar{t}_0 \qquad K_C = \overline{X}_1 \overline{Y} + t_0 \tag{9-16}$$

The logic for write amplifier X_w must also be modified so that the X register circulates upon itself when flip-flop $A = 0$, and the addition is only performed during the word-time that flip-flop $A = 1$.

$$\begin{aligned} J_{Xw} &= A\overline{X}_1\overline{Y}C + A\overline{X}_1 Y\overline{C} + AX_1\overline{Y}\overline{C} + AX_1YC + \overline{A}X_1 \\ &= A\overline{X}_1\overline{Y}C + A\overline{X}_1 Y\overline{C} + X_1\overline{Y}\overline{C} + X_1YC + \overline{A}X_1 \\ K_{Xw} &= A\overline{X}_1\overline{Y}\overline{C} + AX_1\overline{Y}C + A\overline{X}_1YC + AX_1Y\overline{C} + \overline{A}\overline{X}_1 \\ &= \overline{X}_1\overline{Y}\overline{C} + AX_1\overline{Y}C + \overline{X}_1YC + AX_1Y\overline{C} + \overline{A}\overline{X}_1 \end{aligned} \tag{9-17}$$

A three-input serial binary adder. Let us next design an adder which forms the sum of three binary numbers appearing serially least significant bit first in three flip-flops X, Y, and Z. The truth table describing the operation of such an adder, with the assumption that the sum is to be put in a flip-flop X_w, is given in Table 9-6. It can

Table 9-6

		n				$n+1$	
X	Y	Z	C_1	C_2	C_1	C_2	X_w
0	0	0	0	0	0	0	0
1	0	0	0	0	0	0	1
0	1	0	0	0	0	0	1
1	1	0	0	0	0	1	0
0	0	1	0	0	0	0	1
1	0	1	0	0	0	1	0
0	1	1	0	0	0	1	0
1	1	1	0	0	0	1	1
0	0	0	0	1	0	0	1
1	0	0	0	1	0	1	0
0	1	0	0	1	0	1	0
1	1	0	0	1	0	1	1
0	0	1	0	1	0	1	0
1	0	1	0	1	0	1	1
0	1	1	0	1	0	1	1
1	1	1	0	1	1	0	0
0	0	0	1	0	0	1	0
1	0	0	1	0	0	1	1
0	1	0	1	0	0	1	1
1	1	0	1	0	1	0	0
0	0	1	1	0	0	1	1
1	0	1	1	0	1	0	0
0	1	1	1	0	1	0	0
1	1	1	1	0	1	0	1

easily be shown that two carry flip-flops are required to mechanize such an adder. Starting at the top of the table and assuming no incoming carries, we find the first eight entries are exactly those of Table 9-5 for the two-input adder. Four of these eight states cause a carry to be generated and that carry is given the name C_2. The second eight entries in the table cover all the combinations of inputs for X, Y, and Z under the assumption that the carry C_2 is a "one." The last entry of this group of eight indicates that four "ones" may be present simultaneously at the nth binary place. Such a circumstance would require a zero sum digit as shown, and a carry not into the next consecutive bit position but to the one after that. This new carry, which is given the designation C_1, is displayed in the third group of eight entries on the truth table. The largest sum that can occur in a single binary place of a three-input adder is the last entry in Table 9-6 and occurs when all three inputs X, Y, and Z are "one" at a time when the C_2 carry is "one." These three "ones" plus a carry having weight two give a total of five for the sum in that binary position, and this sum is represented by a sum digit of "one" sent to X_w and a new carry two bit positions to the left. Thus the carry generated under this condition has the same significance as the carry generated at the sixteenth entry in this table and no new configuration of carries is required. In particular, it is never possible for C_1 and C_2 to be "one" at the same time, and this redundancy may be used in the design of the adder.

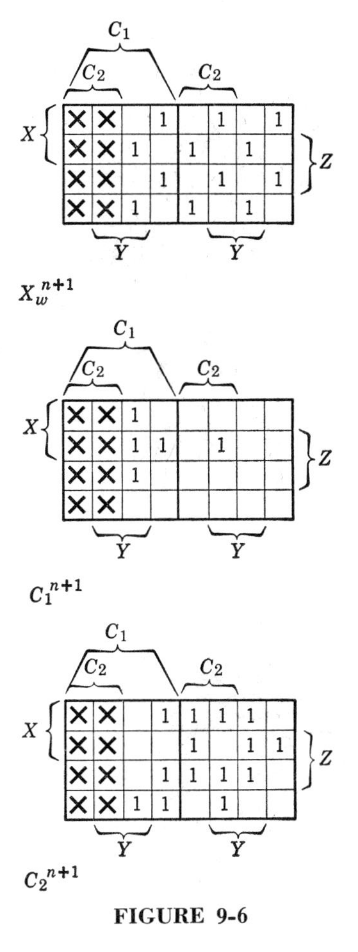

FIGURE 9-6

In the Veitch diagrams of Figure 9-6, the difference equations for amplifier X_w and flip-flops C_1 and C_2 are plotted. The input equations for these circuit elements, derived from the Veitch diagrams, are:

$$J_{Xw} = \overline{K}_{Xw} = \overline{X}\overline{Y}\overline{Z}C_2 + \overline{X}\overline{Y}Z\overline{C}_2 + \overline{X}Y\overline{Z}\overline{C}_2 + X\overline{Y}\overline{Z}\overline{C}_2 + \overline{X}YZC_2 + X\overline{Y}ZC_2 + XY\overline{Z}C_2 + XYZ\overline{C}_2 \quad (9\text{-}18)$$

$$J_{C1} = XYZC_2$$
$$K_{C1} = \overline{X}\overline{Z} + \overline{X}\overline{Y} + \overline{Y}\overline{Z} \quad (9\text{-}19$$

$$J_{C2} = XY\overline{C}_1 + XZ\overline{C}_1 + YZ\overline{C}_1 + \overline{X}\overline{Z}C_1 + \overline{X}\overline{Y}C_1 + \overline{Y}\overline{Z}C_1$$
$$K_{C2} = \overline{X}\overline{Y}\overline{Z} + XYZ \quad (9\text{-}20$$

A one-bit-time, parallel, binary adder. Suppose that we have two twenty-bit binary numbers standing in two flip-flop registers X and Y. We shall now design an adder which forms the sum of these two binary numbers and places them in the Y-register one bit-time after the addend and augend appear in X and Y. If we look at any pair of binary digits in these two registers at the first bit-time, t_1, we can write a truth table which defines the operation of the adder and determines what the sum digit must be in the corresponding Y flip-flop at the next bit-time. This table, which has the same appearance as do the other addition truth tables which we have drawn, is given in Table 9-7. As usual, the sum digit in the ith binary position is a function

Table 9-7

	t_1			t_2
X_i	Y_i	C_i	C_{i+1}	Y_i
0	0	0	0	0
1	0	0	0	1
0	1	0	0	1
1	1	0	1	0
0	0	1	0	1
1	0	1	1	0
0	1	1	1	0
1	1	1	1	1

not only of the corresponding bits of the addend and augend but also of the carry from the previous bit position. The sum bit Y_i at the second bit-time t_2 is thus determined from the values of X_i, Y_i, and C_i at t_1 time. However, note that the carry, C_i, is not a signal that can be obtained from a single flip-flop as X_i and Y_i are. The ith carry is a function of the addend and augend bits in the $(i - 1)$st bit position and of the incoming carry bit to that bit position. The $(i - 1)$st carry, in turn, is a function of bits further toward the least significant end of the adder, and this functional relationship continues down to the least significant end of the words, where C_1, the carry bit into the least significant bit position, is "zero."

This dependence of Y_i at t_2 time on the values of not only X_i and Y_i, but also all X and Y flip-flops with numbers less than i, is indicated in equations 9-21 and 9-22, which are derived directly from Table 9-7.

$$Y_i^{t_2} = (\overline{X}_i\overline{Y}_iC_i + \overline{X}_iY_i\overline{C}_i + X_i\overline{Y}_i\overline{C}_i + X_iY_iC_i)^{t_1}$$

$$= [Y_i(\overline{X}_i\overline{C}_i + X_iC_i) + \overline{Y}_i(\overline{X}_iC_i + X_i\overline{C}_i)]^{t_1} \qquad (9\text{-}21)$$

where

$$
\begin{aligned}
C_i^{t_1} &= (X_{i-1}Y_{i-1} + X_{i-1}C_{i-1} + Y_{i-1}C_{i-1})^{t_1} \\
&= [X_{i-1}Y_{i-1} + (X_{i-1} + Y_{i-1})C_{i-1}]^{t_1} \\
\overline{C}_i^{t_1} &= [(\overline{X}_{i-1} + \overline{Y}_{i-1})(\overline{X}_{i-1} + \overline{C}_{i-1})(\overline{Y}_{i-1} + \overline{C}_{i-1})]^{t_1} \\
C_1 &= 0
\end{aligned}
\tag{9-22}
$$

Note that the equation for Y_i is a difference equation, whereas the expression C_i involves only one bit-time, t_1.

From the application equation 9-21 we can derive the input equations for flip-flop Y_i

$$
\begin{aligned}
J_{Yi} &= \overline{X}_iC_i + X_i\overline{C}_i \\
K_{Yi} &= \overline{\overline{X}_i\overline{C}_i + X_iC_i} \\
&= \overline{X}_iC_i + X_i\overline{C}_i \\
&= J_{Yi}
\end{aligned}
\tag{9-23}
$$

and the specific input equations for particular flip-flops may be written by combining the carries from equation 9-22 with the input equations 9-23. The input equations for the first three Y flip-flops are given in equations 9-24, 9-25, and 9-26. The input equations for flip-flop Y_{20} are indicated in equation 9-27, where it must be understood that C_{20} is a function of all X and Y flip-flops except X_{20} and Y_{20}.

$$J_{Y1} = K_{Y1} = X_1 \tag{9-24}$$

$$C_2 = X_1Y_1$$

$$J_{Y2} = K_{Y2} = \overline{X}_2X_1Y_1 + X_2(\overline{X}_1 + \overline{Y}_1) \tag{9-25}$$

$$C_3 = X_2Y_2 + (X_2 + Y_2)X_1Y_1$$

$$
\begin{aligned}
J_{Y3} = K_{Y3} &= \overline{X}_3[X_2Y_2 + (X_2 + Y_2)X_1Y_1] \\
&\quad + X_3(\overline{X}_2 + \overline{Y}_2)(\overline{X}_2\overline{Y}_2 + \overline{X}_1 + \overline{Y}_1) \\
&= \overline{X}_3X_2Y_2 + \overline{X}_3X_2X_1Y_1 + \overline{X}_3Y_2X_1Y_1 \\
&\quad + X_3\overline{X}_2\overline{Y}_2 + X_3\overline{X}_2\overline{X}_1 + X_3\overline{X}_2\overline{Y}_1 \\
&\quad + X_3\overline{Y}_2\overline{X}_1 + X_3\overline{Y}_2\overline{Y}_1
\end{aligned}
\tag{9-26}
$$

$$C_4 = X_3Y_3 + (X_3 + Y_3)[X_2Y_2 + (X_2 + Y_2)X_1Y_1]$$

. . . .

$$J_{Y20} = K_{Y20} = \overline{X}_{20}C_{20} + X_{20}\overline{C}_{20} \tag{9-27}$$

Logically speaking, this completes the design of the one-bit parallel adder. Practically speaking, many problems still remain. As can be seen from the foregoing equations, the carry logic gets progressively more and more complex in higher and higher bit positions. This gives rise to two difficulties closely related to one another. The first is the

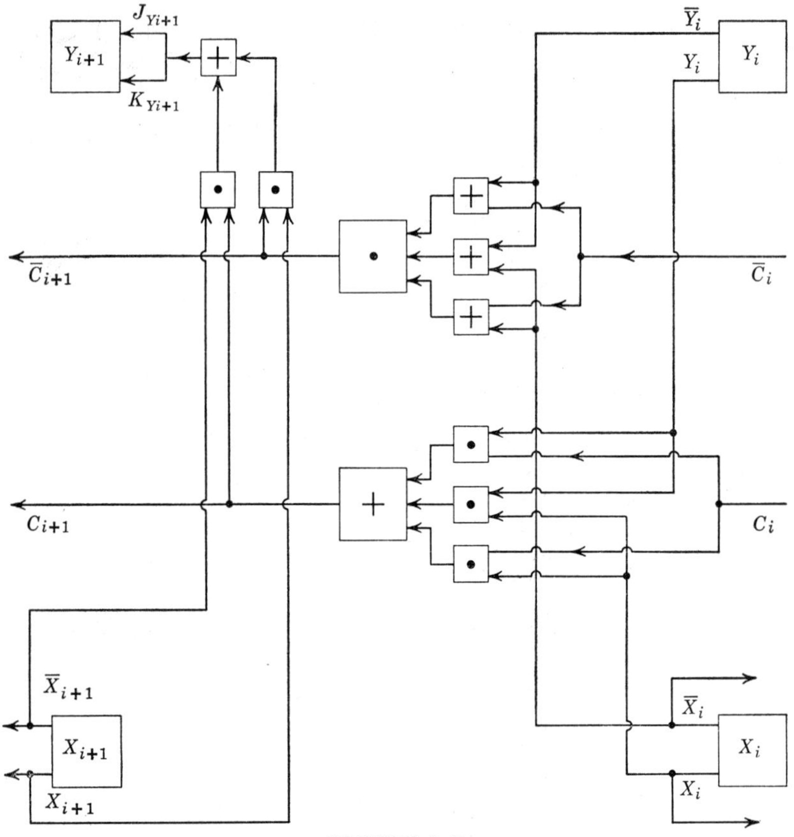

FIGURE 9-7

fact that if C_{20}, for example, were written in simplest minterm form as a function of the thirty-eight flip-flops in the X- and Y-registers, several million diodes would be required to mechanize the resulting equation. The number of components required can be reduced to quite manageable proportions if the circuit designer permits the use of higher-than-second-order expressions, however. The logical designer can then use equation 9-22 to express each carry as a function of previous carries, with the result shown in Figure 9-7, which represents a wiring diagram for one stage of the adder.

Although the arrangement of Figure 9-7 greatly reduces the number of diodes required to mechanize the parallel adder (only 24 diodes per stage or a total of about 450 are required for a twenty-bit adder), another problem still remains. Two levels of gating appear in series in each stage of the adder, so that the carry signal into the twentieth bit position of the sum represents a signal which has passed through about thirty-six "and" and "or" circuits. The connection of so many gates in series is in general not a practical thing to do with conventional decision elements, and it is necessary to introduce booster amplifiers at various points along the path. In addition, as has been mentioned before, there is a delay through each of these gates, and a proportionally longer delay through the combination of thirty-six gates and associated booster amplifiers. This delay is often called the *carry propagation time* and is a limiting factor on the speed with which this particular adder can be made to work. In designing such an adder, the logical and circuit designers must work together to make sure that, if the augend and addend are inserted in the X- and Y-registers at t_1 time, the long series of gates propagates the carry signal to the most significant bit position by t_2 time. Alternatively, the logical designer may arrange to insert the augend and addend in the X- and Y-registers several bit-times before the addition is to take place, thus allowing the carry to propagate while (perhaps) other operations not involving changes in the X- and Y-registers are carried out.

A two-bit-time, parallel, binary adder. The logical designer may take other steps to simplify the logic for a parallel adder by increasing the time taken to perform an addition. He thus can make the rapid operation of gating circuits a less critical factor in design. One example of the decrease in complexity that can be achieved through an increase in addition time will be given here.

The adder to be designed forms a complete sum in two bit-times, and its operation in a particular addition is indicated in Figure 9-8. At t_1 time two numbers X and Y appear in the X- and the Y-registers, and the logical designer provides logic to carry out a partial addition. In this partial addition each pair of bits in the X- and Y-registers are compared with one another without reference to any preceding bits. The sum of this pair of bits is computed, and the logical designer arranges that the sum bit appears in the Y-register at t_2 time while the carry bit is transferred to the next higher bit position in the X-register. In the least significant four bit positions of Figure 9-8, for example, there appear all four possible combinations of two bits in X and Y. In the least significant bit position, the two digits are both "zero" and the sum and carry digits, both "zero," appear in flip-flop Y_1 and

X_2, respectively, at t_2 time. In the second bit position, X is a "zero" while Y is a "one" at t_1 time. The sum of these two bits is a "one," which is inserted in flip-flop Y_2 at t_2 time, and the carry is a "zero" which appears in flip-flop X_3. In the fourth bit position, both X and Y are "one" at t_1 time, and the sum digit, "zero," appears in flip-flop

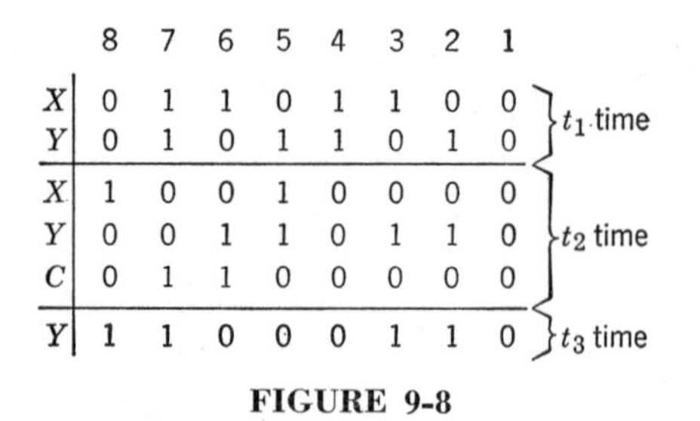

	8	7	6	5	4	3	2	1	
X	0	1	1	0	1	1	0	0	t_1 time
Y	0	1	0	1	1	0	1	0	
X	1	0	0	1	0	0	0	0	t_2 time
Y	0	0	1	1	0	1	1	0	
C	0	1	1	0	0	0	0	0	
Y	1	1	0	0	0	1	1	0	t_3 time

FIGURE 9-8

Y_4 at t_2 time while the carry appears in flip-flop X_5. The same combinations appear in bit positions five through eight at t_1 time, and give rise to the same results at t_2 time.

We now have two new numbers in registers X and Y. Because of the way these numbers were generated from the original numbers, their sum is identical to the sum of the original two numbers. We complete the design of the adder, then, by designing a one-bit-time adder to form the complete sum of X and Y in the Y-register by t_3 time.

It is not immediately apparent that the adder which has just been described is any simpler than the one-bit-time adder previously designed. In fact, it would appear that we require the same one-bit-time parallel adder at t_2 time that we require in the faster adder, and that some additional logic is required at t_1 time as well. However, we shall now show that the simple operation carried out at t_1 time results in a certain redundancy in the X- and Y-registers at t_2 time, and that this redundancy may be used to simplify the logic for the one-bit-time adder.

The redundancy which exists can be expressed by the following formula:

$$(X_iC_i)^{t_2} = 0 \tag{9-28}$$

That is to say, it is impossible for both X_i (the partial carry generated at t_1 time) and C_i (the ordinary binary carry employed at t_2 time to complete the adding process) to be "one" at t_2 time. The truth of this statement may be proved by referring to Figure 9-9a, where it is assumed that X and C in digit position i are both "one" at t_2 time. We can prove that this assumption leads to an inconsistency as follows.

Since X_i is "one" at t_2 time and since it represents the carry from the partial addition of the two digits in X_{i-1} and Y_{i-1} at t_1 time, it follows that X_{i-1} and Y_{i-1} must both have been "one" at t_1 time, and that Y_{i-1} must therefore be "zero" at t_2 time. This is illustrated in Figure 9-9*b*. Furthermore, if C_i is "one" at t_2 time, it must be true that X_{i-1}

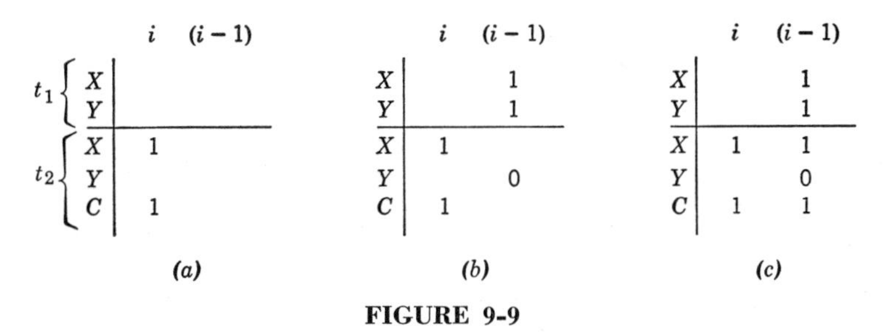

FIGURE 9-9

and C_{i-1} were both "one" at t_2 time, since it would be impossible to get a carry into bit position i otherwise with Y_{i-1} equal to "zero." This last conclusion is indicated in Figure 9-9*c*.

We have thus proved so far that if X and C are both "one" at t_2 time in some bit position, then they must also be "one" in the next less significant bit position at that same bit-time. By the same argument, if they are both "one" in the next less significant bit position, they must both be "one" at the bit position to the right of that. This argument can be continued until we come to the least significant end of the word, where we conclude that both X_1 and C_1 are "one." We know, however, that the least significant carry must be "zero." We therefore conclude that our original assumption was incorrect, and that it is not possible for X_i and C_i both to be unity at t_2 time. Furthermore, we can quickly convince ourselves that there are no other redundancies inherent in the X- and Y-registers at t_2 time by examining the example of Figure 9-8, where all combinations of X, Y, and C appear except those for which X and C are both equal to "one."

We are now ready to begin the detailed design of the adder. The truth table for the partial addition is shown in Table 9-8, and the application equations and input logic for flip-flops Y_i and X_{i-1} are:

$$Y_i^{t_2} = (X_i\overline{Y}_i + \overline{X}_iY_i)^{t_1} \qquad J_{Y_i}{}^{t_1} = K_{Y_i}{}^{t_1} = X_i^{t_1} \tag{9-29}$$

$$X_{i+1}{}^{t_2} = (X_iY_i)^{t_1} \qquad J_{X_{i+1}}{}^{t_1} = (X_iY_i)^{t_1} \qquad K_{X_{i+1}}{}^{t_1} = (\overline{X}_i + \overline{Y}_i)^{t_1}$$

$$J_{X1}{}^{t_1} = 0 \qquad K_{X1}{}^{t_1} = 1 \tag{9-30}$$

We must now write the logic for the complete parallel binary addition which takes place at bit-time t_2, employing the redundancy of equation

9-28 if we can. The equations which define this operation are the same as equations 9-21 and 9-22. They are shown below, and are plotted in the Veitch diagram of Figure 9-10 where the redundancies also appear.

$$Y_i^{t_3} = (\bar{X}_i\bar{Y}_iC_i + \bar{X}_iY_i\bar{C}_i + X_i\bar{Y}_i\bar{C}_i + X_iY_iC_i)^{t_2} \tag{9-31}$$

$$C_i^{t_2} = (X_{i-1}Y_{i-1} + X_{i-1}C_{i-1} + Y_{i-1}C_{i-1})^{t_2} \tag{9-32}$$

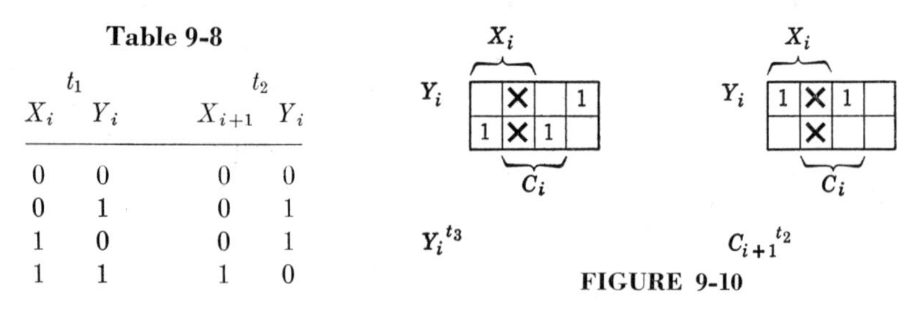

Table 9-8

t_1		t_2	
X_i	Y_i	X_{i+1}	Y_i
0	0	0	0
0	1	0	1
1	0	0	1
1	1	1	0

FIGURE 9-10

Examining the Veitch diagram, we find the simplest forms of the input equations for flip-flop Y_i to be

$$J_{Y_i}{}^{t_2} = K_{Y_i}{}^{t_2} = (X_i + C_i)^{t_2} \tag{9-33}$$

where

$$C_i^{t_2} = (X_{i-1}Y_{i-1} + Y_{i-1}C_{i-1})^{t_2} \tag{9-34}$$

We are now in a position to write the input logic for all the Y flip-flops at t_2 time by substituting particular values of i into equations 9-33 and 9-34.

Since

$$X_1^{t_2} = C_1^{t_2} = 0$$

$$J_{Y1}{}^{t_2} = K_{Y1}{}^{t_2} = 0 \tag{9-35}$$

$$C_2^{t_2} = 0$$

$$J_{Y2}{}^{t_2} = K_{Y2}{}^{t_2} = X_2^{t_2} \tag{9-36}$$

$$C_3^{t_2} = (X_2Y_2)^{t_2}$$

$$J_{Y3}{}^{t_2} = K_{Y3}{}^{t_2} = (X_3 + X_2Y_2)^{t_2} \tag{9-37}$$

$$C_4^{t_2} = (X_3Y_3 + Y_3X_2Y_2)^{t_2}$$

$$J_{Y4}{}^{t_2} = K_{Y4}{}^{t_2} = (X_4 + X_3Y_3 + Y_3X_2Y_2)^{t_2} \tag{9-38}$$

$$C_5^{t_2} = (X_4Y_4 + Y_4X_3Y_3 + Y_4Y_3X_2Y_2)^{t_2}$$

. . . .

$$J_{Y20}{}^{t_2} = K_{Y20}{}^{t_2} = (X_{20} + C_{20})^{t_2} \tag{9-39}$$

In order to complete the design for the adder, we must now combine the equations which are true at t_1 time with those which must be effective at t_2 time. The complete logic is indicated in equations 9-40 through 9-44.

$$\begin{aligned} J_{Y1} &= K_{Y1} = X_1 t_1 \\ J_{X1} &= 0 \qquad K_{X1} = t_1 \end{aligned} \tag{9-40}$$

$$\begin{aligned} J_{Y2} &= K_{Y2} = X_2 t_1 + X_2 t_2 \\ J_{X2} &= X_1 Y_1 t_1 \qquad K_{X2} = \overline{X}_1 t_1 + \overline{Y}_1 t_1 \end{aligned} \tag{9-41}$$

$$\begin{aligned} J_{Y3} &= K_{Y3} = X_3 t_1 + X_3 t_2 + X_2 Y_2 t_2 \\ J_{X3} &= X_2 Y_2 t_1 \qquad K_{X3} = \overline{X}_2 t_1 + \overline{Y}_2 t_1 \end{aligned} \tag{9-42}$$

$$\begin{aligned} J_{Y4} &= K_{Y4} = X_4 t_1 + X_4 t_2 + X_3 Y_3 t_2 + Y_3 X_2 Y_2 t_2 \\ J_{X4} &= X_3 Y_3 t_1 \qquad K_{X4} = \overline{X}_3 t_1 + \overline{Y}_3 t_1 \end{aligned} \tag{9-43}$$

. . . .

$$\begin{aligned} J_{Yi} &= K_{Yi} = X_i t_1 + X_i t_2 + C_i t_2 \\ J_{Xi} &= X_{i-1} Y_{i-1} t_1 \qquad K_{Xi} = \overline{X}_{i-1} t_1 + \overline{Y}_{i-1} t_1 \end{aligned} \tag{9-44}$$

Except for the K input to flip-flop X_1, all of these equations are derived simply by combining the logic of equations 9-29 and 9-30 with that of equations 9-33 and 9-34. The X_1 flip-flop is a special problem and must be reset to "zero" at t_1 time so that it does not contribute a "one" to the least significant bit position of the sum.

Comparing these with equations 9-24 through 9-27, we see that a remarkable simplification has been achieved. One measure of this simplification is the fact that the carry input into the twentieth bit position for this adder requires only about 200 diodes when mechanized as a second-order function, where the carry for the one bit-time adder would require more than 20 million diodes. The input equations to the Y flip-flops exhibit an even more striking simplicity. Note, for example, that nowhere is it necessary to form the complement of the carry C_i.

The problem of mechanizing this two-bit-time parallel adder is much simpler than that for the faster adder described in the previous section. The carry function is still of course somewhat complicated, and because of the delay in complex gating circuits there is still a carry propagation time which limits the speed of addition for a given set of circuit components. However, the carry and therefore the input logic to the Y equations is simpler by an order of magnitude than that for the one-bit-time adder, and this simplicity may be worth the additional bit-time required.

A serial, decimal adder employing the excess-three code. A completely general approach to the design of a serial decimal adder—an approach which will be effective no matter what binary code is chosen for the decimal digits—may be described as follows. Suppose we want to add two decimal numbers X and Y. We first arrange to shift the four (or more) bits representing one decimal digit of X into flip-flops X_1, X_2, X_3, and X_4 and at the same time to shift the corresponding Y digit into four flip-flops Y_1, Y_2, Y_3, and Y_4. We have also available another flip-flop, C, which contains the decimal carry generated from the previous addition of two decimal digits. We now draw up a truth table having 200 entries corresponding to the ten possible digits of X, the ten possible digits of Y, and the two possible values of the carry bit from the previous digit addition. The left-hand part of this table, which represents the state of the registers at a particular bit-time, will contain nine columns: four each for the X and Y digits and one for the carry digit. The right-hand side of the table will contain five columns, four of them representing the code for the sum digit, and the fifth representing the new value of the decimal carry. From this truth table we can write a set of application equations and input equations for the flip-flops used to produce the new sum, and for the carry flip-flop.

For some decimal codes, an approach of this kind may be the only possible way to solve the design problem. However, for many decimal adders, it is possible to determine the decimal sum digit without having all of the bits of both decimal digits available simultaneously. In this section and the one that follows, two very simple adders employing commonly used codes will be designed.

One of the easiest adders to design and to understand operates on numbers represented in the excess-three code shown in Table 9-1. In this code, each decimal digit is represented by a binary number three greater than that decimal digit. The excess-three code has the advantage that, when an *ordinary binary addition* is performed on two decimal digits expressed in this code, their binary sum produces a carry bit from the most significant bit position under exactly those circumstances that a decimal carry is required. This feature is illustrated below

	1983	0 1 0 0	1 1 0 0	1 0 1 1	0 1 1 0
	1824	0 1 0 0	1 0 1 1	0 1 0 1	0 1 1 1
Sum	3807	1 0 0 1	1 0 0 0	0 0 0 0	1 1 0 1
Carry	110	1 0 0 1	1 1 1 1	1 1 1 0	1 1 0

where the two numbers 1983 and 1824, each represented in the excess-three code, are added together as if they were two binary numbers. Examining the carries first, and comparing them with those of the decimal addition shown on the left, we note that when the decimal carry is "zero" the carry entering the least significant bit of a decimal digit is "zero"; and when the decimal carry is "one" the corresponding binary carry is "one." Contrast this with the example below where the same two numbers are added in the same way except that the decimal digits are represented not by an excess-three code but by the ordinary 8-4-2-1 code for decimal digits.

	1983	0 0 0 1	1 0 0 1	1 0 0 0	0 0 1 1
	1824	0 0 0 1	1 0 0 0	0 0 1 0	0 1 0 0
Sum	3807	0 0 1 1	0 0 0 1	1 0 1 0	0 1 1 1
Carry	110	0 0 1 **1**	0 0 0 **0**	0 0 0 **0**	0 0 0

Examining the carries here, we see that two of them are correct as decimal carries, but the third is a "zero" where it should be a "one."

If we look at the four decimal sum digits in each of the above examples, we find of course that they are not correct for either addition. We may summarize by saying that if an ordinary binary adder is to be used as the basis for a decimal adder, we need only provide the logic necessary for correcting the sum digit if we are using an excess-three code, whereas we must modify both the sum logic and the carry logic if we are employing the 8-4-2-1 binary code.

We can best explain the mechanism for the correct decimal carry, and the rules which must be employed to correct the sum digit, by proceeding with the design of this serial, decimal, excess-three adder. Referring to Figure 9-11, suppose we want to form the sum of two

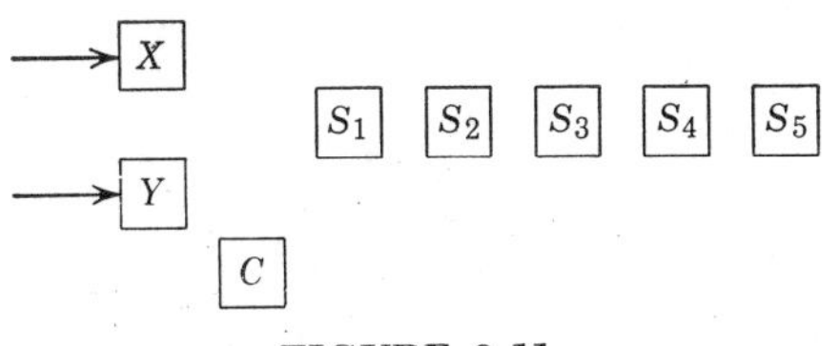

FIGURE 9-11

decimal numbers which are appearing serially least significant bit first in flip-flops X and Y. Suppose further that we provide an ordinary binary adder to add these two numbers, putting their sum, bit by bit, in flip-flop S_1 and using flip-flop C as the carry flip-flop. As each binary bit of the sum is formed in flip-flop S_1, it replaces the last sum bit formed, which is shifted into flip-flop S_2. At the same time, the contents of S_2 shift into S_3, and so on down to flip-flop S_5. The decimal sum is to be read from flip-flop S_5. For example, supposing that the two decimal digits 7 and 5 are to be added, the sequence of events given in Table 9-9 takes place.

Table 9-9

	X	Y	C	S_1	S_2	S_3	S_4	S_5
t_1	0	0	0					
t_2	1	0	0	0				
t_3	0	0	0	1	0			
t_4	1	1	0	0	1	0		
t_1			1	0	0	1	0	

Note that the binary representation of ten (1010), which is the excess-three code for 7, appears serially in flip-flop X from t_1 to t_4 time. Similarly, the binary code for 8 (1000), which is the excess-three representation for 5, appears in flip-flop Y. The sum of these two binary numbers is $10 + 8 = 18$, and this binary sum appears in flip-flops C, S_1, S_2, S_3, and S_4 at the second t_1 time, when the least significant bits of the next two decimal digits are in flip-flops X and Y. At this bit-time, and at every t_1 time, we must correct this sum digit. The correction is carried out on the digit in flip-flops S_1–S_4 at the same time as a shift of one bit position to the right is accomplished. Thus at t_2 time, the corrected decimal digit appears in flip-flops S_2–S_5 and the least significant bit of the new uncorrected sum digit appears in flip-flop S_1. Referring back to Table 9-9, we know that the sum digit 2 must result from adding 7 and 5, and that we therefore must arrange for the formation of the excess-three code for 2 (0101) in S_2–S_5. If at the second t_1 time flip-flops X and Y happen to contain "ones," the operation shown in Table 9-10 must take place.

Table 9-10

	X	Y	C	S_1	S_2	S_3	S_4	S_5
t_1	1	1	1	0	0	1	0	
t_2			1	1	0	1	0	1

In Table 9-11 we show the contents of flip-flops S_2, S_3, S_4, and S_5 at t_2 time as a function of the sum digit which has been generated by the binary addition of the previous two decimal digits. The first entry in the table, for example, represents the configuration which would occur if two decimal zeros were added together. The excess-three code for

Table 9-11

Actual Sum	Binary Sum (Uncorrected Decimal Sum Digit) t_1					Corrected Decimal Sum Digit t_2			
	C	S_1	S_2	S_3	S_4	S_2	S_3	S_4	S_5
0	0	0	1	1	0	0	0	1	1
1	0	0	1	1	1	0	1	0	0
2	0	1	0	0	0	0	1	0	1
3	0	1	0	0	1	0	1	1	0
4	0	1	0	1	0	0	1	1	1
5	0	1	0	1	1	1	0	0	0
6	0	1	1	0	0	1	0	0	1
7	0	1	1	0	1	1	0	1	0
8	0	1	1	1	0	1	0	1	1
9	0	1	1	1	1	1	1	0	0
10	1	0	0	0	0	0	0	1	1
11	1	0	0	0	1	0	1	0	0
12	1	0	0	1	0	0	1	0	1
13	1	0	0	1	1	0	1	1	0
14	1	0	1	0	0	0	1	1	1
15	1	0	1	0	1	1	0	0	0
16	1	0	1	1	0	1	0	0	1
17	1	0	1	1	1	1	0	1	0
18	1	1	0	0	0	1	0	1	1
19	1	1	0	0	1	1	1	0	0

each of these zeros is 0011, and their ordinary binary sum is therefore 0110. The corrected sum digit, zero, which must appear in flip-flops S_2–S_5 at t_2 time, must of course be in the excess-three code again, and must therefore be 0011, as indicated in the right-hand half of the first row of Table 9-11. The remainder of this table shows the other nineteen possible sums formed from the addition of two decimal digits with a decimal carry taken into consideration. Note that when the actual sum is ten or greater the binary sum is sixteen or greater, and a "one" appears in the carry flip-flop, C. *Thus the carry which appears in C at t_1 time is the correct carry to be used with the binary adder logic in forming the next uncorrected decimal sum digit.*

The logic for flip-flops C and S_1 must therefore be the same as the ordinary binary adder logic of equations 9-14 and 9-15. Table 9-11 is a truth table for flip-flops S_2, S_3, S_4, and S_5 and from this table application equations and input logic for these flip-flops may be written. This logic (effective at t_1 time), together with the logic necessary to perform the shifting operations at all bit-times except t_1, is shown in equations 9-45 through 9-48.

$$\begin{aligned} J_{S2} &= t_1(\bar{C}S_3S_4 + CS_1) + \bar{t}_1S_1 \\ K_{S2} &= t_1(\bar{C}\bar{S}_1 + C\bar{S}_3\bar{S}_4) + \bar{t}_1\bar{S}_1 \end{aligned} \tag{9-45}$$

$$\begin{aligned} J_{S3} &= t_1[\bar{C}\bar{S}_2 + C(\bar{S}_2S_4 + S_2\bar{S}_4)] + \bar{t}_1S_2 \\ K_{S3} &= t_1[\bar{C}(\bar{S}_2S_4 + S_2\bar{S}_4) + CS_2] + \bar{t}_1\bar{S}_2 \end{aligned} \tag{9-46}$$

$$\begin{aligned} J_{S4} &= t_1(\bar{C}S_3 + C\bar{S}_3) + \bar{t}_1S_3 \\ K_{S4} &= t_1(\bar{C}S_3 + C\bar{S}_3) + \bar{t}_1\bar{S}_3 \end{aligned} \tag{9-47}$$

$$\begin{aligned} J_{S5} &= t_1\bar{S}_4 + \bar{t}_1S_4 \\ K_{S5} &= t_1S_4 + \bar{t}_1\bar{S}_4 \end{aligned} \tag{9-48}$$

The equations can be reduced in part, but they essentially comprise a complete description of the excess-three adder.

The decimal adder which has been designed in this section requires six flip-flops and imposes a delay of five bit-times between the occurrence of the first bits of an augend and addend and the appearance of the first bit of their sum. A little investigation of the operation of this adder would disclose that the design could be simplified considerably through the elimination of some of these flip-flops, and that the delay time could be similarly shortened. However, rather than modify and improve on the operation of this adder, let us design an 8-4-2-1 code decimal adder, employing the Huffman-Mealy design procedure to insure that we use no more flip-flops than are necessary.

A serial, binary-coded decimal adder. Suppose now that two decimal numbers represented in the 8-4-2-1 decimal code of Table 9-1 appear serially from flip-flops X_1 and X_2. Suppose we begin by attempting to design an adder using the same principle employed in the last section in the design of the excess-three adder. That is to say, we perform an ordinary binary addition on the two numbers appearing serially in X_1 and X_2, storing the binary sum in flip-flop S_1 and the binary carry in flip-flop C. If, for example, we add the two decimal digits 7 and 5, the addition appears as shown in Table 9-12. We hope, then, to correct the binary sum, which appears in flip-flops S_1–S_4 at t_1 time, by performing a shift and correction at t_1 time. We should then be able to obtain the corrected decimal sum digit at t_2 time in S_2–S_5.

Unfortunately, examination of Table 9-12 reveals a difficulty: although in the particular example chosen, the sum is 12 and there should be a decimal carry added to the next pair of decimal digits coming into flip-flops X_1 and X_2, we see that flip-flop C is in the "zero" state at the second t_1 time. We must therefore take some action at t_4 time to insert the proper decimal carry into flip-flop C. Notice that at t_4 time the three least significant bits of the uncorrected binary sum are in flip-flops S_1, S_2, and S_3, and that flip-flops X_1, X_2,

Table 9-12

	X_1	X_2	C	S_1	S_2	S_3	S_4
t_1	1	1	0				
t_2	1	0	1	0			
t_3	1	1	1	0	0		
t_4	0	0	1	1	0	0	
t_1			0	1	1	0	0

and C supply all the information necessary to determine the complete binary sum. Knowing the complete binary sum, we can certainly determine whether or not there will be a decimal carry and insert that carry in flip-flop C. We can also determine the *decimal* sum digit from the same data. Therefore, at t_4 time, instead of carrying out an ordinary binary addition as we did above during bit-times t_1, t_2, and t_3, let us, by examining flip-flops X_1, X_2, C, S_1, S_2, and S_3, generate a decimal carry and the appropriate decimal digit of the sum and place the result in flip-flops C, S_1, S_2, S_3, and S_4. We may then proceed at t_1 time with the binary addition we carried out during the previous t_1 bit-time. This modified mode of operation is illustrated in Table 9-13 for the addition of 7 and 5.

Table 9-13

	X_1	X_2	C	S_1	S_2	S_3	S_4
t_1	1	1	0				
t_2	1	0	1	0			
t_3	1	1	1	0	0		
t_4	0	0	1	1	0	0	
t_1			1	0	0	1	0
t_2					0	0	1
t_3						0	0
t_4							0

The mechanization shown in Table 9-13 permits us to begin reading the least significant bit of the sum digit from flip-flop S_4 at t_1 time, exactly four bit-times after the first bits of the augend and addend digits appear. We may now ask the question: Is it possible to read the sum any earlier than t_1 time? We see at once from Table 9-13 that the answer is yes. At t_4 time the least significant bit of the binary sum (which is always the same as the least significant bit of the decimal sum digit) may be read from flip-flop S_3, with the other bits of the sum following in flip-flop S_3 at t_1, t_2, and t_3 times. We can therefore eliminate flip-flop S_4, and obtain the same result one bit-time earlier from flip-flop S_3.

At first glance, it is not apparent how we can improve on this. Certainly if we look at flip-flop S_2 at t_3 time we shall see the correct least significant bit of the decimal sum digit. The proper third and fourth bits of the decimal sum appear in flip-flop S_2 at t_1 and t_2 times as well. However, at t_4 time flip-flop S_2 contains, not the proper bit of the decimal sum digit, but a bit which is part of the binary sum. Fortunately there is a very simple way around this difficulty. Although we have not yet determined the decimal sum at t_4 time, we have available all the information for generating it. In particular, as was mentioned above, we have in flip-flops X_1, X_2, C, S_1, and S_2 all the data necessary to determine what the entire decimal sum bit should be at that bit-time. We can therefore solve the problem by arranging that the output of the adder is equal to flip-flop S_2 at t_3, t_1, and t_2 times, but that it is some function of X_1, X_2, C, S_1, and S_2, at t_4 time. This mode of operation is indicated in Table 9-14.

Table 9-14

	X_1	X_2	C	S_1	S_2	Output
t_1	1	1	0			
t_2	1	0	1	0		
t_3	1	1	1	0	0	$0 = S_2$
t_4	0	0	1	1	0	$1 = f(X_1, X_2, C, S_1, S_2)$
t_1			1	0	0	$0 = S_2$
t_2					0	$0 = S_2$

The rules defining the operation of the adder may now be summarized as follows. During t_1, t_2, and t_3 times the binary sum of X_1 and X_2 is formed and transferred to S_1, employing flip-flop C as a carry flip-flop. At each of these bit-times, the contents of S_1 are shifted into S_2, and the output of flip-flop S_2 is the output of the adder. At time t_4 the decimal carry is formed in flip-flop C, the two most significant bits of the decimal sum are formed and put in flip-flops S_1 and S_2, and the third bit of the decimal sum digit is read out of the adder directly.

In order to be certain that we do not provide any more states than are absolutely necessary, and to obtain a convenient starting point from which flip-flop equations may be written, let us complete the design of this decimal adder by drawing up a Huffman-Mealy table. We shall assume that a black box contains three flip-flops (C, S_1, and S_2) and has three inputs (t_4, X_1, and X_2). By following carefully the rules which were stated in the previous paragraph, we can draw up Table 9-15. Rather than show the complete derivation of this table, we shall

Table 9-15

Present State				Next State								Output $= z$							
			$t_4 \rightarrow$	0	0	0	0	1	1	1	1	0	0	0	0	1	1	1	1
C	S_1	S_2	$X_1, X_2 \rightarrow$	00	01	10	11	00	01	10	11	00	01	10	11	00	01	10	11
0	0	0	a	a	c	c	e	a	c	c	f	0	0	0	0	0	0	0	1
0	0	1	b	a	c	c	e	a	e	e	g	1	1	1	1	1	0	0	0
0	1	0	c	b	d	d	f	b	e	e	–	0	0	0	0	0	1	1	–
0	1	1	d	b	d	d	f	b	f	f	–	1	1	1	1	1	0	0	–
1	0	0	e	c	e	e	g	c	f	f	–	0	0	0	0	0	1	1	–
1	0	1	f	c	e	e	g	e	–	–	–	1	1	1	1	0	–	–	–
1	1	0	g	d	f	f	h	e	–	–	–	0	0	0	0	1	–	–	–
1	1	1	h	d	f	f	h	f	–	–	–	1	1	1	1	0	–	–	–

	1	2	3	4	5
Rule II:	(a)	(b, d)	(c, e)	(f, h)	(g)
		1 2	2 3	3 2	
		3 2	2 3	3 4	
		. .	. .	. .	
		. .	. .	. .	
		. .	. .	. .	

derive three particular entries on it. Suppose first it is not t_4 time, and that C and S_2 are both "zero" while S_1 is "one." The circuit will then be in state c. If now X_1 and X_2 are both "one," then the sum of X_1, X_2, and C is two. We must therefore put a carry in flip-flop C and a "zero" in flip-flop S_1. At the same time we must shift the "one" in flip-flop S_1 into S_2. We thus go from state c to a state in which C and S_2 are "one" while S_1 is a "zero," i.e., to state f. Therefore in the "next state" entry on line c in the column $t_4 = 0$, $X_1 = X_2 = 1$, we find the entry f. Finally, since it is not t_4 time, the adder output, z, is the same as S_2 and the "output" entry on line c in the column $t_4 = 0$, $X_1 = X_2 = 1$ is "zero."

Next consider the situation when C, S_1, and S_2 are in the states 001. This corresponds to row b. Suppose further that it is t_4 time and that X_1 and X_2 are again both equal to "one." Since it is t_4 time, the two "ones" in flip-flops X_1 and X_2 each have weights of 8, and their sum is 16. The "one" in flip-flop S_2 has a weight 2, making the total sum 18. We therefore require a decimal carry, so that flip-flop C must change to the "one" state. We also require a sum digit of 8, and the first two bits of the code for 8 are "one" and "zero." Therefore we must next go to a state in which C is "one," and S_1 and S_2 are "zero" and "one" respectively—state g. A g is therefore entered on row b in column $t_4 = 1$, $X_1 = X_2 = 1$. The third digit of the code for 8 is also a "zero" and must be the output of the circuit at that bit-time. Therefore, on line b in the output column where $t_4 = 1$ and $X_1 = X_2 = 1$, we find "zero."

Finally let us examine the situation which exists at t_4 time when X_1 is "zero" and X_2 is "one" and the circuit is in the f state. In the f state, C and S_2 are both "one" while S_1 is "zero." The binary sum which has been generated up to this point is therefore either ten or eleven, for the bit in C represents an 8 and the bit in S_2 a 2. (The least significant bit of the sum, which was read out at t_3 time, may have been either "zero" or "one.") Now if $X_2 = 1$, one of the decimal digits being added must be either 8 or 9; and since $X_1 = 0$ the other must be 7 or less. If the digit appearing from X_1 is a 7, and if that from X_2 is a 9, and if there had been a decimal carry from the addition of the previous two decimal digits, then the total binary sum at t_4 time could not be greater than 9, as shown below.

$$\begin{array}{cccc l}
 & & & 1 & = \text{decimal carry} \\
 & 1 & 1 & 1 & = X_1 \\
 & 0 & 0 & 1 & = X_2 \\
\hline
1 & 0 & 0 & 1 &
\end{array}$$

We have thus shown that (1) when the circuit is in state f, the binary sum accumulated by t_4 time is either ten or eleven, and (2) if $X_1 = 0$ and $X_2 = 1$ at t_4 time, the accumulated binary sum cannot be greater than 9. We therefore conclude that X_1 and X_2 cannot be "zero" and "one" at t_4 time with the circuit in state f, and put a dash in the corresponding squares in the "next state" and "output" tables.

With the chart of Table 9-15 complete, we may apply the Huffman-Mealy rules I and II in order to determine whether any superfluous states exist. The application of rule II is indicated at the bottom of Table 9-15, where it is easily seen that no reduction is possible.

We are now ready to derive the logic for this adder. We may assign any code we like to the eight states a through h, and some one particular assignment will undoubtedly result in a smaller number of diodes than any other. As has been indicated before, however, the problem of finding the best assignment is a very difficult one. The assignment which we started out with, and which is indicated at the left-hand side of Table 9-15, is likely to be a fairly simple one because of the simplicity of the binary carry and of the shifting operations involved. Employing techniques already described and discussed, it is possible to draw up Veitch diagrams showing the application equations for each of the three flip-flops involved, and to derive the input logic directly from these Veitch diagrams. That logic together with the logic for the output line z is given below.

$$\begin{aligned} J_C &= X_1X_2 + t_4X_1S_1 + t_4X_2S_1 + t_4X_1S_2 + t_4X_2S_2 \\ K_C &= \bar{t}_4\bar{X}_1\bar{X}_2 + \bar{X}_1\bar{X}_2\bar{S}_1\bar{S}_2 \end{aligned} \qquad (9\text{-}49)$$

$$\begin{aligned} J_{S1} &= X_1X_2C + \bar{t}_4\bar{X}_1\bar{X}_2C + \bar{t}_4\bar{X}_1X_2\bar{C} + \bar{t}_4X_1\bar{X}_2\bar{C} \\ &\quad + \bar{X}_1\bar{X}_2\bar{S}_2C + t_4X_1X_2S_2 + X_1\bar{X}_2\bar{S}_2\bar{C} + \bar{X}_1X_2\bar{C}\bar{S}_2 \end{aligned} \qquad (9\text{-}50)$$

$$K_{S1} = t_4 + \bar{X}_1X_2C + X_1\bar{X}_2C + X_1X_2\bar{C} + \bar{X}_1\bar{X}_2\bar{C}$$

$$\begin{aligned} J_{S2} &= \bar{t}_4S_1 + t_4X_1X_2 + \bar{X}_1\bar{X}_2S_1\bar{C} + t_4X_1C + t_4X_2C \\ K_{S2} &= \bar{S}_1 \end{aligned} \qquad (9\text{-}51)$$

$$\begin{aligned} z &= \bar{t}_4S_2 + t_4S_1\bar{S}_2C + t_4X_1C + t_4X_2C + t_4X_1X_2\bar{S}_2 \\ &\quad + t_4X_1S_1\bar{S}_2 + t_4X_2S_1\bar{S}_2 + \bar{C}S_2\bar{X}_1\bar{X}_2 \end{aligned} \qquad (9\text{-}52)$$

Overflow. When we add two numbers on a scrap of paper we can always contrive to find room for our answer even if that answer contains more digits than do either of the two numbers we are adding. However, in a digital computer the length of a word is limited by the number of memory elements the designer provides for a word (in a parallel computer) or the number of bit-times assigned to a word (in a serial computer). Therefore, if the computer operator is not careful, he may add two large, positive numbers together and obtain an answer smaller than either of them. For example, in a machine having a maximum capacity of five decimal digits per word, the sum of 90,000 and 80,000 would be 70,000 rather than 170,000, for there would be no place to store the "one." When this happens, the arithmetic unit is said to *overflow*.

The computer operator is required to take considerable pains to organize his calculations so that no sums bigger than can be stored in one computer word ever occur, or so that such sums are detected and accounted for when they do occur. The computer designer is often called upon to provide a special facility which will warn the operator when an overflow has taken place. This warning may take different forms in different machines, but it is obvious that its mechanization is fairly simple. The important thing to be done is to observe the carry which results from the addition of the two most significant bits or digits of a word. If the carry is a "one," overflow has occurred; if a "zero," the computer operator may go on. In a serial machine, it is only necessary to look at the contents of the carry flip-flop at the end of the word. In a parallel machine, a special carry circuit (exactly analogous to the other, less significant carry lines) must be formed and the resulting carry stored somewhere at the same bit-time the sum is formed. The use made of this overflow information depends on the desired operation of the computer being designed. It may be made to stop the computer; it is often required to change the sequence of operations carried out by the computer.

Subtraction

A subtracter may be designed in a way very similar to that employed in designing an adder: The subtrahend and minuend must be lined up so that corresponding digits are opposite one another, and a set of rules may then be devised and carried out to produce a difference digit from these digits. The subtracter designed in the next section is an example of this kind of subtracter.

However, the operation of subtraction brings to light a problem we have not yet mentioned: the problem of representing and of operating on negative numbers. After we have designed a very simple binary subtracter, we shall turn our attention to this important matter.

A serial binary subtracter. Suppose we want to modify equations 9-16 and 9-17 which describe the operation of the adder of Figure 9-5 in such a way that the adder becomes an adder-subtracter. Specifically, suppose that instead of the single control flip-flop A which determined whether Y was to be added to X or whether X was to be circulated on itself, we provide two control signals, A and S, which determine what operation is to be carried out as follows:

AS	Operation
00	Circulate X
10	Add Y to X
11	Subtract Y from X
01	(Never occurs)

We shall carry out the design by first designing a subtracter which will work in the framework of Figure 9-5, and then modifying the logic of equations 9-16 and 9-17 to incorporate this subtracter.

The first step in designing the subtracter is to set up a truth table similar to that of Table 9-5. At any bit-time n, X_1 and Y may take on one of the usual four combinations of two binary digits, and these four combinations are listed below.

Time n	X_1	0	1	1	0
	$-Y$	-0	-0	-1	-1
Time $(n + 1)$	X_w	0	1	0	?

The first three entries here are perfectly simple, requiring only that a digit be subtracted from itself or from something larger than itself. The fourth entry, however, requires that "one" be subtracted from "zero" and this is clearly impossible. We can only do this, as we do in ordinary decimal subtraction, by "borrowing" a "one" from the next significant bit position, which has a weight twice that of the bit position we are examining. The "zero" in flip-flop X_1 may then be regarded as part of a binary two, and the "one" may be subtracted from it. The result is shown below, and the borrow is stored in flip-flop C, where it must be examined in connection with the next two bits appearing in X_1 and Y.

Time n	X_1	10
	Y	$-$ 1
Time $(n + 1)$	X_w	1
	C	1

We thus come to the conclusion that subtraction cannot be performed without reference to a quantity called the *borrow*, which is very similar to the carry of addition. The complete truth table for subtraction therefore involves the use of a flip-flop C, and is shown in Table 9-16. The first four entries in this table correspond to the four combinations of X_1 and Y which have already been discussed, where there is no borrow from the previous bit position. The last four entries may be justified as follows. When both X_1 and C are "one," we accomplish the borrow by canceling out X_1, and the new difference and borrow digits are the same as for the entries in the upper half of the table where X_1 and C are both "zero." When C is "one" and X_1 is "zero," we must borrow one from the next stage of the subtraction ($C^{n+1} = 1$), account for the borrow of bit-time n by reducing X_1 from two to "one," and then subtract Y from the reduced X_1.

Comparing Table 9-16 for subtraction with Table 9-5 which defines the operation of addition, we discover a surprising thing: the sum digit and the difference digit are exactly the same function of X_1, Y, and C. Input equation 9-15, which defines the operation of write amplifier X_w in the adder, will thus be unchanged when subtraction is the function

Table 9-16

n			$n+1$	
X_1	Y	C	C	X_w
0	0	0	0	0
1	0	0	0	1
0	1	0	1	1
1	1	0	0	0
0	0	1	1	1
1	0	1	0	0
0	1	1	1	0
1	1	1	1	1

to be performed. The borrow and carry operations are, however, different, and the logic for flip-flop C in subtraction is easily derived from Table 9-16. This logic is given below, where it can be compared with the corresponding logic for addition.

$$J_C = \overline{X}_1 Y \qquad K_C = X_1 \overline{Y} \tag{9-53}$$

The complete logic for the adder-subtracter may now be formed by combining equations 9-14, 9-15, and 9-53, taking into account the significance of signals A and S given above. The complete equations are:

$$\begin{aligned} J_C &= X_1 Y A \overline{S} \bar{t}_0 + \overline{X}_1 Y S \bar{t}_0 \\ K_C &= \overline{X}_1 \overline{Y} A \overline{S} + X_1 \overline{Y} S + t_0 \end{aligned} \tag{9-54}$$

$$\begin{aligned} J_{Xw} &= A\overline{X}_1\overline{Y}C + A\overline{X}_1 Y\overline{C} + AX_1\overline{Y}\overline{C} + AX_1YC + \overline{A}X_1 \\ K_{Xw} &= AX_1Y\overline{C} + AX_1\overline{Y}C + A\overline{X}_1YC + A\overline{X}_1\overline{Y}\overline{C} + \overline{A}\overline{X}_1 \end{aligned} \tag{9-55}$$

Complements. In the last section we devised some rules to mechanize the subtraction of one positive number from another, and designed a circuit which would perform such a subtraction. Suppose, now, that we employ this subtracter to subtract a binary number from "zero." If the circulating register contains a word seven bits long, for example, and if we subtract the number 22 from "zero," the result obtained in the circulating register at the end of the subtraction is:

X_1	0 0 0 0 0 0 0
Y	0 0 1 0 1 1 0
$X_1 - Y$	1 1 0 1 0 1 0

Notice that, when the subtraction is complete, the borrow flip-flop will contain a "one," but that "one" must be discarded because there are no more digits to subtract from. The result shown is called the *two's complement* of the number Y, and may be used to represent the number $-Y$. In this section, we shall examine the properties of complements, show how they may be used in computer design, and discuss the problem of representing and operating on negative numbers in general.

In the discussions which follow, we shall assume that a number in the computer is represented by n significant digits plus a sign digit. (In the example above $n = 6$, and the most significant bit of each word is the sign digit.) The largest number which can be represented is thus $2^n - 1$, and we shall assume that the magnitude of every number considered is less than or equal to this value. Furthermore, because the decimal number system is the more familiar one to work with, we shall discuss the ten's complement of decimal numbers rather than the two's complement of binary numbers, with the understanding that the ideas presented and the conclusions reached are independent of the number base employed.

The ten's complement of a number, like the two's complement, is formed by subtracting that number from zero, making the assumption that we can always borrow a one even though there is no one to be borrowed. For example, if $n = 4$ we can find the ten's complements of 22 and of 1700 by subtracting each from zero as follows:

$$\begin{array}{r} 0\,0\,0\,0\,0 \\ -\;0\,0\,0\,2\,2 \\ \hline 9\,9\,9\,7\,8 \end{array} \qquad \begin{array}{r} 0\,0\,0\,0\,0 \\ -\;0\,1\,7\,0\,0 \\ \hline 9\,8\,3\,0\,0 \end{array}$$

The ten's complement of a number may thus be found by subtracting that number from a power of ten—specifically from 10^{n+1}, for numbers containing n significant digits and a sign digit. Algebraically,

$$\text{ten's complement } Y = 10^{n+1} - Y \qquad (9\text{-}56)$$

In doing this, we see that if we employ zero as the sign digit of a positive number, then 9 will be the sign digit of a negative one.* We can devise a fairly simple *rule* for finding the ten's complement of a decimal number. The rule may be stated as follows:

> Examine each digit one at a time, starting with the least significant. Copy all least significant zeros unchanged into the complement. Subtract the first nonzero digit from 10, and put the difference in the complement. The remaining digits of the complement are found by subtracting the corresponding digits of the original number from 9.

This rule will be employed later on in designing a serial decimal complementer.

By using equation 9-56 as a definition of the ten's complement of a number, we may prove two very simple theorems which explain the usefulness of the complement. The first theorem states that if we take the ten's complement of the ten's complement of a number we obtain the original number again. This may be shown very easily by employing equation 9-56 twice, as follows

$$\text{ten's complement (ten's complement } Y)$$
$$= \text{ten's complement } (10^{n+1} - Y) = 10^{n+1} - (10^{n+1} - Y) = Y$$

This is the equivalent, of course, of saying that the negative of the negative of a number is the number itself.

The second theorem states that, if negative numbers are represented as ten's complements, we can perform addition without regard to the sign of augend or addend, and be sure that the sum will be correct both in magnitude and in sign. For example, if we add 473 to −22, or −473 to +22, or −473 to −22, we obtain the sums shown below

473	00473	−473	99527	−473	99527
− 22	99978	+ 22	00022	− 22	99978
451	00451	−451	99549	−495	99505

The first result is positive and obviously correct. The second and third examples have negative results, as is indicated by the sign digit of 9. Recomplementing these sums, we see that 99549 corresponds to

* Note that the sign digit requires no more than an on-off signal, and can actually be represented by a single bit, 0 or 1, rather than a decimal digit, 0 or 9. In a decimal computer we would therefore need only one bit of storage for a sign. Nevertheless, for clarity we shall continue to use the 9 in these discussions.

the number -451 and 99505 corresponds to the number -495, which are the correct answers to these two operations.

We can easily prove that addition always gives us the correct results for any pair of numbers X and Y, as follows. If the symbols X and Y each represent positive numbers, and we want to perform the addition $X + (-Y)$, the sum will be represented in a computer by

$$X + (\text{ten's complement of } Y) = X + 10^{n+1} - Y = 10^{n+1} + X - Y$$

We must now consider two possibilities:

1. If $X \geqq Y$, then $X - Y \geqq 0$ and the sum will be a positive number with 10^{n+1} added to it. Since 10^{n+1} is a "one" followed by $(n + 1)$ zeros, it will have no effect on the sum or on the sign digit, and the sum will be correct.
2. If $X < Y$, then the sum will be $10^{n+1} - (Y - X)$, which is the ten's complement of the positive number $(Y - X)$.

We must still consider the addition of $(-X)$ to $(-Y)$. The machine representation of this sum will be

$$10^{n+1} - X + 10^{n+1} - Y = 2 \times 10^{n+1} - (X + Y)$$

The 2 here appears to the left of the sign digit and will not actually appear in the machine result. The sum, therefore, will actually be indistinguishable from $10^{n+1} - (X + Y)$, which is the machine representation for the correct sum: $-(X + Y)$.

In the binary number system the two's complement corresponds to the decimal ten's complement. The two's complement of a number Y may be found as follows

$$\text{two's complement } Y = 2^{n+1} - Y \tag{9-57}$$

where n is the number of significant digits in the numbers being employed. The sign digit for a negative number represented as a two's complement is a "one," and three typical additions are now carried out. Note that $n = 5$.

−22	101010	+22	010110	−22	101010
+ 8	001000	− 8	111000	− 8	111000
−14	110010	+14	001110	−30	100010

Another form of complement often used in computer design is closely related to the ten's and two's complements described above. This second kind of complement is called the *nine's* or *one's* complement, and in the decimal system is found simply by subtracting each digit of the number to be complemented from nine. The nine's complement of 22 and of 1,700 are thus 99,977 and 98,299 respectively. Clearly, the nine's complement of a number Y is formed by subtracting that number from $(10^{n+1} - 1)$ instead of simply from 10^{n+1}. Thus

$$\text{nine's complement of } Y = 10^{n+1} - Y - 1 \tag{9-58}$$

and it is apparent that

$$(\text{nine's complement of } Y) + 1 = (\text{ten's complement of } Y) \tag{9-59}$$

It is now possible to devise parallels to the two theorems mentioned above for the ten's complement. The first theorem has an exact counterpart when applied to the nine's complement: the nine's complement of the nine's complement of a number is the number itself. The second theorem must be modified slightly if it is to apply to the nine's complement. It may be stated as follows. We can add numbers without regard to their signs if negative numbers are represented by their nine's complement and if we add "one" to the least significant end of the sum whenever a carry is generated out of the sign digit of the sum. This rule for correcting the sum when an addition is performed using the nine's complement is often called the *end-around carry*. We give three examples of the application of this rule:

473	00473	−473	99526	−473	99526
− 22	99977	+ 22	00022	− 22	99977
451	①00450	−451	99548	−495	①99503
	↳ +1				↳ +1
	00451				99504

The proof that this small correction actually results in the proper answer for all possible numbers X and Y may be carried out as follows. If the symbols X and Y each represent positive numbers, and we want to perform the addition $X + (-Y)$, the sum will be represented in the computer by

$$\begin{aligned} X + (\text{nine's complement of } Y) &= X + 10^{n+1} - Y - 1 \\ &= 10^{n+1} + (X - Y) - 1 \end{aligned}$$

We again must consider two possibilities:

1. If $X \geqq Y$, then $X - Y \geqq 0$ and the sum will be a positive number with 10^{n+1} added to it. This 10^{n+1} will appear as a carry out of

the sign position, and must therefore be added to the sum, forming

$$[10^{n+1} + (X - Y) - 1] + 1$$

which is the correct sum, $(X - Y)$.

2. If $X < Y$, then the sum will be $10^{n+1} - (Y - X) - 1$, which is the nine's complement of the positive number $(Y - X)$.

We again must discuss the addition of $(-X)$ to $(-Y)$. The machine representation of this sum will be

$$10^{n+1} - X - 1 + 10^{n+1} - Y - 1 = 2 \times 10^{n+1} - (X + Y) - 2$$

Now

$$10^n - 1 \geqq X + Y \geqq 0$$

Hence the machine representation for the sum must lie between $2 \times 10^{n+1} - 2 = 19999 \cdots 8$ and $2 \times 10^{n+1} - (10^n - 1) - 2 = 19 \times 10^n - 1 = 18999 \cdots 9$. The "one" at the beginning of each of these numbers (and of all numbers in between) is the carry out of the sign digit position, and must therefore be added to the sum, forming,

$$[2 \times 10^{n+1} - (X + Y) - 2] + 1$$

which is the correct sum, in nine's complement form.

The one's complement bears the same resemblance to the nine's complement as the two's complement does to the ten's complement. The three additions carried out above for the two's complement are repeated below, with negative numbers represented in their one's complement form.

−22	101001	+22	010110	−22	101001
+ 8	001000	− 8	110111	− 8	110111
−14	110001	+14	①001101	−30	①100000
			→ +1		→ +1
			001110		100001

We are now in a position to discuss possible methods for representing positive and negative numbers in a computer, and can state some of the rules which may be employed to carry out arithmetic operations. We have seen that it is possible to represent a negative number by means of a complement; it is also obviously possible to represent such a number by storing its magnitude together with a digit or bit which represents its sign. For example, the number −22 may be stored either in two's complement form (101010) or as magnitude and sign with the most significant bit a "one," indicating that the number is negative (110110). It is possible to design a computer in which negative numbers are represented only by a magnitude and sign, only by means of complements, or else by some combination of the two methods. In Table 9-17 three of these methods are shown, together with the rules

Table 9-17

Method of Representing Negative Numbers	Rules for Addition	Rules for Subtraction
Magnitude and sign	1. Compare signs of numbers to be added.	1. Compare signs of the minuend and subtrahend.
	2. If signs are the same, add the two numbers and affix their common sign to the sum.	2. If signs are different, add the two numbers and affix the sign of the minuend to the sum.
	3. If signs are different, compare the magnitudes of the two numbers, subtract the smaller from the larger, and affix the sign of the larger to the difference.	3. If signs are the same, compare the magnitudes of the two numbers, subtract the smaller from the larger, and affix their *common* sign to the difference if the minuend is larger, the *opposite* sign if the subtrahend is larger.
Complements	Add the two numbers.	Subtract the subtrahend from the minuend, or complement the subtrahend and add the complement to the minuend.
Magnitude and sign in the memory, complements in the arithmetic unit	1. Complement negative numbers as received from the memory.	1. Complement negative numbers as received from the memory.
	2. Add the two numbers.	2. Subtract the subtrahend from the minuend.
	3. If the sum is negative (in complement form), recomplement it and store in the memory as magnitude and negative sign.	3. If the difference is negative (in complement form), recomplement it and store in the memory as magnitude and negative sign.

which must be mechanized for carrying out addition and subtraction. Although the table emphasizes the simplicity of the complement method compared with the other two, it should be noted that the use of complements in computer operations other than addition and subtraction leads to some difficulties which do not occur when one of the other two methods is used. For example, multiplication and division require that certain modifications and corrections be performed when negative operands appear in complemented form; these difficulties are reduced if numbers are represented as magnitude and sign, for multiplications can then always be carried out on positive numbers and the correct sign can be appended to the result. Similarly, input and output operations are usually carried on with numbers represented as magnitude plus sign, for that is the form in which they are most easily understood by the operator. If the machine represents numbers in complement form, it is necessary either for the designer or for the computer user to arrange that negative numbers be recomplemented before they are read out of the computer, and that incoming negative numbers be complemented before they are stored.

A serial, binary-coded decimal complementer. The difficulties arising in the design of a complementer depend upon the code being used and upon the kind of complement desired. It is very easy to see that the one's complement of a binary number may be found very simply by replacing all "zeros" with "ones" and all "ones" with "zeros." Thus, if a binary number is stored in parallel form in a series of flip-flops, the number itself may be read from the Q outputs of these flip-flops, and its one's complement may be read from the $\overline{Q}$ outputs. The nine's complement of a decimal number is similarly very easily found by subtracting each digit from 9; and for some decimal codes (see, e.g., the last three weighted decimal codes and the excess-three code in Table 9-1) the nine's complement of a decimal digit may be found simply by complementing each of the binary digits in the decimal digit.

The ten's and two's complements are, generally speaking, somewhat more difficult to mechanize. Equation 9-59 indicates one possible mechanization: the ten's (or two's) complement of a number may be found by adding unity to the nine's (or one's) complement of that number. Because of the relative ease with which the nine's or one's complement can be found, this mechanization is a natural one for some arithmetic units.

In this section, we shall design a ten's complementer which implements the rule stated in the last section. That is, we shall devise a circuit which examines the decimal digits appearing serially from a flip-flop X, copies all least significant zeros unchanged, finds the ten's complement of the first nonzero digit, and the nine's complement of all subsequent digits. We begin by drawing up Table 9-18 which

Table 9-18

	X				$10 - X$				$9 - X$			
	t_4	t_3	t_2	t_1								
0	0	0	0	0	0	0	0	0	1	0	0	1
1	0	0	0	1	1	0	0	1	1	0	0	0
2	0	0	1	0	1	0	0	0	0	1	1	1
3	0	0	1	1	0	1	1	1	0	1	1	0
4	0	1	0	0	0	1	1	0	0	1	0	1
5	0	1	0	1	0	1	0	1	0	1	0	0
6	0	1	1	0	0	1	0	0	0	0	1	1
7	0	1	1	1	0	0	1	1	0	0	1	0
8	1	0	0	0	0	0	1	0	0	0	0	1
9	1	0	0	1	0	0	0	1	0	0	0	0

shows the 8-4-2-1 code and indicates the ten's and nine's complement of each of the ten decimal digits. Note that the ten's complement of the decimal digit zero is itself zero, satisfying that part of the rule which states that all least significant decimal zeros in the number being complemented must be copied without change. Examination of the table reveals at once that, although the least significant bit of the complement digit depends only on the least significant bit of the digit being complemented, the second bit is not determined until the *most* significant bit of the digit being complemented has appeared. For example, if we read "zeros" at t_1 time and at t_2 time, and if X contains a "zero" at t_3 time, we know that the decimal digit which is to be complemented must be either zero or 8, but we do not know which it is until t_4 time. Since the second least significant bit of the ten's complement will be a "zero" if the decimal digit is zero and a "one" if the decimal digit is 8, we clearly cannot read that bit from the complementer until t_4 time.

Let us therefore plan that the least significant bit of the complement will appear from the complementing circuit at t_3 time, and that the remaining bits will appear at t_4, t_1, and t_2 times. There will thus be a delay of two bit-times between the appearance of a number and the appearance of its complement. If we assume that timing signals for the four bit-times are available to the logical designer, we begin by drawing up a Huffman-Mealy diagram for the complementer as shown in Table 9-19. We assume first that the complementer will be in state a at t_1 time of the first decimal digit in the number to be complemented, and at the beginning of all other least significant digits up to and including the first digit which is not "zero." We then assume that the. complementer has eight states (a through h) corresponding to the eight possible combinations of the binary digits which arrive at t_1, t_2, and t_3 times, and we arrange changes of state so that at t_4 time the complementer is in state a if the three previous digits were 000, in

Table 9-19

Present State	Next State				Output				Remarks			
	Time → t_1	t_2	t_3	t_4	t_1	t_2	t_3	t_4		t_2	t_3	t_4
	X → 0 1	0 1	0 1	0 1	0 1	0 1	0 1	0 1				
a	a b	a c	a e	a i	0 0	0 0	0 0	0 1		0	0 0	0 0 0
b		b d	b f	q i		0 0	1 1	0 0		1	0 1	0 0 1
c			c g	q			0 0	0			1 0	0 1 0
d			d h	s			1 1	1	ten's comple-		1 1	0 1 1
e				s				1	ment			1 0 0
f				s				0				1 0 1
g				s				0				1 1 0
h				i				1				1 1 1
i	i j	i k	i m	q i	0 0	0 0	1 1	0 0		0	0 0	0 0 0
j		j l	j n	q i		0 0	0 0	0 0		1	0 1	0 0 1
k			k o	s			1 1	1			1 0	0 1 0
l			l p	s			0 0	1			1 1	0 1 1
m				s				0				1 0 0
n				s				0	nine's comple-			1 0 1
o				i				1	ment			1 1 0
p				i				1				1 1 1
q	q r	i k			0 0	1 1				0		
r		j l				1 1				1		
s	s t	i k			1 1	0 0				0		
t		j l				0 0				1		

state b if they were 001, etc., down to state h if the three preceding bits were 111. This procedure determines the "next state" entries at t_1, t_2, and t_3 times for state a, b, c, and d. Opposite rows a through h in the "remarks" column at the right of Table 9-19 are indicated the sequence of inputs which must have occurred for the complementer to be in state a through h at t_4 time, as well as at t_3 and t_2 times.

Once a nonzero digit has been read from X, the nine's complement rather than the ten's complement of subsequent digits must be formed. Let us assume for the moment that at t_1 time the complementer is in state i if a nonzero decimal digit has previously been read. We may then add seven more states, j through p, which determine the state of the complementer at t_4 time when a nine's complement of the decimal digit is to be obtained.

We may next enter in the output column at t_3 time the proper least significant output bit for a ten's complement or for a nine's complement digit. Examining Table 9-18, we see that if a ten's complement is to be formed, the least significant bit of the complement digit should be the same as the least significant bit of the original number read from X; whereas if the nine's complement is to be formed, the least significant bit of the complement digit should be the opposite of the least significant bit of the original digit. In the "remarks" column of Table 9-19, we can read, in the t_3 column, the two least significant bits of the number being complemented when the complementer is in states a, b, c, and d (and a ten's complement digit is being formed), and i, j, k, and l (when a nine's complement is desired). The output at t_3 time is thus determined for these states, depending on what complement is being formed and on what the corresponding least significant bit was. For example, the complementer will be in states a and c at t_3 time only if the least significant bit of the digit read from X was a "zero"; and since we are forming the ten's complement, the least significant output bit from the complementer must be "zero" for these states. In states i and k the least significant bit read from X was a "zero" but the nine's complement is being formed; and the outputs at t_3 time in states i and k must therefore both be "one." Note that the output at t_3 time is independent of what is on input line X at that time. That is to say, the least significant bit of the complemented digit is independent of the third significant bit.

Let us next examine the output lines corresponding to each state of the complementer at t_4 time. At that time, we have available all of the information about the decimal digit being complemented: the state of line X determines the most significant bit of this digit, and the state of the complementer (a through p) determines what the three other bits of the previous digit were. Knowing all four bits, and knowing whether a nine's or ten's complement must be formed, we can, of course, determine what the second significant bit of the complemented digit (the output at t_4 time) must be. For example, if the complementer is in state a at t_4 time, the three least significant bits of the digit being complemented must all have been "zero." If X is "zero" at that time, the digit being read was a decimal zero and the second bit of the complemented digit is of course "zero." If $X = 1$ at that time, the decimal digit to be complemented is an 8, and the ten's complement must be decimal 2 (0010). The second least significant bit of the digit 2 is a "one," and the output at t_4 time when $X = 1$ in state a is therefore a "one."

For another example, consider state l. If the complementer is in state l, we are seeking a nine's complement and the three least significant bits of the decimal digits to be complemented are 011. If $X = 0$ at t_4 time, the decimal digit to be complemented is a 3 and its nine's complement is 6 (0110), of which the second significant bit is a "one." If X were "one" at t_4 time when the complementer was in state l, the decimal digit read from X would have been 1011, which is not a permitted combination. We shall assume that no forbidden combination ever occurs, and leave this position blank in Table 9-19. The other entries in the t_4 output column of Table 9-19 are found by analyzing each state just as was done for the two examples above.

We have now arranged to provide the two least significant output bits of the complemented digit. The output at t_2 and at t_1 times are the most significant bit and the next most significant bit respectively of the complemented digit. These two bits must be either 00, 01, or 10, because the combination 11 cannot occur in the 8-4-2-1 code. Furthermore, the outputs at t_1 and t_2 times are independent of the input line X at these two bit-times because these two outputs are determined at t_4 time by one particular decimal input digit, whereas the next two input bits are the least significant part of a subsequent decimal digit. If we assume that the complementer starts in state a and continues in that state as long as a sequence of decimal zeros occurs, then the outputs in state a at t_1 and t_2 times are both "zero." Furthermore, the output at t_2 time when the complementer is in state b is also determined, because the circuit can only be in state b when a decimal zero is to be read out.

When a nine's complement is to be formed we must again provide for the possibility that the two most significant bits of the complement are "zero." Let us do that by entering "zeros" in the output columns of states i and j at t_1 and t_2 times. However, we must also provide for the other possible output combinations: a 01 or a 10. One way of doing this is by providing four more states called q, r, s, and t. States q and r will perform the same kind of operation as do states i and j, except that the outputs from these two states will be 01 at t_1 and t_2 times instead of 00. Similarly, states s and t will have the same effect as do states i and j except that the outputs will be 10 instead of 00. The outputs, and the "next state" entries for these four additions to the table may thus be put down.

We have now explained and derived all entries in Table 9-19 except those which determine the next state at t_4 time. Examining these table entries, we see that they are derived simply by looking at the input decimal digit which corresponds to each state at t_4 time, determining what the corresponding complement should be, and entering a "next state" letter which provides the proper two most significant bits for the desired complement. For example, if a "one" appears at t_4 time when the complementer is in state a, b, i, or j, the decimal digit to be complemented must have been either an 8 or a 9. In either case, the most significant two bits of the complement are 00, and these output bits will be obtained at t_1 and t_2 times if the complementer is set to state i. For another example, if a "zero" appears at t_4 time when the complementer is in state n, the decimal digit to be complemented must have been a 5. The nine's complement of 5 is 4, and the most significant two bits of the digit 4 are 01. These output bits will be obtained at t_1 and t_2 times if the complementer is set to state s. By the use of similar arguments, all "next state" entries in the t_4 column may be determined, and Table 9-19 is completed.

We may now apply rules I and II of the Huffman-Mealy method to eliminate some of the states in Table 9-19. Table 9-20 describes a series of steps which may be taken to eliminate some of the states of Table

Table 9-20

	Eliminate State	By Combining It with State
1	g	f
2	m	f
3	n	f
4	o	h
5	p	h
6	k	d
7	b	i
8	c	j
9	e	d
10	t	j
11	q	h
12	r	d
13	s	l

Table 9-21

Present State	Next State								Output							
Time →	t_1		t_2		t_3		t_4		t_1		t_2		t_3		t_4	
X →	0	1	0	1	0	1	0	1	0	1	0	1	0	1	0	1
a	a	i	a	j	a	d	a	i	0	0	0	0	0	0	0	1
d			j	l	d	h	l				1	1	1	1	1	
f							l								0	
h	h	d	i	d			i		0	0	1	1			1	
i	i	j	i	d	i	f	h	i	0	0	0	0	1	1	0	0
j			j	l	j	f	h	i			0	0	0	0	0	0
l	l	j	i	d	l	h	l		1	1	0	0	0	0	1	

9-19, and the resulting simplified table is shown in Table 9-21. No further simplification is possible. Thirteen of the original twenty states may thus be eliminated and only seven states are necessary.

The design of the complementer may be completed by providing appropriate flip-flops, assigning to each of the seven states a particular code, and deriving the input equations to the flip-flops based on this code. At least three flip-flops must be provided to distinguish between seven different states, and in Table 9-22 the three flip-flops A, B, and C are shown with one particular configuration of these three memory elements assigned to each of the seven states. Note that this code is an arbitrary one, and, although it was chosen with the objective of

Table 9-22

	A	B	C
a	0	1	0
d	1	0	1
f	1	1	0
h	0	0	1
i	0	1	1
j	0	0	0
l	1	0	0

simplifying the logic for the complementer, it is not necessarily the simplest configuration possible—some other code, possibly requiring the use of more than three flip-flops, might easily result in a complementer which is cheaper in terms of diode count and diode and flip-flop costs. The choice of a code, however, makes it possible to transform Table 9-21 into a set of difference equations, to plot these difference equations on a Veitch diagram, and to derive the resulting input logic. The input logic for the code of Table 9-22 is given in equations 9-60 through 9-62, together with equation 9-63 which defines the output from the complementer.

$$J_A = \bar{B}Xt_1 + CXt_2 + \bar{B}Xt_2 + Xt_3$$
$$K_A = Xt_1 + \bar{X}t_2 + Xt_3 \tag{9-60}$$

$$J_B = A\bar{C}\bar{X}t_2 + \bar{A}C\bar{X}t_2 + \bar{A}Xt_3 + Xt_4 + \bar{A}Ct_4$$
$$K_B = CXt_1 + Xt_2 + \bar{C}Xt_3 + At_4 + C\bar{X}t_4 \tag{9-61}$$

$$J_C = \bar{A}Xt_1 + At_2 + BXt_3 + AXt_3 + Xt_4 + \bar{A}\bar{B}t_4$$
$$K_C = BXt_1 + At_2 + BXt_3 + At_4 \tag{9-62}$$

$$\text{Output} = t_1A + t_2\bar{B}C + t_3C + t_4A\bar{B} + t_4\bar{B}C + t_4B\bar{C}\bar{X} \tag{9-63}$$

This completes the design of the decimal ten's complementer. A complementer can be designed in many ways, and it is possible that some other design approach will provide a simpler and cheaper complementer. However, this design based on the Huffman-Mealy method indicates a fairly straightforward approach to design which may be applied to any problem.

Shifting

Although the operations of multiplication and division are somewhat complicated to carry out even with pencil and paper, multiplication and division by ten in the decimal number system are very easy. To multiply a number by ten we simply move its decimal point one place to the right; to divide it by ten, we move the decimal point one place to the left. In shifting the decimal point this way, we are of course taking advantage of the algebraic relationships indicated below, where the series of digits $x_n, x_{n-1}, x_{n-2}, \cdots$ are used to represent a number to the base r.

$$X = x_n r^n + x_{n-1} r^{n-1} + \cdots + x_2 r^2 + x_1 r^1 + x_0 r^0 \qquad (9\text{-}64)$$

$$X r^k = x_n r^{n+k} + x_{n-1} r^{n+k-1} + \cdots + x_2 r^{2+k} + x_1 r^{1+k} + x_0 r^k \qquad (9\text{-}65)$$

where $k = \cdots, -3, -2, -1, 1, 2, 3, \cdots$.

The equations are perfectly general, and when applied to the binary number system they indicate that a movement of the binary point to the right or to the left corresponds to a multiplication or division by some power of two.

The mechanism for performing a shift in a parallel computer is of course very simple. Referring to the parallel register illustrated in Figure 9-2, we see that the general equation for shifting a binary number in the X-register is

$$\begin{aligned} J_{Xi} &= X_{i+1}R + X_{i-1}L \\ K_{Xi} &= \overline{X}_{i+1}R + \overline{X}_{i-1}L \end{aligned} \qquad (9\text{-}66)$$

where R and L are signals which do not occur simultaneously and which are "one" for as many bit-times as the number is to be shifted right or left respectively. If the number in the X-register is a decimal number, the corresponding equations are

$$\begin{aligned} J_{Xi} &= X_{i+4}R + X_{i-4}L \\ K_{Xi} &= \overline{X}_{i+4}R + \overline{X}_{i-4}L \end{aligned} \qquad (9\text{-}67)$$

The shift of a serial number in a circulating register is a different and somewhat more complicated problem. Referring to Figure 9-1, we see that a number in a shifting or circulating register is continually shifting right. However, arithmetic operations on the number in the register are always carried out starting at bit-time t_1, when the least

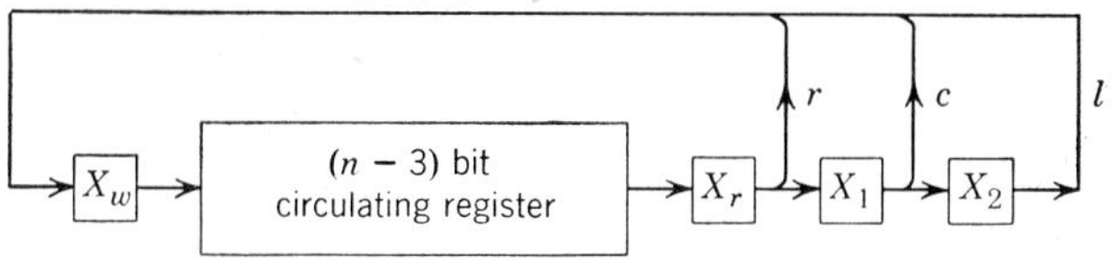

X_w → (n − 3) bit circulating register → X_r → X_1 → X_2 ; r, c, l

t_0	0	0	0	1	1	×		
t_1	×	0	0	0	1	1		
t_2	1	×	0	0	0	1		
t_3	1	1	×	0	0	0		
t_4	0	1	1	×	0	0		Circulate
t_5	0	0	1	1	×	0		
t_0	0	0	0	1	1	×		
t_1	×	0	0	0	1	1	×	
t_2	m	×	0	0	0	1	1	
t_3	1	m	×	0	0	0	1	
t_4	1	1	m	×	0	0	0	Shift left
t_5	0	1	1	m	×	0	0	
t_0	0	0	1	1	m	×	0	
t_1	×	0	0	1	1	m		
t_2	1	×	0	0	1			
t_3	1	1	×	0	0			
t_4	0	1	1	×	0			Shift right
t_5	0	0	1	1	×			
t_0	n	0	0	1	1	×		
t_1	×	n	0	0	1	1		

FIGURE 9-12

significant bit of the number is in the read amplifier. A right or left shift, then, must be interpreted as some operation which leaves the number shifted right or left *at t_1 time.* In Figure 9-12 we show how it is possible to shift a serial number right or left by adding or subtracting delays to or from the normal circulating path. During the first word-time illustrated, the word is shifted through read amplifier X_r and flip-flop X_1; flip-flop X_2 is not employed, and the word appears unmodified at the second t_1 time. During the second word-time, flip-flop X_2 is included in the circulating loop with the result that, at the third t_1 time, the word has been delayed by one bit and appears in the register shifted one bit position to the left. During the third word-time, flip-flops X_1 and X_2 are left out of the circulating loop, so that the number in the register circulates in less time and appears at the fourth t_1 time shifted one bit position to the right from its position the previous t_1 time. A right shift is thus obtained by omitting a delay, and a left shift by inserting one.

The shifting operations of Figure 9-12 bring to light some problems which must be resolved whenever a shifting operation is carried out either in a serial or a parallel machine. When a number is shifted to the left, its most significant digit disappears and a new least significant digit must be supplied. When a number is shifted right, the least significant digit is lost and a new most significant digit must be supplied. In the example of Figure 9-12, the new least significant digit supplied in the left shift operation is indicated by the letter m, and the new most significant digit in the right shift by the letter n. The digit m supplied during a left shift operation must have the significance of a "zero." The most significant digit of a word, however, is generally the sign digit of the number, and therefore cannot in general be treated the same as the other digits.

Table 9-23 indicates the way in which digits are supplied during right and left shifts for positive numbers, and for negative numbers represented as absolute magnitude plus sign, as two's complements, and as one's complements. The following rules may be seen to apply.

Table 9-23

	Absolute Magnitude Plus Sign	Two's Complement	One's Complement
+23	0 0 0 1 0 1 1 1	0 0 0 1 0 1 1 1	0 0 0 1 0 1 1 1
−23	1 0 0 1 0 1 1 1	1 1 1 0 1 0 0 1	1 1 1 0 1 0 0 0
+23(2)	0 0 1 0 1 1 1 **0**	0 0 1 0 1 1 1 **0**	0 0 1 0 1 1 1 **0**
−23(2)	1 0 1 0 1 1 1 **0**	1 1 0 1 0 0 1 **0**	1 1 0 1 0 0 0 **1**
+23(4)	0 1 0 1 1 1 **0 0**	0 1 0 1 1 1 **0 0**	0 1 0 1 1 1 **0 0**
−23(4)	1 1 0 1 1 1 **0 0**	1 0 1 0 0 1 **0 0**	1 0 1 0 0 0 **1 1**
+23(½)	0 **0** 0 0 1 0 1 1	0 **0** 0 0 1 0 1 1	0 **0** 0 0 1 0 1 1
−23(½)	1 **0** 0 0 1 0 1 1	1 **1** 1 1 0 1 0 0	1 **1** 1 1 0 1 0 0
+23(¼)	0 **0 0** 0 0 1 0 1	0 **0 0** 0 0 1 0 1	0 **0 0** 0 0 1 0 1
−23(¼)	1 **0 0** 0 0 1 0 1	1 **1 1** 1 1 0 1 0	1 **1 1** 1 1 0 1 0

(Bits shown in boldface must be supplied)

When numbers are represented as absolute magnitude plus sign, (*a*) in right shifts the sign digit must not change, and a "zero" must be shifted into the next most significant bit position; (*b*) in left shifts the sign digit must not change, and a "zero" must be shifted into the least significant bit position.

If negative numbers are represented in two's complement form, (*a*) in right shifts the sign digit must not change, and a "zero" or a "one" must be shifted into the next most significant bit position, depending on whether the number shifted is positive or negative; (*b*) in left shifts the sign digit must not change, and a "zero" must be shifted into the least significant bit position. An equivalent and simpler statement of these rules is: shift the entire register, including sign digit, right or left as required; in a right shift, replace the sign bit by itself as well as shifting it to the right; and in a left shift, insert a least significant "zero." [A peculiarity of the two's complement system in shifting is that one may obtain a different round-off in shifting positive and negative numbers to the right. For example, in Table 9-23 $(+23)(\frac{1}{2}) = 11$, but $(-23)(\frac{1}{2}) = -12$.]

If negative numbers are represented in one's complement form, (*a*) in right shifts the sign digit must not change, and a "zero" or a "one" must be shifted into the next most significant bit position, depending on whether the number shifted is positive or negative; (*b*) in left shifts the sign digit must not change, and a "zero" or a "one" must be shifted into the least significant bit position, depending on whether the number shifted is positive or negative. An equivalent and simpler statement of these rules is: shift the entire register, including sign digit, right or left as required; in a right shift, replace the sign bit by itself as well as shifting it to the right; and in a left shift, insert the sign bit into the least significant bit position.

It is of course necessary to insure that no number is multiplied by such a large power of r that the product exceeds the capacity of the register in which it is shifted. For example, (23)(8) would be too large a number for the register of Table 9-23.

Multiplication

Multiplication is the most complex operation to be discussed in any detail in this book. It is in some ways the most important operation which must be mechanized, for it is so much more complicated than the operations discussed so far and must often be carried out with such speed that it greatly influences the design of the entire arithmetic unit. Furthermore, its mechanization usually involves facilities for the simultaneous storage, addition, and shifting of several numbers, and these facilities are also employed in the mechanization of the other arithmetic operations. For example, the adder which is used in multiplication may also be employed in mechanizing addition.

If the numbers X and Y are to be multiplied together, and if each is expressed in the form shown in equation 9-64, we can write the following equation for their product:

$$XY = x_0Y + x_1rY + x_2r^2Y + x_3r^3Y + \cdots + x_{n-1}r^{n-1}Y + x_nr^nY \quad (9\text{-}68)$$

In general, Y is an n-digit number, just like X, and the multiplication may be looked upon and is often mechanized as a series of n simplified multiplications, each determining the product of the number Y by a single digit x_i. These subproducts must be shifted with respect to one another (as indicated by the appearance of r^i in each of the terms of equation 9-68) and then added together.

Equation 9-68 may be rewritten in several ways. Very often the multiplication is carried out in a sequence of steps, each of which determines what is called a *partial product.* Each partial product requires a shift, a single-digit multiplication, and an addition, and n of these steps are required to find the product of two n-digit numbers. Two ways of defining partial products and of arriving at the total product are written below. Each is equivalent to equation 9-68.

$$\begin{aligned} P_0 &= x_0Y \\ P_i &= P_{i-1} + x_ir^iY \\ XY &= P_n \end{aligned} \quad (9\text{-}69)$$

$$\begin{aligned} P_0 &= x_0r^nY \\ P_i &= P_{i-1}r^{-1} + x_ir^nY \\ XY &= P_n \end{aligned} \quad (9\text{-}70)$$

A schematic diagram of the equipment necessary to mechanize equations 9-69 and 9-70 is given in Figure 9-13*a* and *b*. The diagrams are general enough that they can represent either serial or parallel multiplication, and several general comments can be made about them. First, each requires three storage registers, one for the multiplier, one for the multiplicand, and one for the partial product. The multiplier and multiplicand registers need be only of normal length; but the partial product register must be double-length to store the final product, which contains twice as many digits as either the multiplier or the multiplicand. Second, these mechanizations of multiplication involve shifting, addition, and one-digit multiplication. Of these, we have discussed all but the last. The methods we have learned previously for adding and shifting must be applied in designing a multiplier, and we need discuss only one new operation—one-digit multiplication. Third and finally, it is evident that one of the principal problems to be faced in designing a multiplier is that of controlling and sequencing the various additions, multiplications, and shifts which must be carried out to obtain the product. We will show how this sequencing may be carried out by designing a complete binary multiplier.

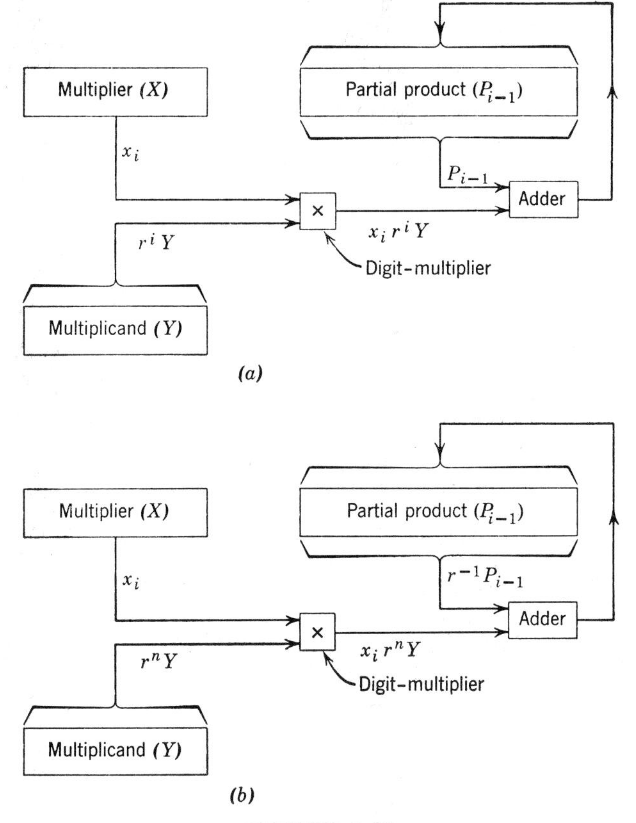

FIGURE 9-13

A serial, binary multiplier. The design of a binary multiplier is simplified by the fact that a one-digit binary multiplication is very easily mechanized. The truth table for the binary multiplication of two digits is given below, and is seen to be the same as the logical operation "and."

x_i \ y_i	0	1
0	0	0
1	0	1

Therefore the multiplication of digit x_i by the number Y may be accomplished simply by storing digit x_i in a flip-flop for a word-time and applying its output to an "and" circuit whose other input is the serial number Y.

Let us use the relationships of equations 9-70 to design a multiplier which operates on two four-bit binary numbers. The circulating registers, amplifiers, and flip-flops necessary to construct this multiplier are indicated in Figure 9-14, which may be seen to correspond in form to

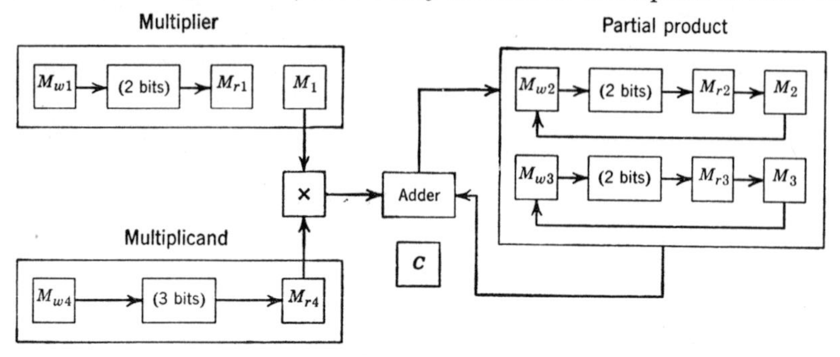

FIGURE 9-14

Figure 9-13*b*. A space bit is provided between the least and most significant bits of every word, so that a five-bit circulating register is required to hold one word. Furthermore, let us assume that at the beginning of the multiplication, the partial product two-word register (which consists of two independent one-word shifting registers) contains "zeros," and the multiplier and the multiplicand are stored in their respective registers. The multiplication is then broken down into a sequence of operations which may be described by the following rules:

a. If the least significant digit in the multiplier register is "one," add the multiplicand register to the left-hand end of the partial product register. Otherwise, go on to step b.

b. Shift the contents of the multiplier register one digit position to the right, putting the former least significant bit into the most significant bit position. This brings the next digit of the multiplier into a position where it can control the next one-digit multiplication. The action of saving the least significant bit in the multiplier register insures that the multiplier is not lost and is still available when the multiplication is complete. (If it is not necessary to save the multiplier, the multiplication unit may be simplified by eliminating one half of the partial product register and shifting the product into the multiplier register, replacing the "used" multiplier digits.)

c. Shift the partial product register one bit position to the right, putting the least significant bit from the left-hand half of the partial product register into the right-hand half.

d. If the previous three rules have been carried out four times, stop. Otherwise, return to step a.

Suppose, for example, we wish to multiply the two binary numbers 5 and 7. Using pencil and paper, we can carry out the multiplication in the following way.

$$
\begin{array}{rrrrrrl}
 & & 0 & 1 & 1 & 1 & = 7 \\
 & & 0 & 1 & 0 & 1 & = 5 \\
\hline
 & & 0 & 1 & 1 & 1 & \\
 & 0 & 0 & 0 & 0 & & \\
0 & 1 & 1 & 1 & & & \\
\hline
1 & 0 & 0 & 0 & 1 & 1 & = 35
\end{array}
$$

The same multiplication carried out according to the sequence of steps described above is shown in Table 9-24. Note that the partial

Table 9-24

(Multiplicand Register contains 0111)

Step	Multiplier * Register	Partial Product Register: Left-Hand Half	Partial Product Register: Right-Hand Half
		0 0 0 0	0 0 0 0
a	0 1 0 1.	0 1 1 1	0 0 0 0
b	1. 0 1 0		
c		0 0 1 1	1 0 0 0
a	1. 0 1 0	0 0 1 1	1 0 0 0
b	0 1. 0 1		
c		0 0 0 1	1 1 0 0
a	0 1. 0 1	1 0 0 0	1 1 0 0
b	1 0 1. 0		
c		0 1 0 0	0 1 1 0
a	1 0 1. 0	0 1 0 0	0 1 1 0
b	0 1 0 1.		
c		0 0 1 0	0 0 1 1
d	stop		

* The dot indicates the position of the least significant bit of the multiplier.

product P_i of equations 9-70 is formed in the partial product register during each step a. The complete product P_n appears in the partial product register after the fourth step a, and the final shift right carried out in the last step c is merely inserted to shift the product to the right-hand-most side of the product register.

In the foregoing rules, the various additions and shifts were described as if they took place sequentially, one at a time. In designing a multiplier, however, the logical designer can speed up the multiplication operation by carrying out shifts and additions simultaneously. In Table 9-25 the multiplication of Table 9-24 is written out in complete detail, with the contents of every register and every flip-flop at every bit-time during the multiplication indicated explicitly. During each word-time in Table 9-25, the three steps defined above are carried out. The multiplication is then complete at the end of the fourth word-time, and a fifth word-time is added to shift the product to the right-

Table 9-25

Word-Time		Multiplier						Partial Product										Multiplicand				
		M_{w1}			M_{r1}	M_1	C	M_{w2}			M_{r2}	M_2	M_{w3}			M_{r3}	M_3	M_{w4}				M_{r4}
1	0	0	1	0	1.	×		0	0	0	0	×	0	0	0	0	×	0	1	1	1	×
	1	1.	0	1	0	1.	0	0	0	0	0	×	0	0	0	0	×	×	0	1	1	1
	2	0	1.	0	1	1.	0	1	0	0	0	×	0	0	0	0	×	1	×	0	1	1
	3	1	0	1.	0	1	0	1	1	0	0	×	0	0	0	0	×	1	1	×	0	1
	4	0	1	0	1.	1.	0	1	1	1	0	×	0	0	0	0	×	1	1	1	×	0
2	0	1.	0	1	0	1.	0	0	1	1	1	×	0	0	0	0	×	0	1	1	1	×
	1	0	1.	0	1	0	0	0	0	1	1	×	1	0	0	0	×	×	0	1	1	1
	2	1	0	1.	0	0	0	1	0	0	1	×	0	1	0	0	×	1	×	0	1	1
	3	0	1	0	1.	0	0	1	1	0	0	×	0	0	1	0	×	1	1	×	0	1
	4	1.	0	1	0	0	0	0	1	1	0	×	0	0	0	1	×	1	1	1	×	0
3	0	0	1.	0	1	0	0	0	0	1	1	×	1	0	0	0	×	0	1	1	1	×
	1	1	0	1.	0	1	0	0	0	0	1	×	1	1	0	0	×	×	0	1	1	1
	2	0	1	0	1.	1	1	0	0	0	0	×	0	1	1	0	×	1	×	0	1	1
	3	1.	0	1	0	1	1	0	0	0	0	×	0	0	1	1	×	1	1	×	0	1
	4	0	1.	0	1	1	1	0	0	0	0	×	1	0	0	1	×	1	1	1	×	0
4	0	1	0	1.	0	1	0	1	0	0	0	×	1	1	0	0	×	0	1	1	1	×
	1	0	1	0	1.	0	0	0	1	0	0	×	0	1	1	0	×	×	0	1	1	1
	2	1.	0	1	0	0	0	0	0	1	0	×	0	0	1	1	×	1	×	0	1	1
	3	0	1.	0	1	0	0	1	0	0	1	×	1	0	0	1	×	1	1	×	0	1
	4	1	0	1.	0	0	0	0	1	0	0	×	1	1	0	0	×	1	1	1	×	0
5	0	0	1	0	1.	0	0	0	0	1	0	×	0	1	1	0	×	0	1	1	1	×
	1	×	0	1	0	1.		0	0	0	1	×	0	0	1	1	×	×	0	1	1	1
	2	1.	×	0	1	0		1	0	0	0	×	1	0	0	1	×	1	×	0	1	1
	3	0	1.	×	0	1		0	1	0	0	×	1	1	0	0	×	1	1	×	0	1
	4	1	0	1.	×	0		0	0	1	0	×	0	1	1	0	×	1	1	1	×	0
6	0	0	1	0	1.	×		0	0	0	1	×	0	0	1	1	×	0	1	1	1	×
	1	×	0	1	0	1.		×	0	0	0	1	×	0	0	1	1	×	0	1	1	1
	2	1	×	0	1	0		1	×	0	0	0	1	×	0	0	1	1	×	0	1	1
	3	0	1.	×	0	1		0	1	×	0	0	1	1	×	0	0	1	1	×	0	1
	4	1	0	1.	×	0		0	0	1	×	0	0	1	1	×	0	1	1	1	×	0
	0	0	1	0	1.	×		0	0	0	1	×	0	0	1	1	×	0	1	1	1	×
	1	×	0	1	0	1.		×	0	0	0	1	×	0	0	1	1	×	0	1	1	1

hand side of the product register. The sixth word-time of Table 9-25 is added merely to show the normal circulating path of numbers in the multiplier and product registers.

Examining the table in some detail, we see that the important features of the shifting and adding operations are as follows:

1. During the course of the multiplication, the multiplier and partial product registers are shifted right by leaving flip-flops M_1, M_2, and M_3 out of the circulating paths.
2. Because flip-flop M_1 is not used in the circulating path, it may be used to hold the multiplier digit which determines whether or not the multiplicand is added into the partial product. Therefore, at every t_0 time during the multiplication, the appropriate multiplier digit is transferred into flip-flop M_1.
3. The shifted partial product appearing at read amplifier M_{r2} must be added to the product of M_1 and the multiplicand, which is read from read amplifier M_{r4}. A carry flip-flop C contains the carry for this addition.
4. The carry flip-flop C must be in the "zero" state at the beginning of each partial product addition, i.e., at each t_1 time.
5. When the left-hand half of the partial product is shifted right, a "zero" must be inserted in its most significant bit position. A "zero" must therefore be written in write amplifier M_{w2} at every t_0 time.
6. When the right-hand half of the partial product register is shifted right, the least significant bit of the left-hand half of the partial product register must be transferred from the left-hand partial product register to the right-hand one. Therefore the contents of M_{r2} must be transferred to M_{w3} at every t_0 time.
7. At every t_0 time the space bit appears in the multiplicand read amplifier. Since this bit may be either a "zero" or a "one," we must be sure not to add the multiplicand to the partial product at this bit time.

The input logic for all the flip-flops and write amplifiers in the multiplier is indicated in Table 9-26. Note that each of the seven points just raised has been accounted for, and that the logic includes all necessary shifts and additions. The table indicates that, in order to write the complete logical equations, we must distinguish six different stages of multiplication: the time before and after multiplication, when all registers must circulate and no additions or shifts are taking place; the four word-times during which the multiplication proper takes place; and the fifth word-time, when the partial products are shifted. We

Table 9-26

	No Multiplication in Progress	First Four Word-Times of Multiplication		Fifth (and Last) Word-Time of Multiplication	
		t_0	$\bar{t}_0$	t_0	$\bar{t}_0$
$J_{Mw1} = \overline{K}_{Mw1}$	M_1	M_{r1}	M_{r1}	M_1	M_1
J_{M1}	M_{r1}	M_{r1}	0	M_{r1}	M_{r1}
K_{M1}	$\overline{M}_{r1}$	$\overline{M}_{r1}$	0	$\overline{M}_{r1}$	$\overline{M}_{r1}$
J_C	–	0	$M_{r2}M_1M_{r4}$	–	–
K_C	–	1	$\overline{M}_{r2}(\overline{M}_1 + \overline{M}_{r4})$	–	–
$J_{Mw2} = \overline{K}_{Mw2}$	M_2	0	$CM_{r2}M_1M_{r4} + \overline{C}\overline{M}_{r2}M_1M_{r4} + C\overline{M}_{r2}(\overline{M}_1 + \overline{M}_{r4}) + \overline{C}M_{r2}(\overline{M}_1 + \overline{M}_{r4})$	0	M_{r2}
$J_{M2} = \overline{K}_{M2}$	M_{r2}	–	–	–	–
$J_{Mw3} = \overline{K}_{Mw3}$	M_3	M_{r2}	M_{r3}	M_{r2}	M_{r3}
$J_{M3} = \overline{K}_{M3}$	M_{r3}	–	–	–	–
$J_{Mw4} = \overline{K}_{Mw4}$	M_{r4}	M_{r4}	M_{r4}	M_{r4}	M_{r4}

distinguish these six stages by providing three flip-flops M_4, M_5, and M_6 which are normally in the 000 state when no multiplication is in progress, but which step through five other states during the five word-times of multiplication. We can, of course, assign any desired sequence of states to this counter. However, by making a judicious

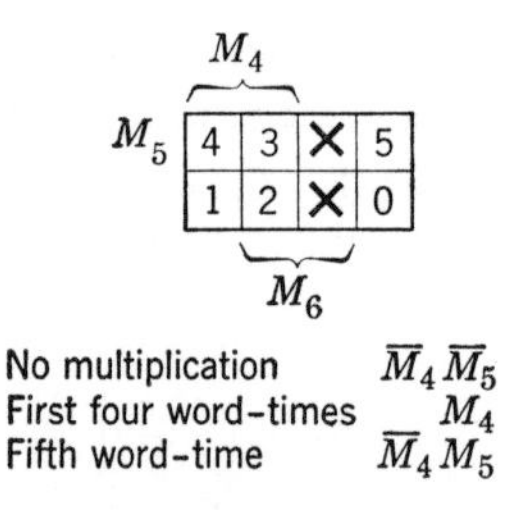

FIGURE 9-15

choice of the six states, we may simplify the logic necessary to distinguish the first four word-times from the fifth, and the fifth from the "no multiplication" state. One such choice of states is illustrated in Figure 9-15, and the truth table which defines the counter's operation is shown in Table 9-27. The logic for these three flip-flops is given in the following text.

Table 9-27

n				$n+1$		
M_4	M_5	M_6	t_4	M_4	M_5	M_6
0	0	0	0	0	0	0
0	0	0	1	0	0	0
1	0	0	0	1	0	0
1	0	0	1	1	0	1
1	0	1	0	1	0	1
1	0	1	1	1	1	1
1	1	1	0	1	1	1
1	1	1	1	1	1	0
1	1	0	0	1	1	0
1	1	0	1	0	1	0
0	1	0	0	0	1	0
0	1	0	1	0	0	0

$$J_{M4} = (\text{Multiply})\, t_4 \qquad K_{M4} = M_5\overline{M}_6t_4 \tag{9-71}$$

$$J_{M5} = M_6t_4 \qquad K_{M5} = \overline{M}_4t_4 \tag{9-72}$$

$$J_{M6} = M_4\overline{M}_5t_4 \qquad K_{M6} = M_5t_4 \tag{9-73}$$

Note that when the counter is in the 000 state, it will remain in that state indefinitely until a logical signal identified as "multiply" sets flip-flop M_4.

We can now write the complete equations for the multiplier, by combining the separate equations of Table 9-26 together, employing the counter to make the necessary distinction between steps.

$$J_{Mw1} = M_1\overline{M}_4 + M_{r1}M_4 \qquad K_{Mw1} = \overline{M}_1\overline{M}_4 + \overline{M}_{r1}M_4 \tag{9-74}$$

$$J_{M1} = M_{r1}\overline{M}_4 + M_{r1}t_0 \qquad K_{M1} = \overline{M}_{r1}\overline{M}_4 + \overline{M}_{r1}t_0 \tag{9-75}$$

$$J_C = M_{r2}M_1M_{r4}\bar{t}_0 \qquad K_C = t_0 + \overline{M}_{r2}\overline{M}_1 + \overline{M}_{r4}\overline{M}_{r2} \tag{9-76}$$

$$\begin{aligned} J_{Mw2} &= M_2\overline{M}_4\overline{M}_5 + M_4\bar{t}_0[CM_{r2}M_1M_{r4} + \bar{C}\overline{M}_{r2}M_1M_{r4} \\ &\quad + C\overline{M}_{r2}(\overline{M}_1 + \overline{M}_{r4}) + \bar{C}M_{r2}(\overline{M}_1 + \overline{M}_{r4})] \\ &\quad + \overline{M}_4M_5\bar{t}_0M_{r2} \\ K_{Mw2} &= \overline{M}_2\overline{M}_4\overline{M}_5 + t_0(M_4 + M_5) + M_4\bar{t}_0[\bar{C}M_{r2}M_1M_{r4} \\ &\quad + C\overline{M}_{r2}M_1M_{r4} + CM_{r2}(\overline{M}_1 + \overline{M}_{r4}) \\ &\quad + \bar{C}\overline{M}_{r2}(\overline{M}_1 + \overline{M}_{r4})] + \overline{M}_4M_5\bar{t}_0\overline{M}_{r2} \end{aligned} \tag{9-77}$$

$$\begin{aligned} J_{M2} &= M_{r2} \\ K_{M2} &= \overline{M}_{r2} \end{aligned} \tag{9-78}$$

$$\begin{aligned} J_{Mw3} &= M_3\overline{M}_4\overline{M}_5 + M_{r2}t_0(M_4 + M_5) + M_{r3}\bar{t}_0(M_4 + M_5) \\ K_{Mw3} &= \overline{M}_3\overline{M}_4\overline{M}_5 + \overline{M}_{r2}t_0(M_4 + M_5) + \overline{M}_{r3}\bar{t}_0(M_4 + M_5) \end{aligned} \tag{9-79}$$

$$\begin{aligned} J_{M3} &= M_{r3} \\ K_{M3} &= \overline{M}_{r3} \end{aligned} \tag{9-80}$$

$$\begin{aligned} J_{Mw4} &= M_{r4} \\ K_{Mw4} &= \overline{M}_{r4} \end{aligned} \tag{9-81}$$

The "don't-care" conditions of the input logic, indicated by dashes in Table 9-26, have been included in equations 9-74 through 9-81 in order to simplify them. For example, since the C flip-flop is usefully employed only during the first four word-times of a multiplication, it is unnecessary to multiply its input equations by M_4. As defined in equations 9-76, flip-flop C operates constantly as a carry flip-flop. However, equations 9-77 show that C enters into the multiplication operation only when M_4 is "one."

Before turning to the problem of designing a decimal multiplier, two comments should be made upon the work just done. First, certain modifications must be made to the logic described above before we have what might be called a practical multiplier. Clearly, some means must be provided for reading new multipliers and multiplicands into the circulating registers from main computer storage; and some method must be provided for reading the product back out of the arithmetic unit. In addition, the multiplier as designed will handle only positive numbers, and in general some provision must be made to multiply numbers which may be either positive or negative. We shall discuss the problem of signed multiplication later in this chapter.

The second remark is concerned with the method of design employed in deriving equations 9-71 through 9-81. The method of design was largely empirical, and required first that a model of the unit be set up (Figure 9-14), that the rules of operation for multiplication be established (Table 9-24), that the bit-time by bit-time operation of the proposed model be examined in great detail (Table 9-25), that the functions of various memory elements in the model be defined for each of the several stages involved in the multiplication operation (Table 9-26), that a counter of some kind be provided to distinguish between the various stages and to control the sequence of operations (Table 9-27), and finally that the control logic and the operation logic be combined. This sequence of steps will be formalized in Chapter 11. It is a far cry from the neat, systematic, and mathematical approach to synthesis embodied in the Huffman-Mealy tables, but it is often the only practical method which can be employed in design.

A binary-coded decimal multiplier. A very nice feature of the serial binary multiplier is the fact that the multiplication of one binary number by a single binary digit is very simple. The multiplication of a serial decimal number by either a binary digit or by a decimal digit is much more complicated, and it is the purpose of this section to indicate how this problem may be solved. We shall not design a complete serial decimal multiplier, mechanizing all the various shifts, additions, and counts necessary, but shall concentrate our attention on the problem of designing a decimal digit-multiplier (see Figure 9-13) whose function it is to multiply a serial decimal number Y by a single decimal digit x_i.

One way of simplifying this problem is to break the digit-multiplier down into two parts, as shown in Figure 9-16. The first part is called

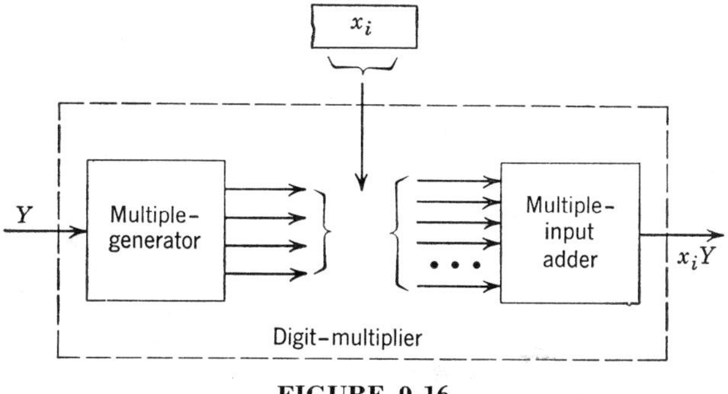

FIGURE 9-16

a *multiple-generator*, and continuously provides on each of its output lines a specified multiple of the number Y fed into it. The second part of the digit-multiplier consists of an adder having several input lines.

Certain of the outputs of the multiple-generator are connected to certain inputs of the many-input adder, where the connections depend on the multiplier digit x_i. In Table 9-28, five different digit-multipliers are indicated, each consisting of a multiple-generator having certain outputs and an adder with a certain number of inputs. The first adder, for example, has only two inputs but requires a multiple-generator having five outputs which provide the serial numbers Y, $3Y$, $5Y$, $7Y$, and $9Y$. If the digit x_i is 8, one input to the adder will be $7Y$ and the other Y. The other multipliers indicated in Table 9-28 require adders and multiple-generators with various combinations of inputs and outputs. The fourth multiplier, for example, requires a three-input adder but only three multiples of Y: Y, $2Y$, and $4Y$. The fifth multiplier employs a four-input adder but requires only two outputs from the multiple-generator: Y and $3Y$.

Table 9-28

Desired Product		Adder Inputs													
		Adder ①		Adder ②		Adder ③			Adder ④			Adder ⑤			
x_i	Y	X_1	X_2	X_1	X_2	X_1	X_2	X_3	X_1	X_2	X_3	X_1	X_2	X_3	X_4
1	Y	Y		Y		Y			Y			Y			
2	Y	Y	Y	$2Y$			$2Y$			$2Y$		Y	Y		
3	Y	$3Y$		$2Y$	Y	Y	$2Y$		Y	$2Y$		Y	Y	Y	
4	Y	$3Y$	Y	$2Y$	$2Y$			$4Y$			$4Y$	Y			$3Y$
5	Y	$5Y$		$5Y$		Y		$4Y$	Y		$4Y$	Y	Y		$3Y$
6	Y	$5Y$	Y	$5Y$	Y		$2Y$	$4Y$		$2Y$	$4Y$			$3Y$	$3Y$
7	Y	$7Y$		$5Y$	$2Y$	Y	$2Y$	$4Y$	Y	$2Y$	$4Y$	Y		$3Y$	$3Y$
8	Y	$7Y$	Y	$8Y$				$8Y$		$4Y$	$4Y$	Y	Y	$3Y$	$3Y$
9	Y	$9Y$		$8Y$	Y	Y		$8Y$	Y	$4Y$	$4Y$		$3Y$	$3Y$	$3Y$
Multiples needed		1, 3, 5, 7, 9		1, 2, 5, 8		1, 2, 4, 8			1, 2, 4			1, 3			

The digit-multiplier to be designed in this section supplies multiples to a three-input adder, and requires a multiple-generator whose outputs are Y, $-Y$, and $4Y$. It is to operate on numbers expressed in the 8-4-2-1 decimal code. Table 9-29 indicates how all nine multiples of Y can be obtained by adding various combinations of these three multiples, and Figure 9-17 shows the digit-multiplier in more detail. The multiple $-Y$ can be obtained from the complementer designed earlier in this chapter. The number Y is, of course, already available at the input to the multiple-generator, although some means will have to be provided for delaying it so that its least significant bit occurs at the same time as the least significant bit of $-Y$ and $4Y$. The $4Y$ multiplier will be designed below. The multiplier digit x_i is held in four flip-flops R_1, R_2, R_3, and R_4, where the least significant bit of the decimal digit is in flip-flop R_1. These four flip-flops will, of course, determine which combination of multiples are connected to the three-input decimal adder during the word-time in which the digit-multiplication is to take place. The inputs to this adder are labeled X_1, X_2, and X_3. If the R_1 to R_4 register should happen to contain the decimal digit 7 and the serial decimal number 0028 appears on the Y line, the multiplication proceeds as shown in Figure 9-18, where the sequence of bits appearing on each of the various input and output lines is indicated.

Table 9-29

Desired Product	Adder Inputs X_1	X_2	X_3
Y		Y	
$2Y$	Y		Y
$3Y$	Y	Y	Y
$4Y$	$4Y$		
$5Y$	$4Y$	Y	
$6Y$	$4Y$	Y	Y
$7Y$	$4Y$	$-Y$	$4Y$
$8Y$	$4Y$		$4Y$
$9Y$	$4Y$	Y	$4Y$

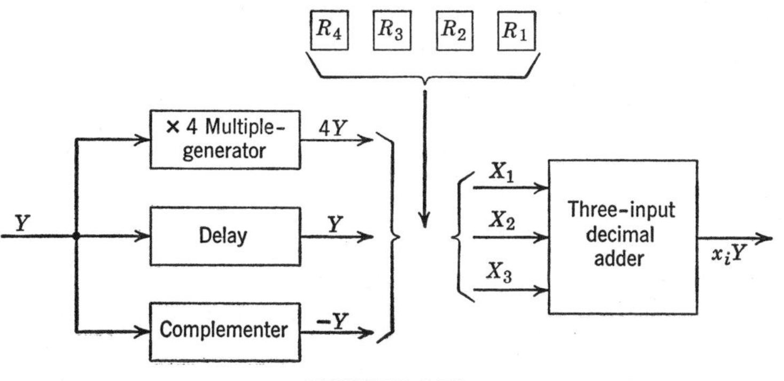

FIGURE 9-17

The bit-time by bit-time operation of the $4Y$ multiple-generator may be described as follows. During the first three bit-times of every decimal digit, the incoming bits from Y will be shifted into a flip-flop C_1,

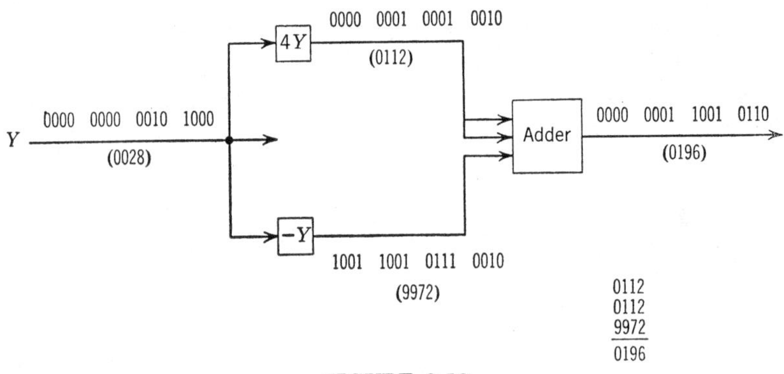

FIGURE 9-18

C_1 is shifted into C_2, C_2 into M_1, and M_1 into M_2. At t_4 time all four bits of the decimal digit will thus be available in Y, C_1, C_2, and M_1. When this digit is multiplied by four, a product digit will be formed along with a carry of 0, 1, 2, or 3 as shown in Table 9-30. Before the

Table 9-30

Digit in $YC_1C_2M_1$ at t_4 Time	Carry Digit	Product Digit
0	0	0
1	0	4
2	0	8
3	1	2
4	1	6
5	2	0
6	2	4
7	2	8
8	3	2
9	3	6

product digit can be read from the multiple-generator, it must be modified by adding to it the carry digit which was formed in the ×4 multiplication of the previous decimal digit. For example, if the latest decimal digit is a 3 and a carry digit of 2 was formed in the previous digit-multiplication, the result will be $(4 \times 3) + 2 = 14$, and a four must be read from the multiple-generator while a one is stored as a carry to be used in modifying the next product digit. We must now determine at which bit-time the least significant bits of the product digits may be read out. First, note from Table 9-30 that all product digits are even before they are modified by the addition of a previous carry. Therefore, if the previous carry is even (0 or 2) the product digit will be even; if odd (1 or 3) the product digit will be odd. Since the least significant bit of a digit in the 8-4-2-1 code determines whether that digit is even or odd, the least significant bit of a product digit may be read from the multiple-generator at any time after the previous carry is determined. However, the next-to-least significant bit of a product digit cannot be determined until t_4 time, when all four bits of the digit to be multiplied are available in Y, C_1, C_2, and M_1. Therefore, the earliest that a product digit can be read from the $4Y$ multiple-generator is t_3, t_4, t_1, and t_2 times for the 1, 2, 4, and 8 bits of that digit.

If we arrange to put the carry digit in flip-flops C_1 and C_2 at every t_4 time, the previously described shifting operation will make the least significant bit of the carry digit (and therefore of the product digit, as shown in the last paragraph) available in M_2 at t_3 time. At t_4 time, the most significant bit of the carry will be in M_2, ready to be combined with the latest digit to be multiplied (in Y, C_1, C_2, and M_1). We can therefore arrange, at that bit-time, for (*a*) the carry digit to be stored in C_1 and C_2, (*b*) the two most significant bits of the product digit to be stored in M_1 and M_2, where they will be shifted out of the multiple-generator during t_1 and t_2 times, and (*c*) the next-to-least significant bit of the product digit to be read directly from the multiple-generator.

The Huffman-Mealy table for this multiple-generator is shown in Table 9-31. The first two columns in the "next state" and "output" parts of this table define its operation during the first three bit-times of every incoming decimal digit, when $t_4 = 0$. Examination of these entries will show that nothing more than a simple shifting operation is taking place, and that the output bit in every entry is equal to M_2. For example, if the multiple-generator is in state *n*, flip-flops C_1, C_2, M_1, and M_2 contain 1101. If it is not t_4 time, output is "one," which is the bit in M_2. Furthermore, if $Y = 1$ at that time, the shifting operation will put the "one" from Y into C_1, C_1 into C_2, C_2 into M_1, and M_1 into M_2 so that these four memory elements should change to 1110, which corresponds to state o.

The right-hand two columns in the "next state" and "output" parts of Table 9-31 determine the operations of the multiple-generator at t_4 time as described previously. (The two columns at the far right of Table 9-31 indicate what product is formed for each of the sixteen states a through p, and may be of some use in developing the table.) For example, if the multiple-generator is in state h at t_4 time, and if there is a "zero" on input line Y, then the bits Y, C_1, C_2, and M_1 are 0011, and the most recent decimal digit is 3. Furthermore, since $M_2 = 1$,

Table 9-31

Present State					Next State				Output Z_4				t_4	
C_1	C_2	M_1	M_2		$t_4Y \rightarrow$ 00	01	10	11	00	01	10	11	$Y \rightarrow$ 0	1
0	0	0	0	a	a	i	a	m	0	0	0	1	0	32
0	0	0	1	b	a	i	a	n	1	1	1	0	2	34
0	0	1	0	c	b	j	b	n	0	0	0	1	4	36
0	0	1	1	d	b	j	b	o	1	1	1	0	6	38
0	1	0	0	e	c	k	c		0	0	0		8	
0	1	0	1	f	c	k	e		1	1	0		10	
0	1	1	0	g	d	l	e		0	0	1		12	
0	1	1	1	h	d	l	f		1	1	0		14	
1	0	0	0	i	e	m	f		0	0	1		16	
1	0	0	1	j	e	m	g		1	1	0		18	
1	0	1	0	k	f	n	i		0	0	0		20	
1	0	1	1	l	f	n	i		1	1	1		22	
1	1	0	0	m	g	o	j		0	0	0		24	
1	1	0	1	n	g	o	j		1	1	1		26	
1	1	1	0	o	h	p	k		0	0	0		28	
1	1	1	1	p	h	p	m		1	1	0		30	

①						②				③				④	
(a	c	e	k	m	o)	(b	d	l	n)	(f	h	j	p)	(g	i)
1	2	1	3	4	3	1	2	3	4	1	2	1	3	2	1
4	3	1	2	1	3	4	3	2	1	1	2	1	3	2	1
1	2	1	4	3	1	1	2	4	3	1	3	4	1	1	3
1	2					2	1								

the decimal carry from the previous digit-multiplication must have been either 2 or 3, depending on what bit was read from M_2 at t_3 time. The new product digit will therefore be either $(4 \times 3) + 2 = 14$ or $(4 \times 3) + 3 = 15$, and the decimal digit to be read out is either 4 (0100) or 5 (0101). The next-to-least significant bit of either of these digits is "zero," and is indicated in the output part of the table on row h under column 10. The two most significant bits of either of these digits are 01, and must be formed in M_1 and M_2. The carry is also 01, and must appear in C_1 and C_2 at the next bit-time. The multiple-generator must therefore change to state 0101 or f, and an f appears in the "next state" part of the table on row h under column 10. The other entries in the table may similarly be derived.

Application of the Huffman-Mealy rules to the completed table is indicated at the bottom of Table 9-31, where it is seen that no simplification is possible. The complete logic for the multiple-generator, assuming that the various states are coded as shown at the left of Table 9-31, is given in equations 9-82 through 9-86.

$$Z_4 = \bar{t}_4 M_2 + t_4 Y \overline{M}_2 + t_4 \overline{C}_1 C_2 M_1 \overline{M}_2 + C_1 C_2 \overline{M}_1 M_2 + \overline{Y}\overline{C}_1 \overline{C}_2 M_2 + \overline{Y}\overline{C}_2 M_1 M_2 + t_4 C_1 \overline{C}_2 \overline{M}_1 \overline{M}_2 \quad (9\text{-}82)$$

$$J_{C1} = Y \qquad K_{C1} = \bar{t}_4 \overline{Y} + \overline{Y}\overline{C}_2 \overline{M}_1 \quad (9\text{-}83)$$

$$J_{C2} = C_1 \bar{t}_4 + t_4 Y + C_1 \overline{M}_1 \qquad K_{C2} = \overline{C}_1 \bar{t}_4 + \overline{C}_1 \overline{M}_1 \overline{M}_2 + t_4 C_1 \overline{M}_1 + t_4 C_1 \overline{M}_2 \quad (9\text{-}84)$$

$$J_{M1} = \bar{t}_4 C_2 + \overline{C}_1 C_2 \overline{M}_2 + t_4 C_1 \overline{C}_2 M_2 \qquad K_{M1} = \bar{t}_4 \overline{C}_2 + \overline{C}_2 \overline{M}_2 + t_4 \overline{Y} M_2 + t_4 \overline{Y}\overline{C}_1 \quad (9\text{-}85)$$

$$J_{M2} = \bar{t}_4 M_1 + \overline{C}_1 \overline{C}_2 M_1 + t_4 C_1 \overline{M}_1 \qquad K_{M2} = \bar{t}_4 \overline{M}_1 + t_4 Y M_1 + \overline{Y}\overline{C}_1 \overline{M}_1 + t_4 M_1 C_1 + \overline{Y}\overline{C}_2 \overline{M}_1 \quad (9\text{-}86)$$

The design of the one-digit multiplier cannot be said to be complete until (*a*) the complementer has been designed; (*b*) a two-bit delay has been provided for the number Y, so that it appears at the input to the adder in synchronism with $4Y$ and $-Y$; (*c*) the three-input adder itself has been designed; and (*d*) the inputs to this adder are written as functions of the multiplier bits in flip-flops R_1 to R_4, and the three multiples $4Y$, Y, and $-Y$. The complementer has already been designed; the delay is easily provided by shifting Y through two flip-flops; the three-input adder is a standard design problem which will not be considered; and the inputs to this adder may be written as follows with reference to Table 9-29. In these equations Z_4 means the output of the $4Y$ multiple-generator; Z_1 refers to the delayed number Y; and Z_{-1} refers to the output of the complementer.

$$X_1 = Z_4 R_4 + Z_4 R_3 + Z_1 R_2 \overline{R}_3 \quad (9\text{-}87)$$

$$X_2 = Z_1 R_1 \overline{R}_3 + Z_1 R_1 \overline{R}_2 + Z_1 \overline{R}_1 R_2 R_3 + Z_{-1} R_1 R_2 R_3 \quad (9\text{-}88)$$

$$X_3 = Z_4 R_4 + Z_1 R_2 \overline{R}_3 + Z_1 \overline{R}_1 R_2 + Z_4 R_1 R_2 R_3 \quad (9\text{-}89)$$

Signed multiplication. The discussion so far has been devoted entirely to the multiplication of positive numbers. The method employed to operate on negative multipliers and multiplicands depends upon the way negative numbers are represented in the computer. If numbers are represented by absolute magnitude and sign, the solution of this problem is very easy: simply perform the multiplication in a straightforward manner on the two operands, using their magnitudes only, and append to the product a positive sign if both operands have the same sign and a negative sign if one operand is positive and the other negative.

If, on the other hand, negative numbers are expressed in complement form in the computer, it is necessary for the logical designer to make allowance for negative operands when designing the multiplier. The correction which must be made is a surprisingly simple one, as will be shown below. We shall only consider corrections made for the ten's complement and two's complement representations of negative numbers, although corrections for the nine's and one's complement are not much more complicated.

Suppose first that the multiplier is positive and the multiplicand negative. It is particularly easy to make the correction in this instance if we remember that, when a negative number expressed in complement form is shifted right (divided by two or ten, in the binary or decimal number systems respectively) a "one" or nine must be shifted into the most significant digit position. (See p. 295.) When we multiply with a negative multiplicand, all partial products are negative and we must observe this shifting rule. For example, the multiplication of -27 by 19 is illustrated below in the binary and decimal number systems where it will be seen that a "one" (or a nine) is inserted as the most significant bit every time a partial product is shifted to the right.

```
1 0 0 1 0 1                 -27         9 7 3
0 1 0 0 1 1                 +19         0 1 9
-----------                            ---------
1 0 0 1 0 1                 multiply    7 5 7
1 1 0 0 1 0 1               shift       9 7 5 7
1 0 0 1 0 1                 multiply    9 7 3
-------------                          ---------
0 1 0 1 1 1 1               add         9 4 8 7
1 0 1 0 1 1 1 1             shift       9 9 4 8 7
1 1 0 1 0 1 1 1 1           shift
1 1 1 0 1 0 1 1 1 1         shift
1 0 0 1 0 1                 multiply
-------------------
0 1 1 1 1 1 1 1 1 1         add
1 0 1 1 1 1 1 1 1 1 1       shift
```

Let us next suppose that the multiplicand is positive but the multiplier is negative, and investigate what special steps must be taken to obtain a proper product when the multiplier is expressed in complement form. If the multiplier has a magnitude X, it will be expressed in the machine as $r^{n+1} - X$. The straightforward multiplication of the multiplier and the multiplicand Y will evidently give the result $r^{n+1}Y - XY$. This result is too large by $r^{n+1}Y$, and that amount must be subtracted from the result of a straightforward calculation. For example, the multiplication of 27 by -19 would be carried out as follows:

0 1 1 0 1 1	+27	0 2 7
1 0 1 1 0 1	−19	9 8 1
0 1 1 0 1 1	multiply	0 2 7
0 0 1 1 0 1 1	shift	0 0 2 7
	multiply	2 1 6
	add	2 1 8 7
0 0 0 1 1 0 1 1	shift	0 2 1 8 7
0 1 1 0 1 1	multiply	2 4 3
1 0 0 0 0 1 1 1	add	2 6 4 8 7
0 1 0 0 0 0 1 1 1	shift	
0 1 1 0 1 1	multiply	
1 0 1 0 1 1 1 1 1	add	
0 1 0 1 0 1 1 1 1 1	shift	
0 0 1 0 1 0 1 1 1 1 1	shift	
0 1 1 0 1 1	multiply	
1 0 0 1 0 1 1 1 1 1 1	add	
− 1 1 0 1 1	subtract $27r^{n+1}$	− 2 7 0
1 0 1 1 1 1 1 1 1 1 1		9 9 4 8 7

This correction process may be simplified a little by noting that the most significant digit of a negative multiplier, the sign digit, is $(r - 1)$. In the normal course of events, this digit is multiplied by the multiplicand and the result is added into the partial product. The most significant digit of the multiplier thus causes the partial product to be increased by $(r - 1)r^nY = r^{n+1}Y - r^nY$, and therefore the partial product is $r^{n+1}Y - XY - (r^{n+1}Y - r^nY) = r^nY - XY$ *before* the product of the sign digit and the multiplicand have been added in. It should be evident, then, that instead of performing the correction by subtracting $r^{n+1}Y$ from the product after the multiplication is complete, we may simply subtract r^nY from the partial product instead of multiplying by the sign digit of the multiplier. In the example just given of the multiplication of 27 by -19, we may substitute the correction shown below for the last multiplication, addition, and correction given there.

0 0 1 0 1 0 1 1 1 1 1		0 2 1 8 7
− 0 1 1 0 1 1	subtract $27r^n$	− 0 2 7
1 0 1 1 1 1 1 1 1 1 1		9 9 4 8 7

It is now possible to state a very simple rule which applies either to a positive or to a negative multiplier, and results in a properly corrected product: always take the product of the most significant digit of the multiplier and the *complement* of the multiplicand, and add to the partial product. A combination of the use of this rule, and of the rules for shifting negative partial products, will result in a proper product no matter what combination of negative and positive operands is employed. As a final example, the multiplication of two negative numbers is indicated below.

1 0 0 1 0 1	−27	9 7 3
1 0 1 1 0 1	−19	9 8 1
1 0 0 1 0 1		9 7 3
1 1 0 0 1 0 1		9 9 7 3
1 1 1 0 0 1 0 1		7 8 4
1 0 0 1 0 1		
		7 8 1 3
0 1 1 1 1 0 0 1		9 7 8 1 3
1 0 1 1 1 1 0 1 1		
1 0 0 1 0 1		
0 1 0 1 0 0 0 1		
1 0 1 0 1 0 0 0 1		
1 1 0 1 0 1 0 0 0 1		
− 1 0 0 1 0 1	subtract $(-27)r^n$	− 9 7 3
0 1 0 0 0 0 0 0 0 0 0 1		0 0 5 1 3

Fast multiplication. Among the most important factors which affect the design of the multiplier and therefore of the arithmetic unit in general is the speed with which multiplication should take place.

The desired speed for multiplication should be based on the speed of addition or subtraction and upon the expected frequency with which multiplication is carried out. For example, if it is expected in the average problem that there will be ten additions and subtractions for every multiplication, a reduction in multiplication time from ten addition times to one addition time may increase the complexity of the multiplication logic by a factor of ten, but cannot even reduce total computing time by factor of two. Of course, the system designer may decide that some such slight increase in speed justifies a large additional expense—but this would depend on the job the system is designed to do.

The simplest way to speed up multiplication is to multiply by more than one multiplier digit at a time. Referring to Figure 9-13, we see that if x_i is a single digit of the multiplier, the total multiplication will require as many addition times as there are digits in the multiplier. (In a serial machine an addition time will be one word-time; in a parallel machine, one or a few bit-times.) If the multiplier digit x_i is taken to be two digits instead of one, the digit-multiplier is correspondingly more complicated and the shifting operations are changed, but the multiplication will now be complete in half as many addition times as there are digits in the multiplier. For binary machines, the design of a multidigit-multiplier based on Figure 9-16 is very easy. If x_i is to be three binary digits, we need only provide a three-input binary adder and a multiple-generator which produces Y, $2Y$, and $4Y$—each of which may be found simply by shifting Y to the left.

The ultimate in multiplication speed obtained this way is indicated in Figure 9-19, where all the multiplier digits are available simultaneously and control the addition of Y into the product. The particular configuration of multiplications, additions, and delays indicated in Figure 9-19 is that employed in the University of Manchester computer. The heart of this multiplier is a group of two-input binary adders (indicated by the dashed circles), each of which forms the sum of x_iY and the delayed sum of other products. A set of flip-flops C_i are required for the carries, and another set P_i are required to provide the necessary delays. The addition is accomplished in two word-times (the time required to read the product out of the multiplier), and obviously about $3n$ memory elements are required for an n-bit binary multiplier.

Another common method employed to speed up multiplication simply requires that the designer recognize that no addition-time is necessary to modify a partial product when a multiplier digit is zero. This obvious fact is most often used to stop the multiplication operations as soon as the last nonzero digit of the multiplier has been used, with the result that a multiplication lasts as many word-times as there

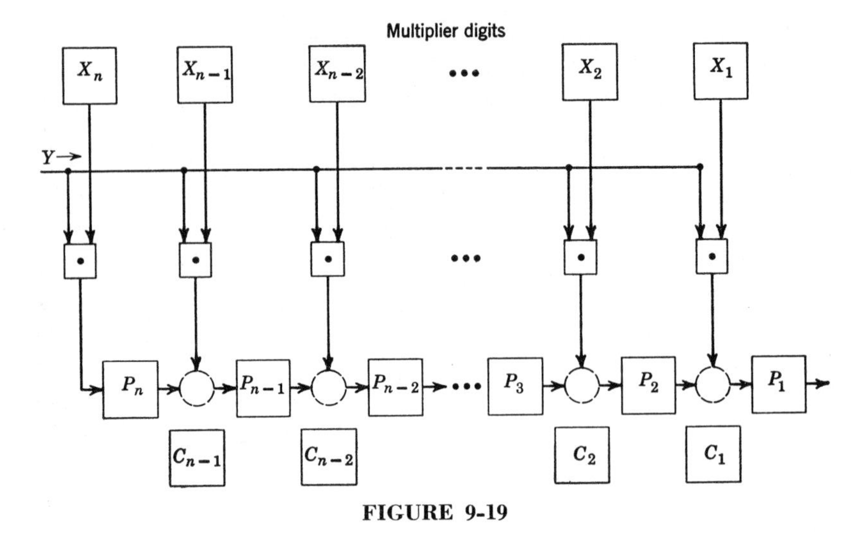

FIGURE 9-19

are significant digits in the multiplier. Thus, in a machine having words ten decimal digits long and requiring one addition-time per multiplier digit, a multiplication of any multiplicand by

0000000243 will require three addition-times
0000014902 will require five addition-times
0079265822 will require eight addition-times.

A fast multiplier of this kind requires either that the designer provide circuits which recognize a string of most significant zeros in the multiplier and stop the multiplication accordingly; or that the computer user specify how many significant multiplier digits should be used for each multiplier, and the designer provide circuits to store and interpret this number properly.

Division

Division may be mechanized in at least three distinct ways: by employing the common and familiar trial-and-error method; by using a "nonrestoring" algorithm; or by making use of an iterative procedure. The trial-and-error method is the one taught in school and used by all of us. In it, a multiple of the divisor is subtracted from the dividend, leaving a remainder. The multiplier is stored as the first digit of the quotient, and a new multiple of the divisor is chosen to subtract from the new remainder. It is characteristic of this process that the multiples are always chosen so that the remainders are positive, and this characteristic makes the process somewhat difficult to mechanize. In a binary divider, for example, it is either necessary to compare the divisor and old remainder before subtraction in order to find out whether the new remainder will or will not be positive, or else to go ahead with the subtraction and then correct the new remainder (if it is negative) by adding back the divisor.

The nonrestoring method eliminates this last step, leaving the remainder negative and then taking appropriate action to correct the quotient. The division of 26930 by 3 is carried out in Table 9-32 using a nonrestoring algorithm. The divisor or shifted divisor D is first subtracted from the dividend V until the difference is negative; as soon as a negative difference is detected, the divisor is shifted to the right and added to the remainder until the sum turns positive. The divisor is again shifted right one digit, and subtracted until the quotient again becomes negative. The process may be repeated indefinitely. The quotient Q is built up by adding some power of ten to the quotient register every time the shifted divisor is subtracted, and subtracting a power of ten every time the shifted divisor is added to the remainder. A binary division may be carried out in precisely the same general manner.

Division may also be accomplished by use of some iterative formula which, when applied repeatedly to an initial quotient, derives a new quotient which is closer to the correct one than was the initial trial. Repeated application of the formula will then eventually result in a number as close to the quotient as is desired. The Harvard computer, Mark IV, employs an iterative divider which may be described as follows. Suppose we want to find the quotient of X and Y. We first shift these two numbers by multiplying or dividing them both by the base r, until the divisor Y is less than one but greater than or equal to r^{-1}. The shifted values of X and Y we call X_1 and Y_1. The desired quotient is then X_1/Y_1. We now find a number p_1 such that $1 < p_1 < 1/Y_1$ and multiply p_1 by X_1 and Y_1, obtaining two new numbers called X_2 and Y_2. Then obviously

$$\frac{X_1}{Y_1} = \frac{X_1 p_1}{Y_1 p_1} = \frac{X_2}{Y_2}$$

We continue in the same way, finding a sequence of numbers p_i such that $1 < p_i < 1/Y_i$ and performing suitable multiplications to form

$$X_{i+1} = p_i X_i \qquad Y_{i+1} = p_i Y_i$$

As the process continues, Y_i approaches one as closely as we desire; and therefore X_i approaches the quotient.

Table 9-32

V	D	δQ	Q
+26930			
	−30000	+10000	10000
− 3070			
	+ 3000	− 1000	9000
− 0070			
	+ 3000	− 1000	8000
+ 2930			
	− 300	+ 100	8100
+ 2630			
	− 300	+ 100	8200
+ 2330			
	− 300	+ 100	8300
+ 2030			
	− 300	+ 100	8400
+ 1730			
	− 300	+ 100	8500
+ 1430			
	− 300	+ 100	8600
+ 1130			
	− 300	+ 100	8700
+ 830			
	− 300	+ 100	8800
+ 530			
	− 300	+ 100	8900
+ 230			
	− 300	+ 100	9000
− 70			
	+ 30	− 10	8990
− 40			
	+ 30	− 10	8980
− 10			
	+ 30	− 10	8970
+ 20			
	− 3	+ 1	8971
+ 17			
	− 3	+ 1	8972
+ 14			
	− 3	+ 1	8973
+ 11			
	− 3	+ 1	8974
+ 8			
	− 3	+ 1	8975
+ 5			
	− 3	+ 1	8976
+ 2			
	− 3	+ 1	8977
− 1			

In a decimal divider, the number p_i may be found by the following rule: subtract Y_i from $0.99999 \cdots$, drop all digits to the right of the first nonzero digit, and add one to the result. The method is illustrated in Table 9-33, where we find the quotient of ten and five. Note one

Table 9-33

i	X_i	Y_i	p_i
1	1.0	0.5	1.4
2	1.4	.7	1.2
3	1.68	.84	1.1
4	1.848	.924	1.07
5	1.97736	.98868	1.01
6	1.9971336	.9985668	1.001
7	1.9991307336	.9995653668	1.0005

advantage to this method of choosing p_i: each multiplication by p_i requires only one single-digit multiplication.

Comparison

Weighted number systems in which the weight of each digit position is r times the weight of the next less significant bit position have the nice property that it is relatively easy to determine which of two numbers is the larger. Starting with the most significant digits of the two numbers, we need only compare them, digit by digit, until we find a digit position in which two digits are different. The larger digit is then in the larger of the two numbers regardless of the magnitude of subsequent digits. A little reflection will disclose that the same rule applies when both numbers are negative and expressed in complement form, for then the number whose absolute magnitude is larger will in complement form be the smaller number. If the signs of the two numbers are different, one can determine which is the larger by looking at the sign digit alone without regard to the other digits of the numbers.

Table 9-34

	Most Significant																Least Significant
$X =$	0	3	7	5	9	0	4	2	6	6	7	1	4	7	1	2	6
$Y =$	0	3	7	6	4	0	4	2	1	1	7	9	2	8	5	2	6
X larger					✓	✓	✓	✓	✓	✓			✓				
Y larger	✓	✓	✓	✓							✓	✓		✓	✓		

← Start here

In a serial digital computer, of course, the least significant digits appear first in time. A serial comparator must therefore be made to remember which of the two numbers being compared contained the bigger digit the last time two digits were unequal. For example, the comparison of two positive decimal numbers is shown in Table 9-34.

In Table 9-35 the Huffman-Mealy table for a binary comparator is shown. The two numbers are assumed to appear serially, least significant bit first, from flip-flops X and Y. Negative numbers are represented in two's complement form, and a signal t_n marks the most significant (sign) digit of the two words. Two outputs, Z_1 and Z_2, are desired which display the result of the comparison at t_n time as follows:

Table 9-35

State	Next State								Outputs: Z_1, Z_2				
$t_n\ X\ Y \rightarrow$	000	001	010	011	100	101	110	111	0–	100	101	110	111
a	a	b	c	a	a	a	a	a	–	00	10	01	00
b	b	b	c	b	a	a	a	a	–	01	10	01	01
c	c	b	c	c	a	a	a	a	–	10	10	01	10

$t_n = 1$		
Z_1	Z_2	
0	0	$X = Y$
1	0	$X > Y$
0	1	$X < Y$

The circuit is assumed to be in state a at the beginning of a word-time, and it remains in state a as long as the two numbers X and Y are equal. As soon as two digits of the two numbers are unequal, the circuit assumes either state b or c, depending on which of the two digits was a "one," Y or X respectively. At t_n time, if X and Y (the sign digits) are equal, the state of the circuit determines which of the two numbers was larger. If the sign digits are unequal, they themselves determine which number was larger. The circuit is then returned to state a.

If states a, b, and c are identified with flip-flops A and B as follows:

	A	B
a	0	0
b	0	1
c	1	0

the flip-flop logic and output logic for the comparator are as follows:

$$Z_1 = \overline{X}Y + YA + \overline{X}A \qquad Z_2 = X\overline{Y} + XB + \overline{Y}B \tag{9-90}$$

$$J_A = \bar{t}_n X\overline{Y} \qquad K_A = t_n + \overline{X}Y \tag{9-91}$$

$$J_B = \bar{t}_n \overline{X}Y \qquad K_B = t_n + X\overline{Y} \tag{9-92}$$

The design of a parallel comparator for two numbers residing in two flip-flop registers is easily written. However, the resulting expression is a function of every memory element in the two registers, and is therefore expensive to mechanize.

BIBLIOGRAPHY

General

Staff of Harvard Computation Laboratory, *Synthesis of Electronic Computing and Control Circuits,* Harvard University Press, Cambridge, 1951.

A. D. Booth and K. H. V. Booth, *Automatic Digital Calculators,* Butterworth Scientific Publications, London, 1953.

R. K. Richards, *Arithmetic Operations in Digital Computers,* D. Van Nostrand Company, New York, 1955.

R. C. Jeffrey, "Arithmetic Analysis of Digital Computing Nets," *J. Association for Computing Machinery,* **3,** no. 4, 360–375 (1956).

J. V. Blankenbaker, "How Computers Do Arithmetic," *Control Engineering,* **3,** no. 4, 93–99 (Apr. 1956).

Number Representation

S. Lubkin, "Decimal Location in Computing Machines," *Mathematical Tables and Other Aids to Computation,* **3,** 44–50 (1948).

R. F. Shaw, "Arithmetic Operations in a Binary Computer," *Review of Scientific Instruments,* **21,** no. 8, 687–693 (1950).

G. S. White, "Coded Decimal Number Systems for Digital Computers," *Proc. I.R.E.,* **41,** no. 10, 1450–1452 (Oct. 1953).

B. M. Gordon, "Adapting Digital Techniques for Automatic Controls," *Electrical Manufacturing,* 136–143 (Nov. 1954); 120–125 (Dec. 1954).

R. W. Murphy, "A Positive-Integer Arithmetic for Data Processing," *I.B.M. J. of Research and Development,* **1,** no. 2, 158–170 (1957).

L. B. Wadel, "Negative Base Number Systems," *I.R.E. Trans. on Electronic Computers,* **EC-6,** no. 2, 123 (1957).

Complete Arithmetic Units

F. C. Williams et al., "Universal High Speed Computers; Serial Computing Circuits," *Proc. of the I.E.E.*, **99,** Part II, 107–123 (1952).

J. R. Stock, "An Arithmetic Unit for Automatic Digital Computers," *Z. angew. Math. und Phys.*, **5,** 168–172 (Mar. 1954).

W. Woods-Hill, "An Electronic Arithmetic Unit," *Electronic Engineering*, 212–217 (May 1955).

R. C. M. Barnes, G. A. Howells, and G. H. Cooke-Yarborough, "Transistor Arithmetic Circuits for an Interleaved Digit Computer," *Proc. of the I.E.E.*, **103,** Part B, Suppl. 3, 371–381 (Apr. 1956).

K. D. Tocher and M. Lehman, "A Fast Parallel Arithmetic Unit," *Proc. of the I.E.E.*, **103,** Part B, Suppl. 3, 520–527 (Apr. 1956).

Addition and Subtraction

R. Townsend, "Serial Digital Adders for a Variable Radix of Notation," *Electronic Engineering*, 410–416 (Oct. 1953).

M. W. Allen, "A Decimal Addition-Subtraction Unit," *Proc. of the I.E.E.*, **103,** Part B, Suppl. 1, 138–145 (Apr. 1956).

A. Weinberger and J. L. Smith, "A One-Microsecond Adder Using One-Megacycle Circuitry," *I.R.E. Trans. on Electronic Computers*, **EC-5,** no. 2, 65–72 (June 1956).

Multiplication

A. A. Robinson, "Multiplication in the Manchester University High Speed Digital Computer," *Electronic Engineering*, 6–10 (Jan. 1953).

G. B. B. Chaplin, R. E. Hayes, and A. R. Owens, "A Transistor Digital Fast Multiplier with Magnetostrictive Storage," *Proc. of the I.E.E.*, **102,** Part B, no. 4, 412–414 (July 1955).

J. E. Robertson, "Two's Complement Multiplication in Binary Parallel Digital Computers," *I.R.E. Trans. on Electronic Computers*, **EC-4,** no. 3, 118 (Sept. 1955).

G. Estrin, B. Gilchrist, and H. J. Pomerene, "A Note on High Speed Digital Multiplication," *I.R.E. Trans. on Electronic Computers*, **EC-5,** no. 3, 140 (Sept. 1956).

Division

Harvard Computation Laboratory Progress Report, Number 18, May-August, 1951.

Square Rooting

E. H. Lenaerts, "Automatic Square Rooting," *Electronic Engineering*, 287–289 (July 1955).

EXERCISES

1. Examine the weighted codes of Table 9-1. Can you devise a different 8-4-2-1 code? A different 5-2-1-1 code?

2. In a "floating-binary" machine, a number N is represented by a coefficient, a, and an exponent, b, chosen so that $N = a \times 2^b$. Suppose that two such numbers are stored in parallel form in two flip-flop registers. Write the rules which must be mechanized to add these two numbers together.

3. Design a parallel decimal counter which employs the 5-1-1-1-1 code of Table 9-1. Write the complete (simplified) logic for two digits (ten memory elements) of this counter. Use J-K flip-flops.

4. Use the Huffman-Mealy method of synthesis to design, (*a*) the two-input binary adder, (*b*) the three-input binary adder.

5. Draw a schematic wiring diagram (similar to that of Figure 9-7) for one stage of the two-bit-time parallel adder of equations 9-40 through 9-44.

6. Use the Huffman-Mealy design procedure to design a serial decimal adder which sums two numbers represented in the excess-three code. The least significant bit of each sum digit should be read out at t_3 time.

7. Use the Huffman-Mealy synthesis procedure to design a decimal adder-subtracter operating on numbers in the 8-4-2-1 code. Start with Table 9-15 and include another input, X_3, which is "one" when addition is to take place, and "zero" when the number from X_2 is to be subtracted from the number from X_1. Simplify the table wherever possible. Assume that the circuit is always in state a when an addition or a subtraction begins.

8. In a certain serial, binary computer, negative numbers are represented in two's complement form. Numbers are always added and subtracted without regard to sign, since the sign of the result is automatically correct as long as no overflow has occurred. State the rules which must be mechanized to detect an overflow during either addition or subtraction. In a computer having a word-length of six bits including sign, for example, overflow occurs when 25 is added to 23, when (-25) is subtracted from 23, and when 25 is subtracted from (-23).

9. Design a serial binary adder which forms the sum of a number, Y, and the contents of a circulating register, X. (Cf. Figure 9-5 and equations 9-16 and 9-17.) The adder is to handle negative numbers as well as positive ones, and a negative number is represented by its *one's* complement. Carry out the design in two ways: first, aim for the least amount of equipment; and second, provide whatever equipment is necessary to mechanize an addition in one word-time.

10. A two-decimal-digit number is stored in the X-register of Figure 9-2. Write the logic which replaces that number by its ten's complement in one bit-time. Use J-K flip-flops, and assume the digits are represented by the

a. 3-3-2-1 code of Table 9-1.
b. 6-3-1-1 code of Table 9-1.

11. Design a circuit which replaces the decimal number in the circulating register of Figure 9-5 by its ten's complement in one word-time. Assume the decimal number is coded in the

a. 3-3-2-1 code of Table 9-1.
b. 6-3-1-1 code of Table 9-1.

12. Two eight-bit binary numbers appear in two flip-flop registers X_1–X_8 and Y_1–Y_8. The least significant bits of the two numbers appear in X_1 and Y_1, and the sign bits in X_8 and Y_8. Negative numbers are expressed as two's complements.

a. Design a one-bit-time subtracter which puts the difference $(Y - X)$ into the Y-register. Write the complete logic for flip-flops Y_1, Y_2, and Y_3, and the general logic for flip-flop Y_i. Use J-K flip-flops.

b. For the subtracter designed in part a, write an equation for a function which will be "one" when the Y-register overflows, i.e., when the difference to be stored in Y is a number whose absolute magnitude is 128 or larger.

13. In a particular computer, decimal numbers are to be represented in *series-parallel* fashion. That is to say, a ten-decimal-digit number is read from the memory in ten bit-times, so that in each bit-time the four bits corresponding to one decimal digit are read. A decimal circulating register thus consists of four ten-bit binary circulating registers operated in parallel. The read amplifiers for these binary registers may be denoted by X_{r1}, X_{r2}, X_{r3}, and X_{r4}, where X_{r1} contains the least significant bit of a decimal digit, and X_{r4} the most at any bit-time. Design an adder, using

a. the 8-4-2-1 code

b. the 5-2-1-1 code of Table 9-1

which places the sum of the number in the circulating register X, and a number from the memory (flip-flops Y_1, Y_2, Y_3, and Y_4), back into the circulating register (write amplifiers X_{w1}, X_{w2}, X_{w3}, and X_{w4}.)

14. Design a $4Y$ multiple-generator which forms the quadruple of a number Y read from a memory in series-parallel form (see Exercise 13). The decimal digits are each expressed in the

a. 8-4-2-1 code

b. 6-3-1-1 code of Table 9-1.

15. The binary multiplier described by equations 9-74 through 9-81 is to be incorporated into a complete arithmetic unit. Flip-flops M_4–M_7 are provided to control the arithmetic unit, determining what operation is to be performed. As before, M_4 is set to "one" when a multiplication is to be carried out, and the logic of equations 9-71 through 9-73 then applies. However, when $M_4 = 0$, the action taken depends on the state of the other control flip-flops, as will be described below. The operands to be used are read from an amplifier X_r, and appear least significant bit first, always starting at t_1 time. Results are to be stored by writing in a write amplifier X_w, and the least significant bit of a number sent to storage must be *in* X_w at t_1 time. It may be assumed that each configuration of flip-flops M_4–M_7 lasts for one word-time, beginning on some t_0 and ending on the following t_4.

M_4	M_5	M_6	M_7	
0	0	0	0	Circulate contents of all registers—do nothing else.
0	0	0	1	Transfer the number from X_r into the multiplier register.
0	0	1	1	Transfer the number from X_r into the multiplicand register.
0	1	0	1	Add the number from X_r to the number in the most significant half of the partial product register, putting the result in that same register.
0	1	1	1	Subtract the number from X_r from the number in the most significant half of the partial product register, putting the result in that same register.

M_4	M_5	M_6	M_7	
0	1	1	0	Transfer the most significant half of the partial product register to write amplifier X_w.
0	0	1	0	Replace the number in the most significant half of the partial product register by the number in the least significant half. Then store zero in the least significant half.
1	–	–	–	Multiply.

Write the logic for this arithmetic unit; i.e., rewrite equations 9-74 through 9-81 so as to carry out these new functions. Add another equation for the input to write amplifier X_w. (Assume that all numbers are positive.)

16. Redesign the binary multiplier (equations 9-71 through 9-81) so that it finds the correct product of two numbers, regardless of their signs. Assume that negative numbers are represented by two's complements, and use the correction procedure developed in this chapter.

17. (*a*) Use the Huffman-Mealy synthesis procedure to design a $3Y$ multiple-generator, to operate on decimal numbers represented in the 8-4-2-1 code. (*b*) Write the equations defining the inputs to a four-input adder which combines Y and $3Y$ to form any multiple between Y and $9Y$. (Cf. column 5 of Table 9-28. The result will be four equations similar in form to equations 9-87 through 9-89.)

18. Derive the rules which make it possible automatically to multiply binary numbers without regard to sign, when negative numbers are represented in *one's* complement form.

19. Write the input equation for the ith delay element (P_i) of the fast multiplier of Figure 9-19. Write, also, the input equations for the carry flip-flop C_i (assumed to be a J-K flip-flop.)

20. Two binary numbers are stored in parallel in two eight-bit registers A_1–A_8 and B_1–B_8. A_1 and B_1 hold the least significant bit of each number, and A_8 and B_8 the sign bit. Negative numbers are represented as two's complements. Write an equation which is true whenever the number in A is greater than or equal to the number in B.

10 / *Error-free computer operation*

The circuit designer does his best, in putting together the circuits that make up a digital computer, to insure long and trouble-free operation. However, no matter how carefully and conservatively he works, the circuit components he uses have a certain probability of failure, and in any given computer one or more of the hundreds present will eventually fail. It is possible for the logical designer to anticipate troubles of this kind and to provide, in the computer logic, certain means for detecting and even correcting some errors. It is the purpose of this chapter to show what the logical designer can do.

However, it can be and has been argued that no amount of self-checking by the computer can eliminate the effect of *all* failures; and if reliable operation is important, it is necessary for the computer *user* to regard the computer's results with a certain amount of scepticism, and to check them before using them. The computer user can himself arrange to carry out any kind of check he desires automatically in some computers, without the necessity for the logical designer to provide any checking equipment at all (see, e.g., the general-purpose computer of Chapter 11). Because of this, and because error-checking equipment is used only moderately in most present-day computers, this chapter will be a short one.

CHECKING INFORMATION TRANSFERS

At any particular instant in a computing system a great deal of information is stored in the computer memory and in the arithmetic and control units. In addition, a word or a few words will, at that instant, be in transit from one place to another, e.g., from input device to memory unit, or from the memory to the arithmetic unit. Transfers of this kind present both an opportunity for circuits to fail and an opportunity for checks to be made on the information being transferred.

One obvious way of providing an automatic check on the transfer of information is to duplicate the transfer, once or many times, and to compare the results of the duplications. Von Neumann has shown that, under suitable conditions, the probability of failure may be reduced as far as is desired by providing enough duplicate equipment. However, the number of duplications required is impractical with conventional components, and his conclusions are not useful in present-day computers. The simple duplication of equipment and comparison of results are certainly justified in some applications (see, e.g., the reservation computer described in Chapter 1), but it is possible to provide simpler methods for error-detection. It is also possible to devise circuits which detect and *correct* simple errors of data transmission.

Error-Detecting Codes

Instead of duplicating transfer operations and equipment, it is possible to attach extra bits to each block of data being transferred in such a way that these bits make it possible to detect and correct certain errors. The simplest kind of error-detection scheme requires that one bit, usually called the *parity* bit, be added to each word. In Table 10-1, for example, each of the sixteen possible four-bit words is augmented by a parity bit P which is chosen so that the total number of "ones" in each word is even. If the original code (A, B, C, D) contained one or three "ones," the parity bit is made "one"; if it contained zero, two, or four "ones," the parity bit is chosen to be "zero." This five-bit word may then be stored and transferred in place of the original four-bit code, and as often as is desired the five bits may be examined to see whether an even number of "ones" are present. Obviously an odd number of errors, each involving the incorrect transmission of a "zero" or a "one," will always result in a code containing an odd number of "ones," which can then immediately be detected as an error. On the other hand, an even number of errors will either result in the original code (e.g., when 10100 changes to 10110 and then back to

10100) or in an incorrect but acceptable code (e.g., if the first two bits change and 10100 becomes 01100).

The precise effect of the addition of this parity bit may be seen in Figure 10-1, where the code of Table 10-1 is plotted on a Veitch diagram. Note that each of the sixteen code signals is separated from its nearest neighbor by an empty square. That is to say, if we examine any particular code symbol, the change of any one bit in that symbol corresponds to a move on the Veitch diagram from a numbered square to a blank square. The blank squares, of course, correspond to words containing an odd number of "ones" and therefore to errors in transmission.

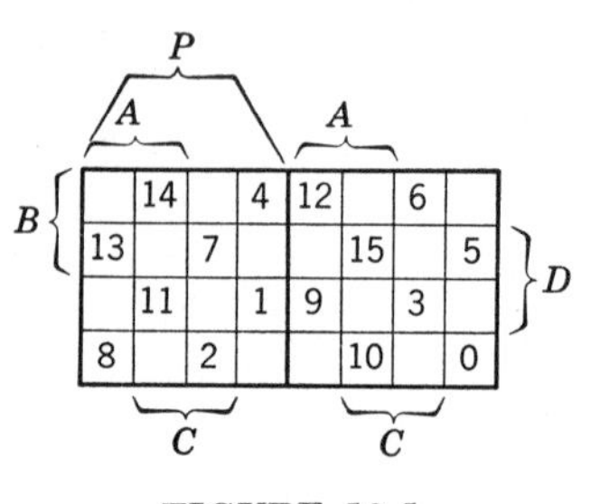

FIGURE 10-1

We may detect the presence of an error in this code by examining the five bits either all at once or sequentially—either in parallel or in series fashion. For parallel error-detection, we can derive a circuit

Table 10-1

	A	B	C	D	P
0	0	0	0	0	0
1	0	0	0	1	1
2	0	0	1	0	1
3	0	0	1	1	0
4	0	1	0	0	1
5	0	1	0	1	0
6	0	1	1	0	0
7	0	1	1	1	1
8	1	0	0	0	1
9	1	0	0	1	0
10	1	0	1	0	0
11	1	0	1	1	1
12	1	1	0	0	0
13	1	1	0	1	1
14	1	1	1	0	1
15	1	1	1	1	0

whose output will be "one" whenever five flip-flops A, B, C, D, and P contain a forbidden code simply by writing an equation for the blank squares in Figure 10-1. For serial error-detection, we require a counter to count "ones" as the number appears serially, and we must examine the counter after the last bit of the number has passed, and give an alarm if an odd number of "ones" has been counted. Since all we have to know is whether the count was even or odd, we need only supply the least significant bit of a binary counter. If we are reading the number to be checked serially from X, and if the space bit between words is identified by the signal t_0, then we can reset an error flip-flop E at t_0, and count "ones" by means of the following logic

$$J_E = X\bar{t}_0 \qquad K_E = X + t_0 \tag{10-1}$$

If $E = 1$ at t_0 time, the last word transferred was in error.

Although we have only discussed the addition of a parity bit which makes the number of "ones" in a word even, we could obviously have chosen the opposite convention and made the number of "ones" odd. The choice between even and odd parity should be based on the effect the most probable kind of error will have on the checking operation. For example, if an error in transmission is likely to cause all bits in a word to be transferred as "zeros," then odd parity is obviously called for. On the other hand, if an error is likely to generate a long sequence of "ones," and if a word (including the parity bit) contains an odd number of digits, even parity is to be preferred.

The addition of a parity bit obviously does no more than introduce some redundancy into a code. Whenever redundancies or "forbidden combinations" are possible, the circuits which transmit and store them may be checked by noting whether any such combinations are present. In Chapter 6, for example, we designed a circuit which would give an alarm whenever an incorrect four-bit code was read from a source of 8-4-2-1 coded decimal digits. Such a check is useful and is widely employed. However, it obviously does not in general detect *any* one-bit error as the parity check does.

Error-Correcting Codes

When an error is detected by any means, the designer must decide what action the computer should take. If a parity error is discovered in the course of transferring data from one place to another, what should be done? Should the computer be stopped, and an "error" light be turned on? Should the transfer automatically be repeated, in the hopes that the error is of a transient kind which will not occur again? Or should some other actions be taken? In this section we shall discuss special codes which make it possible for the computer to detect and *correct* certain errors automatically. These codes are generally known as Hamming codes.

It is often surprising to the student to learn that it is possible to devise a code which is able to correct itself. However, a glance at the Veitch diagram of Figure 10-2 will show how this is possible. There, three bits A, B, and C are used to convey one of two messages, a or b. The code for a is 000 and that for b is 111, as indicated by boldface type in Figure 10-2. Note that a *one-bit* error in either of these messages changes the code to a new combination associated with the message from which it changed. Thus a single change in a (000) results in one of the three codes (100, 010, 001) indicated by the lightface a's in Figure 10-2, and a single change in b (111) results in one of three *different* codes (011, 101, 110). Thus if the message a is sent and the first bit gets garbled in transmission, it will be received as 100. By assuming that only one error occurred, we can deduce that the original code cannot have been b but must have been a.

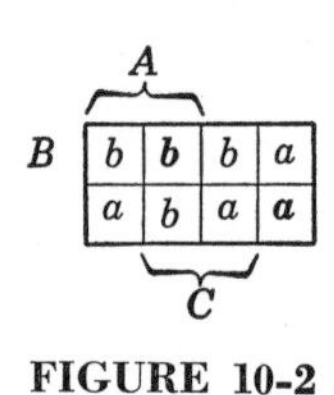

FIGURE 10-2

This general idea can be extended to effect the detection and correction of codes where more than one error in transmission occurs. In Figure 10-3*a*, for example, four bits are used to convey one of two messages. If a single error occurs in transmitting this code, the received code can be identified with the original message as before. If two errors occur, the code will be one of those identified by the numeral 2, and such a code can be interpreted as a double error though it will not be possible to identify it with one of the original messages. In Figure 10-3*b*, five bits are used and permit the detection and correction of single or double errors.

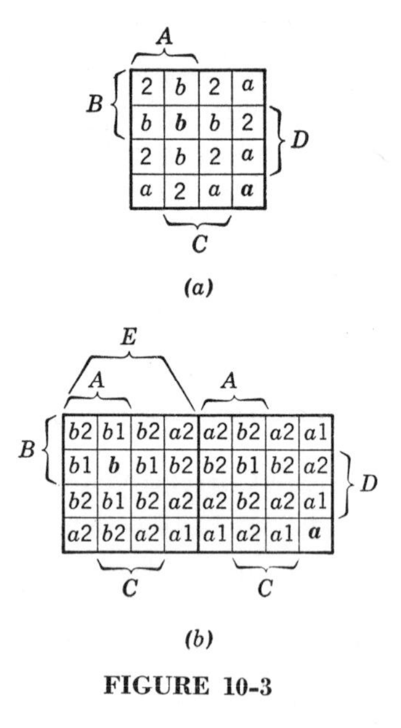

FIGURE 10-3

We shall now show how it is possible to derive codes which make *single*-error-correction possible on messages of any given length. Suppose we want to transfer words having m significant bits, employing a code which will detect and correct any one-bit error of transmittal. We do this by adding k check bits to each word, where the message codes, including check bits, are chosen as follows.

1. Enough checking bits must be supplied so that a k bit *checking number* (whose derivation will be described below) has enough states

to identify any of the $(m + k)$ bit positions, or to indicate that no error has occurred. Since a k bit number can represent 2^k different states, we must have

$$2^k \geq m + k + 1 \qquad (10\text{-}2)$$

Values of k necessary for different values of m are given in Table 10-2.

Table 10-2

m	k_{min}
1	2
2–4	3
5–11	4
12–26	5
27–57	6

2. The $(m + k)$ bit positions in the code are numbered from 1 to $(m + k)$, starting with the least significant bit. The k checking bits are labeled $P_0, P_1, P_2, \cdots, P_{k-1}$, and are inserted in the bit positions numbered 1, 2, 4, 8, $\cdots$, 2^{k-1} respectively. The other m bits—the message bits—may be inserted in any order between checking bit positions.

3. The check bits $P_0, P_1, P_2, \cdots, P_{k-1}$ are chosen in such a way that they serve as parity checks for certain bit positions in the word. P_0 is chosen so that there are an even number of "ones" in bit positions 1, 3, 5, 7, 9, 11, $\cdots$ of each word. P_1 is chosen so that there are an even number of "ones" in bit positions 2, 3, 6, 7, 10, 11, 14, 15, $\cdots$ of each word. Similarly, P_2 checks positions 4, 5, 6, 7, 12, 13, 14, 15, 20, $\cdots$ and P_3 positions 8, 9, 10, 11, 12, 13, 14, 15, 24, 25, $\cdots$.

The code for a four-bit message, for example, requires three check bits (Table 10-2). In Table 10-3 a seven-bit code is shown which was derived using the rules above. The three check bits are labeled P_0, P_1, and P_2, and are inserted in bit positions 1, 2, and 4. A, B, C, and D are then inserted in the remaining positions. P_0 is chosen so as to provide even parity for positions 1, 3, 5, and 7. Similarly, P_1 maintains even parity for positions 2, 3, 6, and 7, and P_2 for positions 4, 5, 6, and 7. The resulting code is plotted on a Veitch diagram in Figure 10-4, where it will be seen that every code position is at least three "moves" from every other one.

The error-detecting and correcting operation is carried out by deriving a k bit *checking number* as follows. The least significant bit of the checking number is found by carrying out a parity check on bits

Table 10-3

	Bit Position						
	7	6	5	4	3	2	1
Message	A	B	C	P_2	D	P_1	P_0
0	0	0	0	0	0	0	0
1	0	0	0	0	1	1	1
2	0	0	1	1	0	0	1
3	0	0	1	1	1	1	0
4	0	1	0	1	0	1	0
5	0	1	0	1	1	0	1
6	0	1	1	0	0	1	1
7	0	1	1	0	1	0	0
8	1	0	0	1	0	1	1
9	1	0	0	1	1	0	0
10	1	0	1	0	0	1	0
11	1	0	1	0	1	0	1
12	1	1	0	0	0	0	1
13	1	1	0	0	1	1	0
14	1	1	1	1	0	0	0
15	1	1	1	1	1	1	1

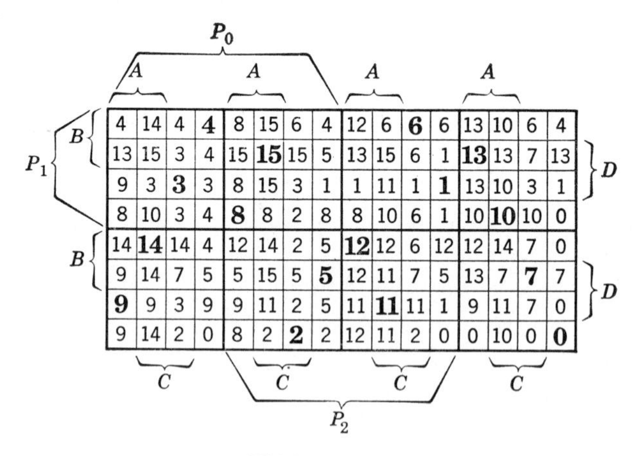

FIGURE 10-4

1, 3, 5, 7, 9, ···, and writing a "zero" if the check is correct and a "one" if it is incorrect. The next bit of the checking number is made "zero" if the P_1 parity check (on bits 2, 3, 6, 7, 10, 11, 14, 15, ···) is correct, and a "one" if it is not. The other bits of the checking number are found in a similar way by carrying out the remaining parity checks.

If the check number is zero (all parity checks correct), the received code was one of the allowable ones, and it is presumed that no error of transmission has occurred. If the check number is not zero, it is interpreted as the number of the bit position which is in error, and that bit position must be changed.

In the example of Table 10-3, suppose that message six (0110011) is transmitted, and is received as 0110111, with an error in the third bit position. Carrying out the P_0, P_1, and P_2 parity check, we find

		Checking Number
P_0: (1, 3, 5, 7) = (1, 1, 1, 0).	Odd parity	1
P_1: (2, 3, 6, 7) = (1, 1, 1, 0).	Odd parity	1
P_2: (4, 5, 6, 7) = (0, 1, 1, 0).	Even parity	0

The checking number is thus 011, indicating that the third bit position is in error.

It should now be apparent why this ingenious coding method operates correctly: parity bit P_i is chosen so that it checks the parity of that set of bit positions whose checking numbers contain a "one" in the (2^i)th position.

We conclude this section by designing a simple error-correcting circuit for messages expressed in the code of Table 10-3. The equipment configuration is shown in Figure 10-5, where it will be assumed that the number to be checked appears serially at the output of X_1 during seven consecutive bit-times. A space bit separates adjacent words. A counter register (T_2, T_1, and T_0) counts from zero to seven and is in the zero state ($T_2 = T_1 = T_0 = 0$) when the space bit is read from X_1.

During every word-time, flip-flops C_0, C_1, and C_2 carry out the P_0, P_1, and P_2 parity checks, respectively, on the number from X_1. Meanwhile, the number from X_1 is shifted into and through the shift register shown. When the space bit is in X_1, the checking number for the previous message is in C_0, C_1, and C_2, and is shifted into C_3, C_4, and C_5. During the next word-time, the word to be corrected is read from A_r, starting with the least significant bit (bit position 1) when T_2, T_1, T_0 contain (0, 0, 1). At every bit-time the number in (C_5, C_4, C_3) is reduced by unity, so that whenever (C_5, C_4, C_3) contain (0, 0, 1),

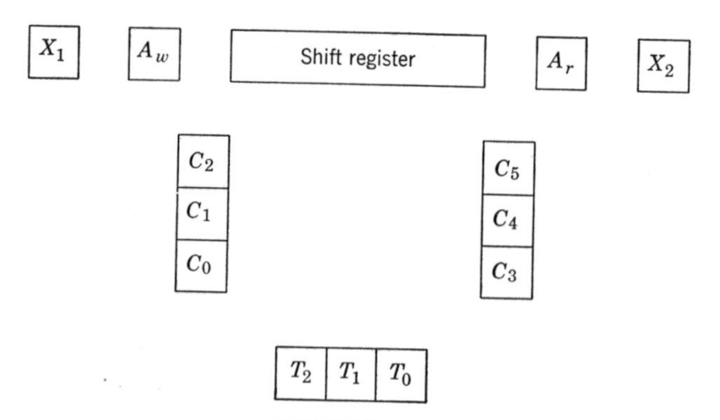

FIGURE 10-5

the bit in A_r is incorrect and should be reversed. The (C_5, C_4, C_3) counter is designed (see Table 10-4) so that it remains in the (0, 0, 0) state when it is once set there.

Table 10-4

n			$n+1$		
C_5	C_4	C_3	C_5	C_4	C_3
1	1	1	1	1	0
1	1	0	1	0	1
1	0	1	1	0	0
1	0	0	0	1	1
0	1	1	0	1	0
0	1	0	0	0	1
0	0	1	0	0	0
0	0	0	0	0	0

$$J_{C5} = 0 \qquad K_{C5} = \bar{C}_3\bar{C}_4$$

$$J_{C4} = \bar{C}_3 C_5 \qquad K_{C4} = \bar{C}_3$$

$$J_{C3} = C_4 + C_5 \qquad K_{C3} = 1$$

An example of the operation of this circuit is indicated in Table 10-5, where it will be seen that an incorrect message (0110111) is read from X_1, least significant bit first. One word-time later, the corrected message is read from X_2.

Table 10-5

T_2	T_1	T_0	X_1	A_w	C_2	C_1	C_0	C_5	C_4	C_3	A_r	X_2
0	0	0										
0	0	1	1		0	0	0					
0	1	0	1	1	0	0	1					
0	1	1	1	1	0	1	1					
1	0	0	0	1	0	0	0					
1	0	1	1	0	0	0	0					
1	1	0	1	1	1	0	1					
1	1	1	0	1	0	1	1					
0	0	0		0	0	1	1	0	0	0		
0	0	1			0	0	0	0	1	1	1	
0	1	0						0	1	0	1	1
0	1	1						0	0	1	1	1
1	0	0						0	0	0	0	0
1	0	1						0	0	0	1	0
1	1	0						0	0	0	1	1
1	1	1						0	0	0	0	1
0	0	0						0	0	0		0

The complete input logic for the error-correcting circuit described above is given in equations 10-3 through 10-10.

$$J_{T0} = K_{T0} = 1 \tag{10-3}$$

$$J_{T1} = K_{T1} = T_0 \tag{10-4}$$

$$J_{T2} = K_{T2} = T_0 T_1 \tag{10-5}$$

$$\left.\begin{aligned} J_{Ci} &= (T_0 + T_1 + T_2) X_1 T_i \\ K_{Ci} &= X_1 T_i + \overline{T}_0 \overline{T}_1 \overline{T}_2 \end{aligned}\right\} \quad i = 0, 1, 2 \tag{10-6}$$

$$\begin{aligned} J_{Aw} &= X_1 \\ K_{Aw} &= \overline{X}_1 \end{aligned} \tag{10-7}$$

$$\begin{aligned} J_{C3} &= \overline{T}_0 \overline{T}_1 \overline{T}_2 C_0 + (T_0 + T_1 + T_2)(C_4 + C_5) \\ K_{C3} &= 1 \end{aligned} \tag{10-8}$$

$$\begin{aligned} J_{C4} &= \overline{T}_0 \overline{T}_1 \overline{T}_2 C_1 + (T_0 + T_1 + T_2) \overline{C}_3 C_5 \\ K_{C4} &= \overline{C}_3 \end{aligned} \tag{10-9}$$

$$\begin{aligned} J_{C5} &= \overline{T}_0 \overline{T}_1 \overline{T}_2 C_2 \\ K_{C5} &= \overline{C}_3 \overline{C}_4 \end{aligned} \tag{10-10}$$

Note that this general circuit configuration is effective for messages of any length; to increase the message length one need only provide more counting and checking flip-flops, and lengthen the shift register.

CHECKING INFORMATION MANIPULATIONS

When equipment is constructed to change information according to prescribed rules, the possibility of an incorrect change arises. Some of the checking methods mentioned earlier in this chapter are applicable to this checking problem, and are used in existing computers. The duplication of arithmetic units, memories, and even complete computers, with comparisons of results at appropriate points, is fairly common. Forbidden combinations may also provide a check under some circumstances. For example, the addition of two decimal numbers should of course generate a sum in which the combinations 1010, 1011, 1100, 1101, 1110, and 1111 do not appear, and a partial check on the correctness of addition can be obtained by checking the sum for forbidden combinations.

There is, however, a different class of error-detecting schemes which can be used to check operations in which some rule (e.g., the addition or multiplication tables) is used to transform one or more words into one or more different words. These schemes are indicated diagrammatically in Figure 10-6, and may be described as follows. Each oper-

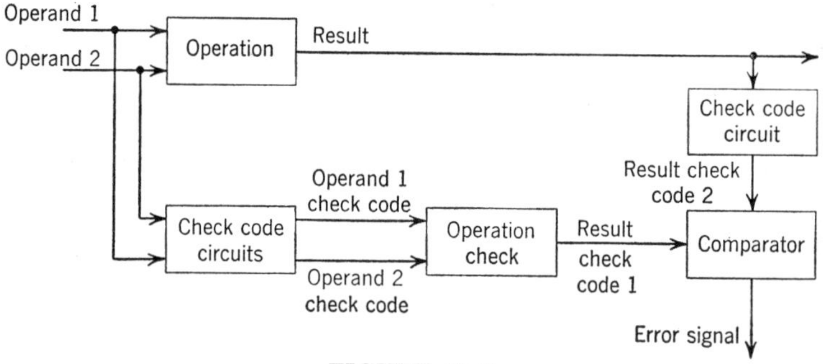

FIGURE 10-6

and is separately operated on to form an operand check code. The operand check codes are then manipulated to form one result check code, and the result of the operation is manipulated to form another result check code. The two result check codes are then compared, and their agreement is an indication that the operation was performed correctly.

The usefulness of such a scheme is of course dependent on the amount of equipment necessary to carry out the checking operations. If the equipment were too complex, it would be easier to duplicate the operation itself and to compare two results directly. However, several checking systems of this kind have been described and are practical. One of these will be briefly discussed.

Suppose that a check code x_c is formed from a number x by dividing that number by three and making the remainder equal to x_c. It can be shown that for the operation of addition, if $x + y = z$, then $(x_c + y_c)_c = z_c$; and for multiplication, if $xy = z$, then $(x_c y_c)_c = z_c$. For example, with $x = 25$ and $y = 17$

$$
\begin{array}{rlrl}
x &= 25 & x_c &= 1 \\
y &= 17 & y_c &= 2 \\
x + y &= 42 & x_c + y_c &= 3 \\
(x + y)_c &= 0 & (x_c + y_c)_c &= 0 \\
xy &= 425 & x_c y_c &= 2 \\
(xy)_c &= 2 & (x_c y_c)_c &= 2
\end{array}
$$

That this simple technique works for any numbers x and y may be seen by expressing x and y as numbers *to the base three.* The least significant digit of the base-three representation of any number x is x_c, for it is obviously the remainder when the number is divided by three (just as the least significant digit of a decimal number is the remainder when that number is divided by ten). The operation of adding or multiplying two base-three numbers thus gives rise to a least significant digit (the result check z_c) which is determined only by the least significant digits of the two operands. Using the numbers of the example above to illustrate this point,

$$
\begin{array}{rlrl}
(25)_{10} &= (221)_3 & x_c &= 1 \\
(17)_{10} &= (122)_3 & y_c &= 2 \\
(25)_{10} + (17)_{10} &= (1120)_3 & z_c &= 0
\end{array}
$$

The mechanization of this checking system requires that circuits be devised to form the check digits from the operands, and to add and multiply them together.

In concluding this chapter, it seems only fair to say something in defense of computer reliability. Forrester has pointed out that if reliability is measured by failures per unit time, a computer may not compare favorably with a human; but if the measure is failures per unit *operation*, the computer is far superior to the human. Because of this, the system designer often goes to a great deal of trouble to minimize the necessity for human intervention, so that the fallible operator has the least opportunity to interfere with what is, after all, a remarkably reliable machine—a digital computer.

BIBLIOGRAPHY

R. M. Bloch, R. V. D. Campbell, and M. Ellis, "Logical Design of the Raytheon Computer," *Mathematical Tables and Other Aids to Computation*, **3,** no. 24, 286–293, 317 (1948).

R. W. Hamming, "Error Detecting and Error Correcting Codes," *Bell System Technical J.*, **29,** no. 2, 147–160 (Apr. 1950).

J. W. Forrester, "Digital Computers: Present and Future Trends," *Review of Electronic Digital Computers*, American Institute of Electrical Engineers, New York, 111 (Feb. 1952).

A. A. Auerbach et al., "The Binac," *Proc. I.R.E.* **40,** no. 1, 12–28 (1952).

Panel Discussion, "Redundancy Checking for Small Digital Computers," *Proc. Eastern Joint Computer Conference*, 56–57, 1954.

P. Elias, "Error-Free Coding," *Trans. of the I.R.E., Symposium on Information Theory*, 29–37, 1954.

I. S. Reed, "A Class of Multiple-Error-Correcting Codes and the Decoding Scheme," *Trans. of the I.R.E., Symposium on Information Theory*, 38–49, 1954.

D. Slepian, "A Class of Binary Signalling Alphabets," *Bell System Technical J.*, **35,** no. 1, 203–234 (1956).

D. A. Huffman, "A Linear Circuit Viewpoint on Error-Correcting Codes," *I.R.E. Trans. on Information Theory*, **IT-2,** no. 3, 20–28 (Sept. 1956).

J. Von Neumann, "Probabilistic Logics and the Synthesis of Reliable Organisms from Unreliable Components, "*Automata Studies*, Princeton University Press, Princeton, 1956.

S. P. Lloyd, "Binary Block Coding," *Bell System Technical J.*, **36,** no. 2, 517–535 (1957).

EXERCISES

1. Write an expression which is "one" only when flip-flops A, B, C, D, and P of Table 10-1 contain a forbidden code.

2. Plot the two-out-of-five decimal code of Table 9-1 on a Veitch diagram, and show that it is a single-error detecting code.

3. Devise a single-error correcting code for a two-bit message ($m = 2$). Plot the resulting code on a Veitch diagram, and indicate on the diagram all the code combinations formed by any one-bit error in a permitted code. (Cf. Figure 10-4.) What is the significance of the remaining blank squares?

4. Suppose that *two* errors occur in transmission of a message which is fed into the correction circuit of Figure 10-5. What will the circuit do to this incorrect code?

5. Add another parity bit P to the code of Table 10-3, chosen to make the *total* number of "ones" in each message even. Plot the resulting code on a Veitch diagram, and show that it is a single-error correcting, double-error detecting code. State the rules which must be followed in interpreting a message in this new code.

6. Show how the codes of Table 10-3 and Exercise 3 can be used as double-error detecting codes instead of single-error correcting codes.

7. A six-bit, positive binary number is read serially from a flip-flop X. Design a sequential circuit having one input (X) and two outputs (Z_1 and Z_2). Z_1 and Z_2 must be "zero" during bit-times one through five, and on the sixth bit-time must represent the remainder formed when the number from X is divided by three. (*Hint:* 1, 4, and 16 have remainders of one when divided by three; 2, 8, and 32 have remainders of two.)

11 / *The control unit: completing computer design*

INTRODUCTION

In Chapter 6, it was pointed out that, as a practical matter, the design of a digital computer is usually carried out by breaking the computer down into a number of different functional parts each of which may be designed independently from the others. In the past few chapters, we have investigated the design of the computer memory, its input and output equipment, and its arithmetic unit. There remains the problem of designing and providing the circuits which cause these various parts of the computer to communicate with one another and to operate together in the proper way. The co-ordinating circuits are usually referred to as the control unit and it will be the purpose of this chapter to illustrate the features of typical control units and to show what the designer must do to make a complete, operating computer out of a number of pieces which have been independently designed.

A method which can be used to tie such pieces together has been described at the end of Chapter 6; that method, however, was necessarily very general and does not provide a coherent approach to the total design problem. In this chapter we shall suggest such an approach, and provide two examples of its use. The first is the complete design of a small and simple general-purpose computer. The general-purpose computer is the most widely employed kind of computer in existence today and is a very powerful tool for carrying out long sequences of arithmetic calculations and logical decisions. The computer designed here is so simple that it would not be a practical computer to construct for use. However, its design will illustrate most of the important characteristics of any general-purpose digital computer. More complicated computers require a considerable extension of the ideas presented here, but an extension in size rather than in kind.

The second computer to be designed in this chapter is intended to do one particular job and is called a limited-purpose (or special-purpose) computer. Special-purpose computers are designed to carry out a single calculation or a very limited set of calculations. They do not have the flexibility inherent in a general-purpose computer, whose function may be completely changed by modifying nothing more than the information contained within the computer memory. However, for a given clock-pulse rate, it is usually possible to fashion a special-purpose computer which executes a given calculation faster than a general-purpose computer could, and with the same precision but with much less equipment. The difficulty is that every new application requires the design of a completely new computer. As logical design and engineering costs are brought down in the future, it may be economically feasible to design special-purpose computers for limited applications even though only a few computers of each kind are required. The limited-purpose computer described in this chapter is not proposed for a practical application, but is merely intended to illustrate what can be done.

THE DESIGN OF A GENERAL-PURPOSE COMPUTER

Functional Description

In Chapter 1, we described a digital computer as a device capable of carrying out a sequence of operations on numbers expressed in digital form, where each operation was one of four kinds: an input, an output, an arithmetic operation, or a sequence-determining operation. A general-purpose computer is a digital computer which has the sequence of operations stored within its own memory (along with the data to be operated on), and which is able to carry out a wide variety of different operations including at least one of each of the four kinds mentioned above. In this section we shall describe a particular and very simple general-purpose computer from the point of view of the user of this computer, and in the following section we shall give the complete design of this computer.

A general-purpose computer, like an automobile or a radio or a desk calculator, may be operated by a person who does not understand why it functions, and who would not be able to design such a machine himself. However, the designer of such a machine must understand very clearly just what the computer will look like to the person who is using it. The computer discussed here has a magnetic drum memory containing 128 words arranged in eight channels of 16 words per channel. Each word consists of ten binary digits plus a sign bit, and each word is separated from the next word by a space bit. The words are numbered from 0 to 127, and each memory space may hold either a number to be operated on by the machine or else an *instruction.* An instruction (also called a *command* or an *order*) is a computer word which identifies which of a limited number of operations the computer is to carry out, and the computer functions by executing a sequence of these instructions, one at a time. By putting a suitable sequence in the computer memory, the user can make the computer perform any complicated sequence of operations.

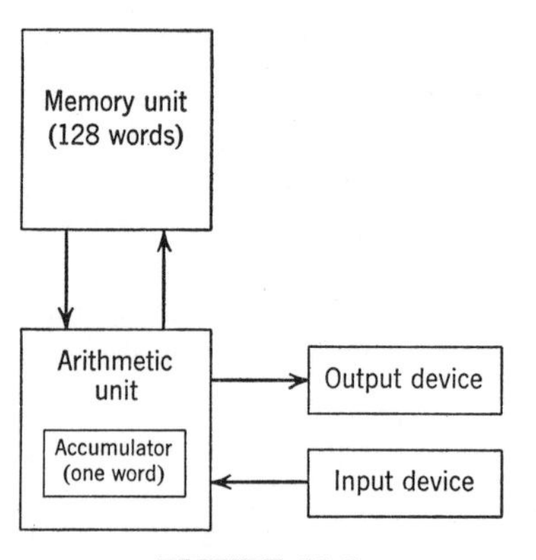

FIGURE 11-1

Our computer's instructions are carried out by the machine in exactly the same sequence as they are stored in the memory. That is to say, when the machine has executed an instruction stored in location n, it next obeys the instruction in location $(n + 1)$. One of the instructions, however, is of the sequence-determining type referred to above, and provides an exception to this rule. When this instruction (the jump command) is performed, it may cause the computer to locate its next instruction anywhere in the memory. The computer therefore does not necessarily always carry out the same cycle of instructions over and over again, but may do one sequence or another depending on how the computation goes.

Our computer has eight different instructions, each of which will be described in more detail below. They enable the computer user to transfer information from the drum into an arithmetic unit where addition and subtraction can take place; to transfer data back to the memory from this arithmetic unit; to transfer information from the arithmetic unit to an output device; and to transfer incoming data from an input device into the arithmetic unit. The arithmetic unit contains one word of storage called the accumulator, and each of the instructions modifies or employs the number stored in this accumulator.

From the user's point of view, the computer has the appearance shown in Figure 11-1.

An instruction, as has been indicated before, occupies one word of storage and therefore consists of ten bits plus the sign bit. The sign bit has no significance as far as the execution of an instruction is concerned. Three of the ten remaining bits identify which of the eight possible instructions is to be carried out. The other seven represent the address of some drum storage location referred to by the command. It is important to note (and sometimes very difficult for the beginner to remember) that each instruction has associated with it two memory locations. One is the memory location in which it is stored; the other is the memory location to which the operation carried out by the instruction refers.

We shall now describe the operations carried out by each of the eight instructions. Remember that, in these descriptions, the letter n refers to the address contained in the instruction, and not the address in which the instruction is stored.

1. *Add. Extract the number in storage location n from the memory and add it to the number in the accumulator.* Suppose, for example, that the instruction in storage location 42 is "add the contents of location 56 to the accumulator." If storage location 56 contains the number 83, and the accumulator contains the number 18 before the addition command is carried out, the computer adds 83 to 18 and leaves the result—the number 101—in the accumulator. Table 11-1a illustrates this operation. Note that the add command does not modify the contents of storage location 42 or of storage location 56. As far as the computer user is concerned, only the accumulator is modified.
2. *Subtract. Subtract the number in storage location n from the number in the accumulator and leave the result in the accumulator.* The subtract operation is illustrated in Table 11-1b, where the computer is made to subtract 83 from 18.
3. *Transfer. Transfer the number in the accumulator to storage location n, where it will replace the number previously in that location.* In Table 11-1c the function of the transfer command is indicated.
4. *Transfer clear. Transfer the number in the accumulator to storage location n, replacing the number previously there. At the same time, store the number zero in the accumulator.* The operation of this command is shown in Table 11-1d. Note that it is the only command which modifies the contents of *two* storage locations: the accumulator and word n.
5. *Shift right. Multiply the number in the accumulator by one-half, i.e., shift it to the right one bit position.* This command is illustrated in

Table 11-1

	Storage Location	Contents of Storage Location	
		Before Execution of Instruction in Storage Location 42	After Execution of Instruction in Storage Location 42
(a)	42	Add 56	Add 56
	56	83	83
	Accumulator	18	101
(b)	42	Subtract 56	Subtract 56
	56	83	83
	Accumulator	18	−65
(c)	42	Transfer 56	Transfer 56
	56	83	18
	Accumulator	18	18
(d)	42	Transfer Clear 56	Transfer Clear 56
	56	83	18
	Accumulator	18	0
(e)	42	Shift Right	Shift Right
	Accumulator	18	9

Table 11-1e. Note that the seven address bits in the shift right command, which were used in the first four instructions to specify a location in the memory, are unused in the shift right instruction (and in the input and output instructions which will be described below).

6. *Jump. If the number in the accumulator is greater than or equal to zero, execute the instruction in storage location n next. If the number in the accumulator is less than zero, proceed as usual, obtaining the next instruction from the location whose address is one greater than the address of the jump command itself.* This is the sequence-determining command mentioned before. It permits the user to carry out different sequences of operations, depending on the results obtained from arithmetic and other operations.

7. *Input. Add the binary digit at the input station to the least significant end of the accumulator and move the input station along to its next input position.* The input device will be assumed to be a paper tape reader which reads a tape having only one hole in every digit position.

8. *Output. Transfer the least significant binary digit in the accumulator into the output station, and move the output device along one hole position.* The output device will be assumed to be a paper tape punch

which either punches or does not punch a hole when stimulated, depending on whether a "one" or a "zero" is sent to it.

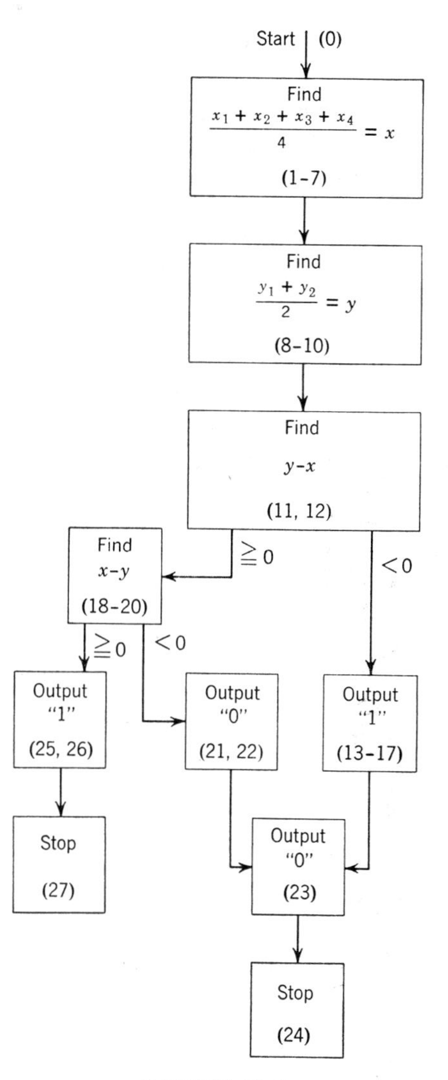

FIGURE 11-2

To illustrate the use of these instructions, and the way they can be combined to carry out complicated sequences of operations, consider the following problem. Six numbers are stored in the computer. If the average of the first four (identified as x_1, x_2, x_3, and x_4) is greater than the average of the last two (identified as y_1 and y_2), punch out the digits 10; if the average of the last two is greater than the average of the first four, punch out 00; and if the two averages are exactly equal, punch out the single digit 1. In Figure 11-2 a *program*, or plan for the solution of the problem, is shown, indicating a sequence of operations which will give the desired result. In Table 11-2 this program is *coded*, or translated into a sequence of machine instructions, and the resulting *routine*, when entered into the computer memory, will punch out the specified results depending on what numbers are present in storage locations 28 through 33. The numbers in parentheses in Figure 11-2 indicate the location of instructions which carry out the corresponding operations. The reader should study this program and code until he is certain that he understands them, for a complete comprehension of the working of the computer is necessary in what follows. (Note that the "stop" operation is performed by executing a jump instruction whose address is the same as the location of the jump instruction; the computer therefore does not really stop, but rather continually repeats one operation. This method of stopping the computer is used because there is no stop instruction among the commands.)

This computer is, of course, not a very powerful one. It has a very small memory and a relatively inflexible order code. In particular, the multiplication instruction is missing, and since multiplication is a very important operation, this is a serious handicap. In order to carry out multiplication on this computer, it would be necessary to write a special multiplication program, using the basic instruction code and carrying out the multiplication as a series of additions and shifts.

Aside from the fact that this computer does not have some of the important kinds of commands common to many general-purpose machines, it is nevertheless illustrative of the type. It differs in capacity, not in kind, from the ordinary general-purpose digital computer, and is therefore a suitable object for study.

Designing the Computer

As was indicated in Chapter 1, the logical designer begins work with a description of the functions to be performed by the system he is to design. The description may be as specific as that given in the above paragraphs, or it may be of a much more general nature. Starting with the functional description, the designer carries out the following steps.

1. *Establishment of a framework.* From the functional description, the designer decides what basic functional units must be provided. If

Table 11-2

Storage Location	Contents of Storage Location		Remarks
0	Transfer clear	127	Puts zero in accumulator
1	Add	28	Forms $\frac{x_1 + x_2 + x_3 + x_4}{4} = x$ in accumulator
2	Add	29	
3	Add	30	
4	Add	31	
5	Right shift		
6	Right shift		
7	Transfer clear	127	Clears accumulator, and puts x in 127
8	Add	32	Forms $\frac{y_1 + y_2}{2} = y$ in accumulator
9	Add	33	
10	Right shift		
11	Subtract	127	Forms $y - x$ in accumulator
12	Jump	18	
13	Transfer clear	127	$y - x < 0$. Therefore output "1"
14	Add	34	
15	Output		
16	Transfer clear	127	
17	Jump	23	
18	Transfer clear	127	
19	Subtract	127	Puts $0 - (y - x) = x - y$ in accumulator
20	Jump	25	
21	Transfer clear	127	
22	Output		$x - y < 0$. Therefore output "0"
23	Output		Output "0"
24	Jump	24	Stops the computer
25	Add	34	$x - y \geq 0$ *and* $y - x \geq 0$. Therefore $x = y$
26	Output		
27	Jump	27	Stops the computer
28		x_1	Data
28		x_2	
30		x_3	
31		x_4	
32		y_1	
33		y_2	
34		1	Least significant "1" for output

the computer is very complicated, these units may be large and complex pieces of equipment (e.g., buffer, error-checking equipment, memory unit, arithmetic unit, etc.). In subsequent steps these units will be independently designed and then joined together. If the computer is simple, the basic functional units may be more elementary (e.g., memory-selection equipment, arithmetic register, adder, complementer, etc.) and the design can proceed more directly. In either event, these functional units provide a framework which can be filled out in subsequent steps.

2. *Provision of a word description.* The designer next works out a word description showing how the basic functional units he has initially specified must operate together in order to make the system perform as desired. This word description goes over every operation of the computer in detail, setting forth the source and destination of all information transfers, and defining what operations are to take place at each step. In carrying out this step, the designer puts flesh on the framework by providing specific memory elements at appropriate places in the basic functional units.

3. *Formation of truth tables.* By going over the word description, the designer can translate it into a series of truth tables of the kind described in the last part of Chapter 6 (e.g., Tables 6-18 and 6-19), which describe the bit-time by bit-time operation of the functional units. These truth tables will in general incorporate logical circuits designed by use of subsidiary truth tables, as the decimal error-detecting circuit was incorporated in Tables 6-18 and 6-19. Examination of these truth tables will indicate what additional signals must be supplied in order to distinguish the various states and conditions of the machine from one another. These states and conditions, or the signals representing them, are added to the tables.

4. *Derivation of input equations.* The difference equations and/or input equations for all flip-flops, writing amplifiers, and output devices may now be written in terms of flip-flop outputs, read amplifier outputs, and computer inputs, by combining the truth tables of step 3. These equations may then be simplified as much as possible by the application of the various techniques which have been described.

5. *Revision and rearrangement.* The complete computer design must now be reviewed to determine whether a rearrangement of terms in the input equations or of the functional parts of the computer or both will lead to simplifications. At this point, such a rearrangement may lead to a completely new design of the whole computer, for the designer may decide to return to step 1 and provide a different framework for the computer. The redesign is repeated as long as the logical designer

has time, or feels that further simplification and improvement are possible.

We shall now carry out these steps for the design of the computer described in the first section of this chapter. For lack of space we shall not, however, conduct as extensive an amount of revision and rearrangement as should be done for a computer which would actually be constructed.

Framework. The basic functional units for our simple computer are shown in Figure 11-3. These are the main memory, the accumulator, the adder and subtracter, a coincidence register (called the C-register) to hold the address of whatever operand is to be read from the memory, a next-instruction register (called the N-register) to hold the address of the next instruction to be carried out, a sector address channel in which is permanently recorded the sector number of the word next to be read from the memory, an origin pulse channel containing

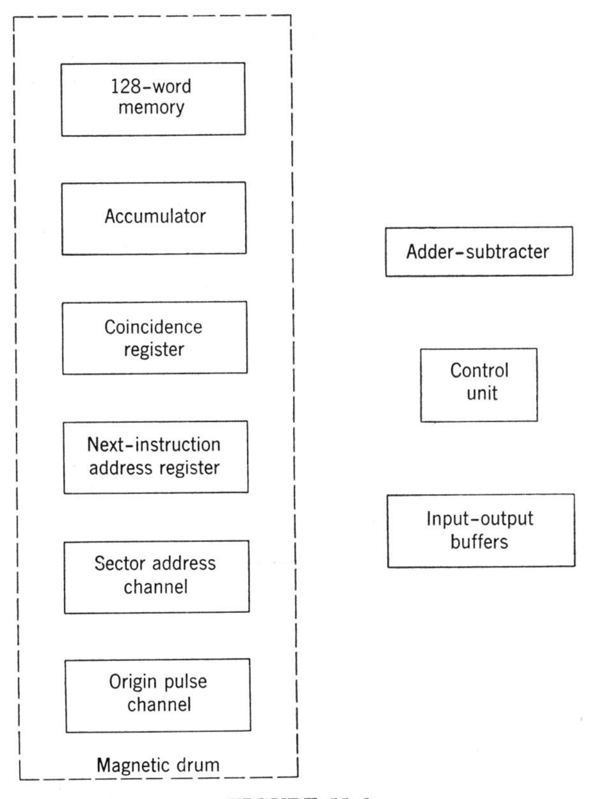

FIGURE 11-3

one pulse which will be used to synchronize the bit-time counter, a control unit, and a buffer for the input-output devices. As is indicated in the diagram, the accumulator, coincidence register, and next-instruction address register are all circulating registers on the drum.

Before beginning the word descriptions, it is necessary to add some details to the framework of Figure 11-3, and these details are shown by the solid lines of Figure 11-4. (The dashed lines in this figure repre-

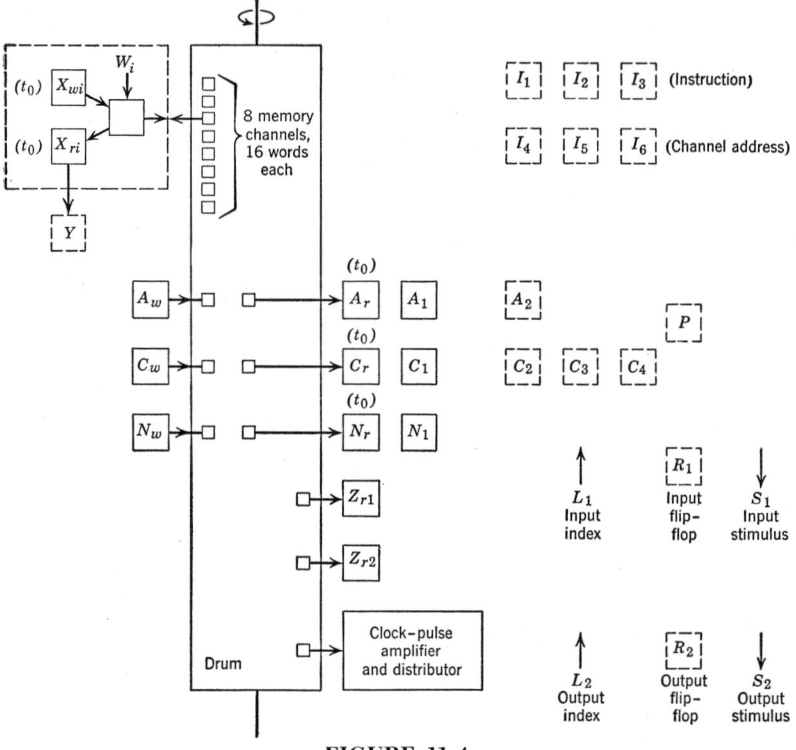

FIGURE 11-4

sent elements which will be added and described later.) The 128-word memory is shown divided into eight channels of 16 words each. Each channel has a read and a write amplifier (X_{wi} and X_{ri}), and an input line W_i controls the writing operation. Each of the three circulating registers contains a write amplifier (A_w, C_w, and N_w), a read amplifier (A_r, C_r, and N_r), and another flip-flop (A_1, C_1, and N_1) which is normally in the circulating path. All amplifiers are of the no-delay type. The extra flip-flop A_1 is clearly required in the accumulator, for we shall eventually have to provide a method for shifting the contents of that register to the right by leaving one storage element out of the circulating path. The need for the other two flip-flops (C_1 and N_1) is not so clear, but we are providing them with the understanding that they may be removed later if found unnecessary. Finally, read amplifiers Z_{r1} and Z_{r2} provide the means for reading the sector address and origin pulse respectively. There must of course be a clock-pulse channel on the drum as well, and circuits must be provided for amplifying and shaping these pulses, and for distributing them throughout the computer.

The C- and N-registers, together with flip-flops C_4, I_1, I_2, I_3, I_4, I_5, I_6, and P (whose functions will be described later) comprise what is usually called the *control unit.*

Word description. We may now proceed with a word description of the computer. Each instruction is carried out in two parts. During the first part, the instruction is read from the memory into the control unit where it can be identified; during the second, the instruction just read is carried out. We begin the word description by showing how this first part (which is the same for all instructions) is accomplished, and then we shall discuss the mechanization of the second part for each of the eight different commands.

1. *Reading in the next instruction.* Suppose the computer has just completed carrying out one instruction and is ready to perform the next one. Suppose further that we have previously arranged for the address of this next instruction to be stored in the N-register of Figure 11-4. The first step in carrying out this new instruction must be to look in the memory for the word whose address is in the N-register, to withdraw this word from the memory, and to store it where it can be referred to. This word is, of course, the next instruction to be executed and we know that it consists of two parts: an instruction code containing three bits (identifying one of eight different commands) and an address containing seven bits (identifying one of 128 different storage locations). The instruction will be read from the memory least significant bit first as indicated in Figure 11-5, where it will be seen that the first four bits to emerge are the last four bits of the operand address and represent a sector number. The next three bits contain the channel address for the operand, and the last three bits are the code for the instruction itself.

In order to extract this new instruction from its memory location, it is necessary to obtain coincidence between the sector number of the next-instruction address and the sector number of the sector next to be read from the memory, as was indicated in the chapter on memory units. It is also necessary to select the proper channel to read the new command from. We must therefore read the address of the new instruction from the N-register, extract the channel number from it, and continually compare the sector number from it with the current sector number read from Z_{r1} until coincidence occurs. As the channel address is read, it is transferred to flip-flops I_4, I_5, and I_6, and their contents subsequently determine which of the eight read amplifiers X_{ri} is connected to the read flip-flop Y. This transfer takes place during the

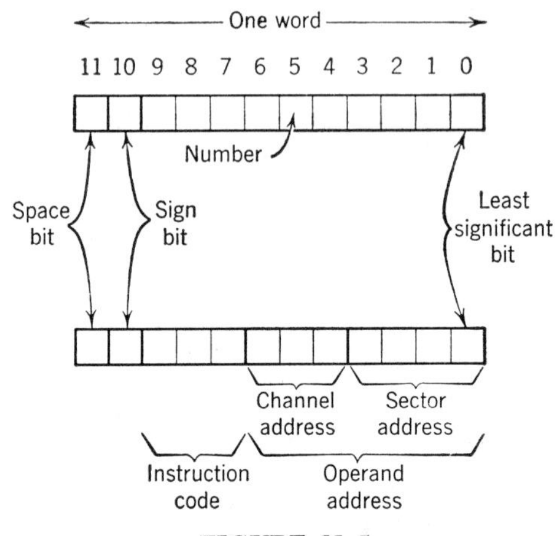

FIGURE 11-5

first word-time of the read instruction phase. During subsequent word-times, and until coincidence is obtained between the sector number in the N-register and the sector address from read amplifier Z_{r1}, the computer does nothing. As soon as coincidence occurs, however, the new instruction must be transferred to the control unit where it can be used. The three bits which represent the instruction code are then transferred to flip-flops I_1, I_2, and I_3. The entire instruction is, at the same time, transferred into the C-register, where the address portion will be available later if this particular command requires that data be read from or sent to the memory.

During the same word-time as the instruction is being read from the memory, it is necessary to increase the number in the N-register by unity so that the next time a command is withdrawn from the memory it will be withdrawn from a storage location whose address is one greater than the address of the previous command.

At the end of the word-time during which the instruction is read from the memory the computer goes into the second phase of operation, and the next step to be taken will depend on the contents of flip-flops I_1, I_2, and I_3, i.e., on what the newly received instruction is. We therefore continue this word description of the computer by detailing exactly what must take place for each of the eight different commands.

2. *Add.* If the instruction flip-flops I_1–I_3 contain the code for add, the computer must immediately begin to search the memory for the operand whose address is a part of the command just read from the memory. The procedure is exactly the same as the one carried out when the new instruction was read. However, the address to be sought is now in the C-register instead of the N-register, and the first step to be carried out is the transfer of the channel address from this C-register into flip-flops I_4, I_5, and I_6. At the same time, and for as long as may be necessary until coincidence occurs, the sector number in the coincidence register must be compared with the sector address from read amplifier Z_{r1}. When coincidence finally occurs, and the operand is read from the memory through the appropriate read amplifier and flip-flop Y, it must immediately be added to the number circulating in the accumulator and the sum must be stored in the accumulator. As soon as this coincidence word-time is over, the computer must return to the instruction read-in phase, and addition is complete.

3. *Subtract.* The subtract operation begins exactly the same as did the add operation, with the transfer of the operand channel address into flip-flops I_4–I_6 and a search for coincidence between the sector address read from Z_{r1} and the sector number in the coincidence register. The only difference between subtraction and addition, of course, is that the new operand must be *subtracted* from the accumulator during the coincidence word-time rather than added to it. We shall use the binary adder-subtracter already designed in Chapter 9, where we saw that an adder and subtracter differ only in the logic for the carry flip-flop. (Negative numbers will be represented in complement form in this computer, so that the adder-subtracter functions without regard to the sign of the operands.) At the end of the coincidence word-time, subtraction is complete.

4. *Transfer.* The transfer instruction begins with a search for coincidence exactly as did the add and subtract instructions. During the coincidence word-time, however, the number in the accumulator is transferred through write amplifier X_{wi} into the appropriate channel in the memory. At the same time, the proper write line W_i must be stimulated so that writing current gets to the correct head. At the end of this coincidence word-time, the command is complete.

5. *Transfer clear.* The transfer clear instruction operates in exactly the same way as does the transfer instruction, except that during the

coincidence word-time a string of zeros must be transferred into the accumulator.

6. *Shift right.* The shifting operation requires no access to the memory and may therefore always be accomplished in the word-time after the shift instruction appears in flip-flops I_1–I_3. A shift operation is carried out, of course, by eliminating one bit of delay in the accumulator loop. Since negative numbers are represented in two's complement form, the sign bit can be shifted right, but must be replaced by itself.

7. *Jump.* It will be remembered that this command causes the address of the next instruction to be changed to the address *contained in* the jump instruction, if the contents of the accumulator are greater than or equal to zero. If the contents of the accumulator are less than zero, nothing happens and the computer goes on as if this instruction had not occurred. It is therefore necessary to extract the sign bit from the accumulator at the beginning of the operation of this command, and to use it to determine whether the N-register will remain unchanged (if the accumulator is negative and the sign bit "one") or whether the contents of the coincidence register should replace the number in the N-register (if the accumulator contains a number greater than or equal to zero, so that the sign bit is "zero").

8. *Input.* The input tape reader must be connected to an input buffer flip-flop R_1. Furthermore, we shall assume that it supplies an index signal L_1, and that it moves from one input position to another when stimulated on line S_1. When S_1 is energized, the index signal immediately becomes "zero," and the tape reader moves from one reading position to the next one. When it reaches the point where the next hole position is opposite the reading station and the next bit of information is in flip-flop R_1, the index line L_1 changes to a "one," indicating that the computer may read the contents of R_1 and add it to the accumulator. The designer must therefore arrange to wait until L_1 is "one," and then to add the bit in R_1 into the accumulator, at the same time supplying a signal to line S_1. As soon as the addition word-time is complete, the computer may go on to seek the next command. (It would be possible for the computer to begin the input command by stimulating line S_1 and then waiting for L_1 to be "one" before reading R_1. This would, however, *always* require that the computer wait for the reader to move from one hole position to the next. Using the sequence of events described above, the computer need never wait this long a time unless two input commands are carried out by the computer in a time interval shorter than the mechanical operating time of the tape reader.)

9. *Output.* The output punch moves from one hole position to the next, punching a hole or not depending on whether the output buffer flip-flop R_2 contains a "one" or a "zero," whenever its input line S_2 is stimulated. Index line L_2 becomes a "zero" immediately after S_2 is energized, and does not return to "one" until the punch is at the next hole position and a new output bit may be inserted in R_2. The designer must therefore arrange that the computer wait until L_2 becomes "one," whereupon he can transfer the least significant bit of the accumulator to R_2 and simultaneously stimulate line S_2. As soon as S_2 has been stimulated, the computer may go on to seek the next command.

Truth tables. Before translating and expanding the word description into truth tables from which the complete input logic may be derived, it is necessary to establish in some detail the timing of words on the drum. To begin with, we shall assume the existence of a counter which changes state every clock-pulse time and which has twelve states labeled t_0, t_1, t_2, $\cdots$, t_{11}. The counter will be synchronized with the information on the drum in such a way that it is in the t_0 state whenever the least significant bit of the accumulator is in read amplifier A_r, and since a computer word contains twelve bits (see Figure 11-5), each of the twelve states of the counter corresponds to one bit in a word. The fact that the counter is in the t_0 state when the least significant bit of the accumulator is in A_r is indicated in Figure 11-4 by a "t_0" printed adjacent to the A_r symbol; and it will be seen that the same notation is used to show that the least significant bits of words written in the memory, read from the memory, in the C-register, or in the N-register, are in amplifiers X_{wi}, X_{ri}, C_r, and N_r, respectively, at t_0 time. With this timing arrangement understood, it is possible to tell when any given bit of a word will be in any position. For example, we know that the least significant bit of the channel address (bit 4 in Figure 11-5) may be read from N_r at t_4 time, and from N_1 at t_5 time. Similarly, the three instruction code bits of a command (bits 7, 8, and 9 of Figure 11-5) will appear in amplifier X_{ri} at t_7, t_8, and t_9 times, and in flip-flop Y at t_8, t_9, and t_{10} times.

We can now make the following stipulations about read amplifiers Z_{r1} and Z_{r2}: the four sector-address bits will be recorded on the drum so that they can be read from Z_{r1} at t_0, t_1, t_2, and t_3 times; a single "one" will be recorded so that it appears in Z_{r2} at some arbitrary t_0 time.

The truth tables can now be constructed as follows:

1. *Reading in the next instruction.* The logic indicated in Table 11-3 shows the coincidence search which goes on during the first part of the

read-instruction phase. Flip-flops C_2 and C_3 are used to compare the next-instruction address in register N with the sector address in read amplifier Z_{r1}. From Figure 11-5 we see that the next-instruction sector address may be read during bit-times t_0 through t_3 from read amplifier

Table 11-3. Read New Instruction. $C_3 = 0$

	C_2		C_3		I_4		I_5		I_6	
	J	K	J	K	J	K	J	K	J	K
t_{11}	0	1	0	1						
t_0	$N_r\bar{Z}_{r1} + \bar{N}_rZ_{r1}$	0	"	0						
t_1	"	"	"	"						
t_2	"	"	"	"						
t_3	"	"	"	"						
t_4	0	"	"	"	N_r	$\bar{N}_r$				
t_5	"	"	"	"	"	"	I_4	$\bar{I}_4$		
t_6	"	"	"	"	"	"	"	"	I_5	$\bar{I}_5$
t_7	"	"	"	"	0	0	0	0	0	0
...	...	...	...		...		...		...	
t_{11}	0	1	$\bar{C}_2$	1	"	"	"	"	"	"
t_0	$N_r\bar{Z}_{r1} + \bar{N}_rZ_{r1}$	0	0	0	"	"	"	"	"	"
t_1	"	"	"	"	"	"	"	"	"	"
t_2	"	"	"	"	"	"	"	"	"	"
t_3	"	"	"	"	"	"	"	"	"	"
t_4	0	"	"	"	N_r	$\bar{N}_r$	"	"	"	"
t_5	"	"	"	"	"	"	I_4	$\bar{I}_4$	"	"
t_6	"	"	"	"	"	"	"	"	I_5	$\bar{I}_5$
t_7	"	"	"	"	0	0	0	0	0	0
...	...	...	...		...		...		...	
t_{11}	0	1	$\bar{C}_2$	1	"	"	"	"	"	"
t_0	$N_r\bar{Z}_{r1} + \bar{N}_rZ_{r1}$	0	0	0	"	"	"	"	"	"

This cycle repeats until coincidence occurs, and $C_3 = 1$ at some t_0 time.

N_r. C_2 is turned on if any of these four bits from N_r differ from any of the four bits identifying the sector next to be read from the memory. At every t_{11} time flip-flop C_3 is turned on if and only if C_2 is off, i.e., if all four of the bits from N_r were equal to the four from Z_{r1}. The word-time during which $C_3 = 1$ is the coincidence word-time. If, however, coincidence did not occur, flip-flop C_3 remains in the "zero" state and the comparison process is repeated.

Meanwhile, the channel address must be transferred from the N-register into channel address flip-flops I_4, I_5, and I_6. This address must, of course, be in place during the word-time that the instruction is being read from the memory. It would be quite possible to read these bits from the N-register during the first word-time that the instruction is being sought, and then never again. However, a moment's reflection will show that if J_{I4}, for example, could be equal to N_r *only* during the first word-time of Table 11-3, it would be necessary to distinguish that first word-time from all subsequent ones which precede coincidence. It is obviously much simpler, and although superfluous does absolutely no harm, to allow the channel address to shift into I_4, I_5, and I_6 every word-time until coincidence occurs. The channel address is shifted into I_4, I_5, and I_6 starting at I_4, and the shift takes place during bit-times t_4, t_5, and t_6, which correspond to the positions of the channel number in a word (see Figure 11-5).

Once coincidence has occurred, the new instruction must be read from the memory through flip-flop Y into the C-register, and the three-bit code for the command itself must be shifted into control flip-flops I_1, I_2, and I_3. The C-register will then contain the entire instruction, and will be used for finding sector coincidence in those commands which require the location of an operand. The logic for taking care of these two functions is indicated in the first four columns of Table 11-4.

Table 11-4. Read New Instruction. $C_3 = 1$

	I_1		I_2		I_3		C_w		N_w		A_2		C_3	
	J	K	J	K	J	K	J	K	J	K	J	K	J	K
t_{11}									N_1	$\overline{N}_1$			$\overline{C}_2$	1
t_0									"	"	1	0	0	0
t_1							Y	$\overline{Y}$	$\overline{N}_1A_2 + N_1\overline{A}_2$	$N_1A_2 + \overline{N}_1\overline{A}_2$	0	$\overline{N}_1$	"	"
t_2							"	"	"	"	"	"	"	"
t_3							"	"	"	"	"	"	"	"
t_4							"	"	"	"	"	"	"	"
t_5							"	"	"	"	"	"	"	"
t_6							"	"	"	"	"	"	"	"
t_7							"	"	"	"	"	"	"	"
t_8	Y	$\overline{Y}$					"	"	"	"	"	"	"	"
t_9	"	"	I_1	$\overline{I}_1$			"	"	"	"	"	"	"	"
t_{10}	"	"	"	"	I_2	$\overline{I}_2$	"	"	"	"	"	"	"	"
t_{11}	0	0	0	0	0	0	"	"	"	"			$\overline{C}_2$	1
t_0	0	0	0	0	0	0	C_1	$\overline{C}_1$	N_1	$\overline{N}_1$			0	0

During this same coincidence word-time, it is also necessary to add unity to the address in the N-register, so that the next instruction will be read from the correct place in the memory. The logic for this addition is also shown in Table 11-4, where it can be seen that a new flip-flop A_2 is used as a carry flip-flop. The addition is begun by putting a "one" into A_2 at t_0 time. When the addition is complete, the number in the N-register must be retained, and therefore during the second phase of all instructions the N-register circulates on itself.

2. *Add.* We are now ready to discuss the mechanization of the commands themselves. As has been mentioned before, the add, subtract, and the two transfer commands all require that an operand be extracted from the memory or transferred to the memory. This operation, common to all four commands, is very similar to the operation of locating

Table 11-5. Search for Operand Address. $C_3 = 0$. (Occurs for Add, Subtract, Transfer, and Transfer Clear Commands Only)

	C_2		C_3		I_4		I_5		I_6	
	J	K	J	K	J	K	J	K	J	K
t_{11}	0	1	0	1						
t_0	$C_r\bar{Z}_{r1} + \bar{C}_rZ_{r1}$	0	"	0						
t_1	"	"	"	"						
t_2	"	"	"	"						
t_3	"	"	"	"						
t_4	0	"	"	"	C_r	$\bar{C}_r$				
t_5	"	"	"	"	"	"	I_4	$\bar{I}_4$		
t_6	"	"	"	"	"	"	"	"	I_5	$\bar{I}_5$
t_7	"	"	"	"	0	0	0	0	0	0
...	...	...	...		...		...		...	
t_{11}	0	1	$\bar{C}_2$	1	"	"	"	"	"	"
t_0	$C_r\bar{Z}_{r1} + \bar{C}_rZ_{r1}$	0	0	0	"	"	"	"	"	"
t_1	"	"	"	"	"	"	"	"	"	"
t_2	"	"	"	"	"	"	"	"	"	"
t_3	"	"	"	"	"	"	"	"	"	"
t_4	0	"	"	"	C_r	$\bar{C}_r$	"	"	"	"
t_5	"	"	"	"	"	"	I_4	$\bar{I}_4$	"	"
t_6	"	"	"	"	"	"	"	"	I_5	$\bar{I}_5$
t_7	"	"	"	"	0	0	0	0	0	0
...	...	...	...		...		...		...	
t_{11}	0	1	$\bar{C}_2$	1	"	"	"	"	"	"
t_0	$C_r\bar{Z}_{r1} + \bar{C}_rZ_{r1}$	0	0	0	"	"	"	"	"	"

This cycle repeats until coincidence occurs, and $C_3 = 1$ at some t_0 time.

the next instruction, as described in Table 11-3. However, the operand address is contained in the C-register where the next instruction address was contained in the N-register. The logic shown in Table 11-5 therefore differs from the logic of Table 11-3 only in that the address being sought comes from read amplifier C_r instead of read amplifier N_r. It must be remembered in particular that the operation shown in Table 11-5 is carried out during only four of the eight instructions.

In Table 11-6 the logic for the add instruction is shown, where it is understood that the number to be added into the accumulator has been

Table 11-6. Add. $C_3 = 1$

	C_3		A_w		A_2	
	J	K	J	K	J	K
t_{11}	$\bar{C}_2$	1	A_1	$\bar{A}_1$		
t_0	0	0	"	"	0	1
t_1	"	"	$\bar{A}_1\bar{A}_2Y + \bar{A}_1A_2\bar{Y} + A_1\bar{A}_2\bar{Y} + A_1A_2Y$	$\bar{A}_1\bar{A}_2\bar{Y} + \bar{A}_1A_2Y + A_1\bar{A}_2Y + A_1A_2\bar{Y}$	A_1Y	$\bar{A}_1\bar{Y}$
t_2	"	"	"	"	"	"
t_3	"	"	"	"	"	"
...	...		...		...	
t_{10}	"	"	"	"	"	"
t_{11}	$\bar{C}_2$	1	"	"	"	"
t_0	0	0	A_1	$\bar{A}_1$		

located and is being read from Y while the operations shown in the table take place. This is indicated by the notation $C_3 = 1$. The addition logic is shown at the input to write amplifier A_w and is the same logic as was derived in Chapter 9. The operand, entering sequentially from the Y flip-flop, is added to the contents of the accumulator from flip-flop A_1, and A_2 is used as the carry flip-flop. When the addition is complete, the accumulator must be made to circulate on itself so that A_1 is subsequently transferred into A_w.

In Figure 11-6 the sequence of events which take place during an add command are shown with the object of clarifying the coincidence timing. In the example shown, the previous command was completed during sector 4, the add command itself is stored in sector 8, and the operand of the add command is in sector 13. The waveforms for coincidence flip-flops C_2 and C_3 are indicated, together with that of a flip-flop P whose function will be discussed later.

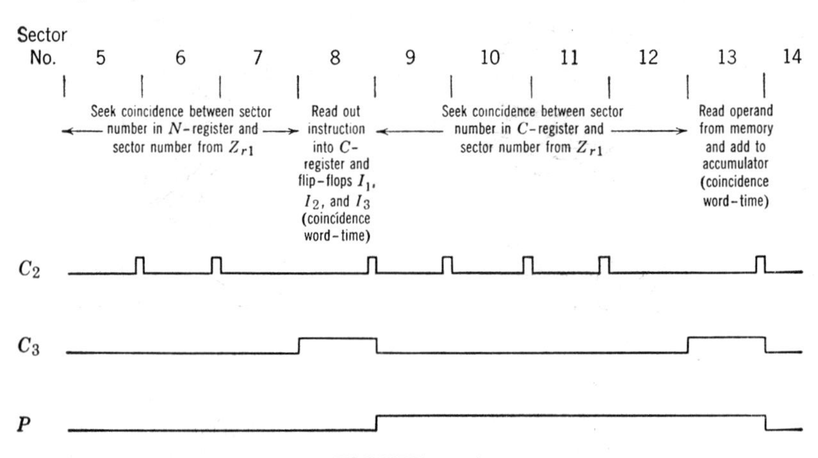

FIGURE 11-6

3. *Subtract.* The subtract operation is illustrated in Table 11-7, where it is seen that, since addition and subtraction differ only in the way the carry is formed, the only change in logic from Table 11-6 is a change in the carry flip-flop A_2.

Table 11-7. Subtract. $C_3 = 1$

	C_3		A_w		A_2	
	J	K	J	K	J	K
t_{11}	$\bar{C}_2$	1	A_1	$\bar{A}_1$		
t_0	0	0	"	"	0	1
t_1	"	"	$\bar{A}_1\bar{A}_2Y + \bar{A}_1A_2\bar{Y} + A_1\bar{A}_2\bar{Y} + A_1A_2Y$	$\bar{A}_1\bar{A}_2\bar{Y} + \bar{A}_1A_2Y + A_1\bar{A}_2Y + A_1A_2\bar{Y}$	$\bar{A}_1Y$	$A_1\bar{Y}$
t_2	"	"	"	"	"	"
t_3	"	"	"	"	"	"
...	...		...		...	
t_{10}	"	"	"	"	"	"
t_{11}	$\bar{C}_2$	1	"	"	"	"
t_0	0	0	A_1	$\bar{A}_1$		

4 and 5. *Transfer and transfer clear.* The transfer and transfer clear commands are illustrated in Tables 11-8 and 11-9, where again only the logic for the word-time during which $C_3 = 1$ is shown. For each command the memory write amplifiers X_{wi} must be connected to the accumulator read amplifier A_r and the address of the particular channel to be written in is used to stimulate the appropriate write line W_i. The

two commands differ only in the fact that zero is written in the A-register during the transfer clear command, whereas the accumulator merely circulates on itself during the transfer command.

Table 11-8. Transfer. $C_3 = 1$

	C_3		X_{wi}		W_i	A_w	
	J	K	J	K		J	K
t_{11}	$\bar{C}_2$	1	A_r	$\bar{A}_r$	0	A_1	$\bar{A}_1$
t_0	0	0	"	"	$(I_4 - I_6)_i$	"	"
t_1	"	"	"	"	"	"	"
t_2	"	"	"	"	"	"	"
t_3	"	"	"	"	"	"	"
...	...		...		...	...	
t_{10}	"	"	"	"	"	"	"
t_{11}	$\bar{C}_2$	1	"	"	"	"	"
t_0	0	0	"	"	0	"	"

Table 11-9. Transfer Clear. $C_3 = 1$

	C_3		X_{wi}		W_i	A_w	
	J	K	J	K		J	K
t_{11}	$\bar{C}_2$	1	A_r	$\bar{A}_r$	0	A_1	$\bar{A}_1$
t_0	0	0	"	"	$(I_4 - I_6)_i$	0	1
t_1	"	"	"	"	"	"	"
t_2	"	"	"	"	"	"	"
t_3	"	"	"	"	"	"	"
...	...		...		...	...	
t_{10}	"	"	"	"	"	"	"
t_{11}	$\bar{C}_2$	1	"	"	"	"	"
t_0	0	0	"	"	0	A_1	$\bar{A}_1$

6. *Shift right.* The shift right command, illustrated in Table 11-10, takes place in one word-time and requires no coincidence. As was indicated in the word description, the right shift can be carried out by omitting a memory element from the accumulator circulating register, and by replacing the shifted sign bit by itself. In Table 11-10 it can be seen that A_1 is left out of the register for a word-time (while A_r shifts directly into A_w). At t_{11} time the sign bit will be in A_1 as well as in A_w (since A_r is the input to both of these elements), and the logic shown will then insert the sign bit from A_1 into the sign bit position.

Table 11-10. Shift Right

	A_w		A_1	
	J	K	J	K
t_{11}	A_1	$\bar{A}_1$	A_r	$\bar{A}_r$
t_0	A_r	$\bar{A}_r$	"	"
t_1	"	"	"	"
t_3	"	"	"	"
...	...		...	
t_{10}	"	"	"	"
t_{11}	A_1	$\bar{A}_1$	"	"
t_0	"	"	"	"

7. *Jump.* The operation of the jump command is indicated in Table 11-11. A new flip-flop, C_4, must be provided to read the sign bit from accumulator flip-flop A_1 at t_{11} time. During the subsequent word-time, nothing unusual happens if the sign bit was a "one," indicating that the accumulator contained a negative number. In particular, the N-register is allowed to circulate on itself again and the next instruction will be obtained from the usual place. If, on the other hand, the

Table 11-11. Jump

	C_4		N_1	
	J	K	J	K
t_{11}	A_1	$\bar{A}_1$	N_r	$\bar{N}_r$
t_0	0	0	$C_4N_r + \bar{C}_4C_r$	$C_4\bar{N}_r + \bar{C}_4\bar{C}_r$
t_1	"	"	"	"
t_2	"	"	"	"
t_3	"	"	"	"
...	...		...	
t_{10}	"	"	"	"
t_{11}	A_1	$\bar{A}_1$	"	"
t_0	"	"	N_r	$\bar{N}_r$

number in the accumulator was positive or zero, the C_4 flip-flop will contain a "zero" during this word-time and the logic shown will transfer the contents of the C-register into the N-register. The next instruction to be carried out will therefore be obtained from the storage location whose address is part of the jump instruction just executed.

8. *Input.* Table 11-12 contains the logic for the input command. In the first place, it must be remembered that the input flip-flop R_1 may not be read until the input tape reader indicates that it is ready by providing a "one" on the L_1 line. The L_1 signal may appear at any time during a word, and is read every t_0 time by C_4. That flip-flop is used to control the S_1 line and thus move the tape reader from

Table 11-12. Input

	C_4		A_2		A_w		S_1
	J	K	J	K	J	K	
t_{11}					A_1	$\bar{A}_1$	
t_0	L_1	$\bar{L}_1$	L_1R_1	$\bar{L}_1 + \bar{R}_1$	"	"	
t_1	0	0	0	$\bar{A}_1$	$\bar{A}_1A_2 + A_1\bar{A}_2$	$A_1A_2 + \bar{A}_1\bar{A}_2$	C_4
t_2	"	"	"	"	"	"	"
t_3	"	"	"	"	"	"	"
...	...		...		...		...
t_{10}	"	"	"	"	"	"	"
t_{11}	0	0	"	"	"	"	"
t_0	L_1	$\bar{L}_1$	L_1R_1	$\bar{L}_1 + \bar{R}_1$	"	"	"
t_1	0	0	0	$\bar{A}_1$	"	"	"
t_2	"	"	"	"	"	"	"
t_3	"	"	"	"	"	"	"
...	...		...		...		...
t_{10}	"	"	"	"	"	"	"
t_{11}	"	"	"	"	"	"	"
t_0	L_1	$\bar{L}_1$	L_1R_1	$\bar{L}_1 + \bar{R}_1$	"	"	"
t_1	0	0	0	$\bar{A}_1$	"	"	"

its present hole position to the next one, where a new input bit may be read for the next input command. At the same time the combination of a "one" on line L_1 and a "one" in the input flip-flop R_1 is used to set the carry flip-flop A_2 at t_0 time. R_1 is thus added to the accumulator in exactly the same way that unity was added to the address of the present command during the read-instruction operation.

9. *Output.* The similar but simpler operation of the output command is shown in Table 11-13. Here the output index line L_2 is used to turn on flip-flop C_4 and thus indicate the proper time for transferring the least significant bit of the accumulator into the R_2 or output flip-flop. C_4 then drives S_2, and the punch moves on to the next hole position after punching a hole or not depending on whether R_2 contains a "one" or a "zero."

Table 11-13. Output

	C_4		R_2		S_2
	J	K	J	K	
t_{11}			0	0	
t_0	L_2	$\bar{L}_2$	0	0	C_4
t_1	0	0	C_4A_1	$C_4\bar{A}_1$	"
t_2	"	"	0	0	"
t_3	"	"	"	"	"
...	...		...		...
t_{10}	"	"	"	"	"
t_{11}	"	"	"	"	"
t_0	L_2	$\bar{L}_2$	0	0	"
t_1	0	0	C_4A_1	$C_4\bar{A}_1$	"
t_2	"	"	"	"	"
t_3	"	"	"	"	"
...	...		...		...
t_{10}	"	"	"	"	"
t_{11}	"	"	"	"	"
t_0	L_2	$\bar{L}_2$	0	0	"
t_1	0	0	C_4A_1	$C_4\bar{A}_1$	"

Input equations. We are now ready to begin collecting together the bits and pieces of logic required for each flip-flop and write amplifier. As a first step, we review Tables 11-3 through 11-13 and collect together in Table 11-14 the logic required for each individual flip-flop. In writing the entries in this table, two conventions have been followed which invite comment.

1. Any and all bit-time signals are indicated by the symbol t_i, just as if there were a flip-flop for each one. We shall subsequently design a bit-time counter and assign logical values to these timing signals.
2. The logic for some flip-flops is specified during every bit-time of every command. For others, logic is specified for some commands but not for all of them. If the logic is unspecified under certain conditions, it means that the flip-flop may assume any state under those conditions. Referring to Table 11-14, we see, for example, that flip-flop A_2 must perform certain definite functions during the instruction read-in operation and during the add, subtract, and input commands. However, its output is not used during the transfer, jump, output, and shift instructions, and its input logic may have any value (and is therefore not specified in Table 11-14) for these commands. Write amplifier A_w,

Table 11-14

Flip-Flop	Instruction	J	K
A_1	all	A_r	$\bar{A}_r$
A_2	(read in)	t_0	$\bar{N}_1\bar{t}_0$
	add	$C_3\bar{t}_0A_1Y$	$C_3t_0 + C_3\bar{t}_0\bar{A}_1\bar{Y}$
	subtract	$C_3\bar{t}_0\bar{A}_1Y$	$C_3t_0 + C_3\bar{t}_0A_1\bar{Y}$
	input	$L_1R_1t_0$	$(\bar{L}_1 + \bar{R}_1)t_0 + \bar{A}_1\bar{t}_0$
A_w	(read in)	A_1	$\bar{A}_1$
	add and subtract	$A_1\bar{C}_3 + A_1t_0C_3$ $+ \bar{t}_0C_3$ (Add logic)	$\bar{A}_1\bar{C}_3 + \bar{A}_1t_0C_3$ $+ \bar{t}_0C_3$ (Add logic)
	transfer clear	$\bar{C}_3A_1$	$\bar{C}_3\bar{A}_1 + C_3$
	right shift	$A_r\bar{t}_{11} + A_1t_{11}$	$\bar{A}_r\bar{t}_{11} + \bar{A}_1t_{11}$
	input	$t_0A_1 + \bar{t}_0(\bar{A}_1A_2 + A_1\bar{A}_2)$	$t_0\bar{A}_1 + \bar{t}_0(A_1A_2 + \bar{A}_1\bar{A}_2)$
	all others	A_1	$\bar{A}_1$
C_1	all	C_r	$\bar{C}_r$
C_2	(read in)	$(t_0 + t_1 + t_2 + t_3)$ $(N_r\bar{Z}_{r1} + \bar{N}_rZ_{r1})$	t_{11}
	add, subtract, transfer, and transfer clear	$(t_0 + t_1 + t_2 + t_3)$ $(C_r\bar{Z}_{r1} + \bar{C}_rZ_{r1})$	t_{11}
	all others	0	t_{11}
C_3	(read in), add, subtract, transfer, and transfer clear	$\bar{C}_2t_{11}$	t_{11}
	all others	0	t_{11}
C_4	(read in)	$t_{11}A_1$	$t_{11}\bar{A}_1$
	input	t_0L_1	$t_0\bar{L}_1$
	output	t_0L_2	$t_0\bar{L}_2$
C_w	(read in)	$C_1\bar{C}_3 + YC_3$	$\bar{C}_1\bar{C}_3 + \bar{Y}C_3$
	all others	C_1	$\bar{C}_1$
I_1	(read in)	$\bar{t}_{11}C_3Y$	$\bar{t}_{11}C_3\bar{Y}$
	all others	0	0
I_2	(read in)	$\bar{t}_{11}C_3I_1$	$\bar{t}_{11}C_3\bar{I}_1$
	all others	0	0

Table 11-14 (CONTINUED)

Flip-Flop	Instruction	J	K
I_3	(read in)	$\bar{t}_{11}C_3I_2$	$\bar{t}_{11}C_3\bar{I}_2$
	all others	0	0
I_4	(read in)	$(t_4 + t_5 + t_6)\overline{C}_3N_r$	$(t_4 + t_5 + t_6)\overline{C}_3\overline{N}_r$
	add, subtract, transfer, and transfer clear	$(t_4 + t_5 + t_6)\overline{C}_3C_r$	$(t_4 + t_5 + t_6)\overline{C}_3\overline{C}_r$
I_5	(read in), add, subtract, transfer, and transfer clear	$(t_4 + t_5 + t_6)\overline{C}_3I_4$	$(t_4 + t_5 + t_6)\overline{C}_3\bar{I}_4$
I_6	(read in), add, subtract, transfer, and transfer clear	$(t_4 + t_5 + t_6)\overline{C}_3I_5$	$(t_4 + t_5 + t_6)\overline{C}_3\bar{I}_5$
N_1	jump	$C_4N_r + \overline{C}_4C_r$	$C_4\overline{N}_r + \overline{C}_4\overline{C}_r$
	(read in) and all others	N_r	$\overline{N}_r$
N_w	(read in)	$\overline{C}_3N_1 + t_0C_3N_1 + \bar{t}_0C_3(\overline{N}_1A_2 + N_1\overline{A}_2)$	$\overline{C}_3\overline{N}_1 + t_0C_3\overline{N}_1 + \bar{t}_0C_3(N_1A_2 + \overline{N}_1\overline{A}_2)$
	all others	N_1	$\overline{N}_1$
R_1	all	from input device	
R_2	output	$t_1C_4A_1$	$t_1C_4\overline{A}_1$
	all others	0	0
S_1	input	C_4	
	all others	0	
S_2	output	C_4	
	all others	0	
X_{wi}	transfer and transfer clear	A_r	$\overline{A}_r$
W_i	transfer and transfer clear	$C_3(I_4 - I_6)_i$	
	all others	0	
Y	(read in), add and subtract	$\sum_{i=0}^{7} X_{ri}(I_4 - I_6)_i$	$\sum_{i=0}^{7} \overline{X}_{ri}(I_4 - I_6)_i$

on the other hand, has a very definite function to perform during every operation, and there are thus no undefined conditions in its input logic.

The reader should study Table 11-14 with some care, and make sure that he understands the purpose of all entries.

One important fact emerges clearly from the partial equations of Table 11-14: it is necessary to distinguish, in the logic, between the time a command is being carried out and the time a new command is being sought in the memory or is being read from the memory. We therefore provide a new flip-flop P whose logic is illustrated in Table 11-15. Flip-flop P will be in the "zero" state during the instruction

Table 11-15. Read Instruction Flip-Flop

Instruction	P J	P K
(read in)	$t_{11}C_3$	0
add, subtract, transfer, transfer-clear	0	$t_{11}C_3$
shift right, jump	0	t_{11}
input, output	0	$t_{11}C_4$

read-in phase, and in the "one" state during the time an instruction is being carried out. It will thus invariably be turned on at the end of the word-time during which a new command has been read into the C-register; i.e., when $C_3 = 1$ at t_{11} time. It must be turned off (indicating that a new command should be read in) at the end of the coincidence word-time in the add, subtract, and transfer commands, at t_{11} time during the shift right and jump commands, and at t_{11} when $C_4 = 1$ during input and output commands. The waveform from P during an add command is shown in Figure 11-6.

We are now ready to choose logical expressions for the timing signals t_0 to t_{11}, and to assign a binary code to each of the commands. We want to assign codes in such a way as to simplify the logic which results, and we therefore begin by looking through the logic of Table 11-14 to discover which timing signals are most often used. We discover that t_0 and $\bar{t}_0$ are used most frequently and that t_{11}, $\bar{t}_{11}$, $t_0 + t_1$

$+ t_2 + t_3$, $t_4 + t_5 + t_6$, and t_1 are all used at least once. The bit-time counter must consist of at least four flip-flops, and since the four can distinguish between sixteen different states while we need to distin-

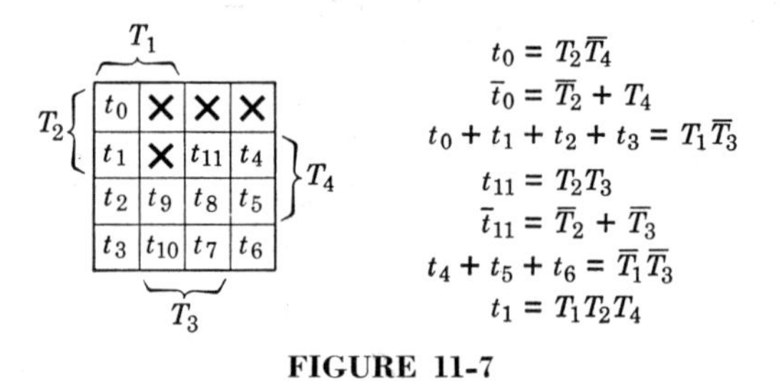

FIGURE 11-7

guish only twelve, we are left with four redundant combinations. In Figure 11-7 a Veitch diagram indicates how the redundant combinations were chosen, and shows the logic which defines the various bit-timing signals.

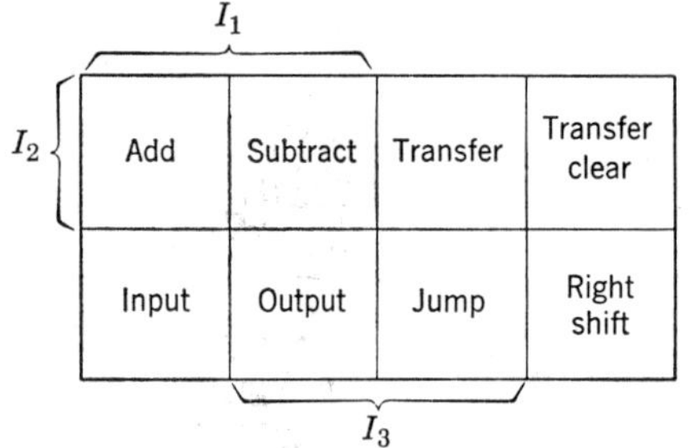

Add = $I_1 I_2 \overline{I}_3$
Subtract = $I_1 I_2 I_3$
Transfer = $\overline{I}_1 I_2 I_3$
Transfer clear = $\overline{I}_1 I_2 \overline{I}_3$
Right shift = $\overline{I}_1 \overline{I}_2 \overline{I}_3$
Jump = $\overline{I}_1 \overline{I}_2 I_3$
Output = $I_1 \overline{I}_2 I_3$
Input = $I_1 \overline{I}_2 \overline{I}_3$

Add + Subtract = $I_1 I_2$
Transfer + Jump + Output = $\overline{I}_1 I_3 + \overline{I}_2 I_3$
Add + Subtract + Transfer + Transfer clear = I_2
Right shift + Jump + Input + Output = $\overline{I}_2$
Transfer + Transfer clear = $\overline{I}_1 I_2$

FIGURE 11-8

The instruction code must be chosen in very much the same way. Inspection of Table 11-14 shows that certain of the commands appear in combinations which should be and can be conveniently combined on a Veitch diagram. The code selection is indicated in Figure 11-8 together with logic which defines the various combinations.

We are now ready to combine the information in Table 11-14 and the timing and instruction logic of Figures 11-7 and 11-8, to form the complete logical description of the computer. The description is given by equations 11-1 through 11-27. This set of equations is a first attempt at a complete description of the computer, and it is hoped that certain rearrangements and simplifications can be effected before computer construction begins. The equations are, therefore, written out in a form which will make them easy to understand and to identify with the work of Table 11-14. However, wherever any obvious simplification is possible, it has been carried out. For example, equation 11-2 for flip-flop A_2 has been simplified by introducing the "unspecified" conditions mentioned above. Since A_2 is used only during the add, subtract, and input commands, it is only necessary to insure that the different input logic terms for these three commands remain separate from one another. Thus, combining the add and subtract functions for J_{A2} in Table 11-14 with the codes for add and subtract shown in Figure 11-7, we would write

$$J_{A2} = \cdots + PI_1I_2\bar{I}_3C_3t_0A_1Y + PI_1I_2I_3C_3\bar{t}_0\bar{A}_1Y + \cdots$$

However, we now note that it is only necessary to distinguish add, subtract, and input from one another, so that we may write

$$J_{A2} = \cdots + PI_2\bar{I}_3C_3\bar{t}_0A_1Y + PI_3C_3\bar{t}_0\bar{A}_1Y + \cdots$$

This change, of course, has the effect of making A_2 act like a carry flip-flop during the transfer, output, and jump commands, but since the output of A_2 is not used during these commands, no harm is done.

The same kind of simplification is used in writing the inputs for C_4, I_4, I_5, and I_6 shown below.

From Table 11-14:

$$\begin{aligned} J_{A1} &= A_r \\ K_{A1} &= \bar{A}_r \end{aligned} \tag{11-1}$$

$$\begin{aligned} J_{A2} &= \bar{P}T_2\bar{T}_4 + PI_2\bar{I}_3C_3(\bar{T}_2 + T_4)A_1Y + PI_3C_3(\bar{T}_2 + T_4)\bar{A}_1Y \\ &\quad + P\bar{I}_2T_2\bar{T}_4L_1R_1 \\ K_{A2} &= \bar{P}\bar{N}_1(\bar{T}_2 + T_4) + PI_2\bar{I}_3[C_3T_2\bar{T}_4 + C_3(\bar{T}_2 + T_4)\bar{A}_1\bar{Y}] \\ &\quad + PI_3[C_3T_2\bar{T}_4 + C_3(\bar{T}_2 + T_4)A_1\bar{Y}] \\ &\quad + P\bar{I}_2[T_2\bar{T}_4\bar{L}_1 + T_2\bar{T}_4\bar{R}_1 + (\bar{T}_2 + T_4)\bar{A}_1] \end{aligned} \tag{11-2}$$

$$\begin{aligned}
J_{Aw} = {} & \bar{P}A_1 + PI_1I_2[A_1\bar{C}_3 + A_1T_2\bar{T}_4C_3 + (\bar{T}_2 + T_4)C_3(\bar{A}_1\bar{A}_2Y \\
& + \bar{A}_1A_2\bar{Y} + A_1\bar{A}_2\bar{Y} + A_1A_2Y)] \\
& + P\bar{I}_1I_2\bar{I}_3\bar{C}_3A_1 + P\bar{I}_1\bar{I}_2\bar{I}_3[A_r(\bar{T}_2 + \bar{T}_3) + A_1T_2T_3] \\
& + PI_1\bar{I}_2\bar{I}_3[T_2\bar{T}_4A_1 + (\bar{T}_2 + T_4)(\bar{A}_1A_2 + A_1\bar{A}_2)] \\
& + P(\bar{I}_1I_3 + \bar{I}_2I_3)A_1 \\
K_{Aw} = {} & \bar{P}\bar{A}_1 + PI_1I_2[\bar{A}_1\bar{C}_3 + \bar{A}_1T_2\bar{T}_4C_3 \\
& + (\bar{T}_2 + T_4)C_3(\bar{A}_1A_2Y + A_1\bar{A}_2Y + A_1A_2\bar{Y} + \bar{A}_1\bar{A}_2\bar{Y})] \\
& + P\bar{I}_1I_2\bar{I}_3(\bar{A}_1 + C_3) + P\bar{I}_1\bar{I}_2\bar{I}_3[\bar{A}_r(\bar{T}_2 + \bar{T}_3) + \bar{A}_1T_2T_3] \\
& + PI_1\bar{I}_2\bar{I}_3[T_2\bar{T}_4\bar{A}_1 + (\bar{T}_2 + T_4)(A_1A_2 + \bar{A}_1\bar{A}_2)] \\
& + P(\bar{I}_1I_3 + \bar{I}_2I_3)\bar{A}_1
\end{aligned} \tag{11-3}$$

$$\begin{aligned} J_{C1} &= C_r \\ K_{C1} &= \bar{C}_r \end{aligned} \tag{11-4}$$

$$\begin{aligned} J_{C2} &= \bar{P}T_1\bar{T}_3(N_r\bar{Z}_{r1} + \bar{N}_rZ_{r1}) + I_2P(C_r\bar{Z}_{r1} + \bar{C}_rZ_{r1}) \\ K_{C2} &= T_2T_3 \end{aligned} \tag{11-5}$$

$$\begin{aligned} J_{C3} &= (\bar{P} + I_2)\bar{C}_2T_2T_3 \\ K_{C3} &= T_2T_3 \end{aligned} \tag{11-6}$$

$$\begin{aligned} J_{C4} &= (\bar{P} + \bar{I}_1)T_2T_3A_1 + P\bar{I}_3T_2\bar{T}_4L_1 + PI_1I_3T_2\bar{T}_4L_2 \\ K_{C4} &= (\bar{P} + \bar{I}_1)T_2T_3\bar{A}_1 + P\bar{I}_3T_2\bar{T}_4\bar{L}_1 + PI_1I_3T_2\bar{T}_4\bar{L}_2 \end{aligned} \tag{11-7}$$

$$\begin{aligned} J_{Cw} &= \bar{P}(C_1\bar{C}_3 + YC_3) + PC_1 \\ K_{Cw} &= \bar{P}(\bar{C}_1\bar{C}_3 + \bar{Y}C_3) + P\bar{C}_1 \end{aligned} \tag{11-8}$$

$$\begin{aligned} J_{I1} &= \bar{P}(\bar{T}_2 + \bar{T}_3)C_3Y \\ K_{I1} &= \bar{P}(\bar{T}_2 + \bar{T}_3)C_3\bar{Y} \end{aligned} \tag{11-9}$$

$$\begin{aligned} J_{I2} &= \bar{P}(\bar{T}_2 + \bar{T}_3)C_3I_1 \\ K_{I2} &= \bar{P}(\bar{T}_2 + \bar{T}_3)C_3\bar{I}_1 \end{aligned} \tag{11-10}$$

$$\begin{aligned} J_{I3} &= \bar{P}(\bar{T}_2 + \bar{T}_3)C_3I_2 \\ K_{I3} &= \bar{P}(\bar{T}_2 + \bar{T}_3)C_3\bar{I}_2 \end{aligned} \tag{11-11}$$

$$\begin{aligned} J_{I4} &= \bar{T}_1\bar{T}_3\bar{C}_3(\bar{P}N_r + PC_r) \\ K_{I4} &= \bar{T}_1\bar{T}_3\bar{C}_3(\bar{P}\bar{N}_r + P\bar{C}_r) \end{aligned} \tag{11-12}$$

$$\begin{aligned} J_{I5} &= \overline{T}_1\overline{T}_3\overline{C}_3I_4 \\ K_{I5} &= \overline{T}_1\overline{T}_3\overline{C}_3\overline{I}_4 \end{aligned} \tag{11-13}$$

$$\begin{aligned} J_{I6} &= \overline{T}_1\overline{T}_3\overline{C}_3I_5 \\ K_{I6} &= \overline{T}_1\overline{T}_3\overline{C}_3\overline{I}_5 \end{aligned} \tag{11-14}$$

$$\begin{aligned} J_{N1} &= P\overline{I}_1\overline{I}_2I_3(C_4N_r + \overline{C}_4C_r) + (\overline{P} + I_1 + I_2 + \overline{I}_3)N_r \\ K_{N1} &= P\overline{I}_1\overline{I}_2I_3(C_4\overline{N}_r + \overline{C}_4\overline{C}_r) + (\overline{P} + I_1 + I_2 + \overline{I}_3)\overline{N}_r \end{aligned} \tag{11-15}$$

$$\begin{aligned} J_{Nw} &= \overline{P}[\overline{C}_3N_1 + T_2\overline{T}_4C_3N_1 \\ &\qquad + (\overline{T}_2 + T_4)C_3(\overline{N}_1A_2 + N_1\overline{A}_2)] + PN_1 \\ K_{Nw} &= \overline{P}[\overline{C}_3\overline{N}_1 + T_2\overline{T}_4C_3\overline{N}_1 \\ &\qquad + (\overline{T}_2 + T_4)C_3(N_1A_2 + \overline{N}_1\overline{A}_2)] + P\overline{N}_1 \end{aligned} \tag{11-16}$$

$$\begin{aligned} J_{R2} &= PI_1\overline{I}_2I_3T_1T_2T_4C_4A_1 \\ K_{R2} &= PI_1\overline{I}_2I_3T_1T_2T_4C_4\overline{A}_1 \end{aligned} \tag{11-17}$$

$$S_1 = PI_1\overline{I}_2\overline{I}_3C_4 \tag{11-18}$$

$$S_2 = PI_1\overline{I}_2I_3C_4 \tag{11-19}$$

$$\begin{aligned} J_{Xwi} &= A_r \\ K_{Xwi} &= \overline{A}_r \end{aligned} \qquad i = 0, 1, 2, \cdots, 7 \tag{11-20}$$

$$W_i = P\overline{I}_1I_2C_3(I_4 - I_6)_i \qquad i = 0, 1, 2, \cdots, 7 \tag{11-21}$$

$$\begin{aligned} J_Y &= \sum_{i=0}^{7} X_{ri}(I_4 - I_6)_i \\ K_Y &= \sum_{i=0}^{7} \overline{X}_{ri}(I_4 - I_6)_i \end{aligned} \tag{11-22}$$

From Table 11-15:

$$\begin{aligned} J_P &= T_2T_3C_3 \\ K_P &= T_2T_3(I_2C_3 + \overline{I}_1\overline{I}_2 + I_1\overline{I}_2C_4) \end{aligned} \tag{11-23}$$

From Figure 11-7:

$$\begin{aligned} J_{T1} &= T_3T_4 + Z_{r2} \\ K_{T1} &= \overline{T}_2\overline{T}_4\overline{Z}_{r2} \end{aligned} \tag{11-24}$$

$$\begin{aligned} J_{T2} &= T_1\overline{T}_4 + Z_{r2} \\ K_{T2} &= \overline{T}_3T_4\overline{Z}_{r2} \end{aligned} \tag{11-25}$$

$$J_{T3} = \bar{T}_1\bar{T}_4\bar{Z}_{r2}$$
$$K_{T3} = T_2 + Z_{r2} \tag{11-26}$$

$$J_{T4} = T_1 + T_3 + Z_{r2}$$
$$K_{T4} = (\bar{T}_2\bar{T}_3 + T_1T_3 + T_2T_3)\bar{Z}_{r2} \tag{11-27}$$

Note that the Z_{r2} signal occurs at t_0 time, and sets the bit-time counter into the (1101) or t_1 state.

These equations must now be simplified as much as possible. In practice, of course, the designer tries to simplify as he goes along and major simplifications are not usually discovered at this stage of the game. However, certain logical reductions are usually apparent, and may be found with little trouble. For example, in equations 11-2

$$C_3T_2\bar{T}_4 + C_3(\bar{T}_2 + T_4)\bar{A}_1\bar{Y} = C_3T_2\bar{T}_4 + C_3\bar{A}_1\bar{Y}$$

and in equations 11-3

$$P\bar{I}_1I_2\bar{I}_3\bar{C}_3A_1 + P(\bar{I}_1I_3 + \bar{I}_2I_3)A_1 = P\bar{I}_1I_2\bar{C}_3A_1 + P(\bar{I}_1I_3 + \bar{I}_2I_3)A_1$$

Other minor simplifications are possible, and the reader should try his hand at finding them.

Revision and rearrangement. As has been indicated before, a comprehensive review of the design at this point may lead the designer to propose a radically different framework for his computer and thus to a repetition of all the steps we have gone through in arriving at equations 11-1 through 11-27. We shall not attempt such a complete redesign, but shall indicate how three major types of modification can lead to major simplifications.

1. *Adding memory elements.* As was pointed out in Chapter 6, the circuit designer is always able to set up some criterion which will make it possible to equate one memory element to a certain number of decision elements, and thus will make it possible to decide which of two equipment configurations having different numbers of logical elements is cheaper. With this criterion in mind, the logical designer can evaluate the effect of adding memory elements to cut down on the need for decision elements, and vice versa.

For the computer we have designed, there are several places where the addition of memory elements may result in a net saving. An obvious possibility is the bit-time counter. The twelve bit-times t_0 to t_{11} are distinguished by means of flip-flops T_1 to T_4, whose twelve states were chosen so as to permit the cheapest possible identification of bit-times or combinations thereof. However, if more flip-flops are provided, the additional redundancies will make it possible to represent these bit-times and combinations with fewer decision elements, and because the bit-time logic is used in most of the input equations this may result in a net saving. For example, if one more flip-flop were added to the counter, it would be possible to represent t_0 and $\bar{t}_0$ by a single letter rather than by the combinations $T_2\bar{T}_4$ and $(\bar{T}_2 + T_4)$.

2. *Removing memory elements.* The removal of memory elements may similarly result in a net simplification. It is of course possible to apply the Huffman-Mealy methods of Chapter 6 to the computer with the objective of eliminating possible states and thus perhaps eliminating one or more memory elements. If the application of the Huffman-Mealy method is too difficult or cumbersome, there are other steps which can be taken. It is necessary for the designer to review the original decisions he made in setting up a framework for the computer; e.g., he should review at this point the necessity for including flip-flops N_1 and C_1 in the N- and C-registers, which were added quite arbitrarily in the beginning. Still another approach requires that the designer examine the completed design with the objective of discovering whether there are two memory elements whose functions do not overlap. It may then be possible to eliminate one of them at little or no cost in decision elements. In our simple computer, flip-flop C_4 can be eliminated in this way by using I_4 to carry out its function. Note that C_4 is only used during t_{11} time of read-in, and during the jump, input, and output commands. I_4, on the other hand, is only employed during $\bar{t}_{11}$ of read-in, and during the add, subtract, and transfer commands. We could thus replace C_4 everywhere it appears in the input equations by I_4, and change the I_4 input logic to

$$\begin{aligned} J_{I4} &= \bar{P}\bar{T}_1\bar{T}_3\bar{C}_3N_r + \bar{P}C_3T_2T_3A_1 + PI_2\bar{T}_1\bar{T}_3\bar{C}_3C_r \\ &\quad + P\bar{I}_1\bar{I}_2T_2T_3A_1 + P\bar{I}_2\bar{I}_3T_2\bar{T}_4L_1 \\ &\quad + PI_1\bar{I}_2I_3T_2\bar{T}_4L_2 \\ K_{I4} &= \bar{P}\bar{T}_1\bar{T}_3\bar{C}_3\bar{N}_r + \bar{P}C_3T_2T_3\bar{A}_1 \\ &\quad + PI_2\bar{T}_1\bar{T}_3\bar{C}_3\bar{C}_r + P\bar{I}_1\bar{I}_2T_2T_3\bar{A}_1 \\ &\quad + P\bar{I}_2\bar{I}_3T_2\bar{T}_4\bar{L}_1 + PI_1\bar{I}_2I_3T_2\bar{T}_4\bar{L}_2 \end{aligned} \qquad (11\text{-}28)$$

3. *Rearranging the functions of memory elements.* A certain amount of simplification can usually be accomplished by rearranging and changing certain of the functions carried out by various flip-flops. For example, consider the J input to the write amplifier A_w of equations 11-3 and Table 11-14. In equation 11-29, part of this equation is written

in a way which emphasizes the logic common to the add, subtract, and input commands.

$$
\begin{array}{lll}
(\text{add} + \text{subtract}) & & \\
 & C_3[A_1t_0 + \bar{t}_0(\bar{A}_1\bar{A}_2Y + \bar{A}_1A_2\bar{Y} + A_1\bar{A}_2\bar{Y} + A_1A_2Y)] \\
+\ (\text{input}) & [A_1t_0 + \bar{t}_0(\qquad \bar{A}_1A_2 \quad + A_1\bar{A}_2 \qquad)] \\
+\ (\text{output}) & [A_1t_0 + \bar{t}_0(\qquad\qquad A_1 \qquad\qquad)]
\end{array}
\tag{11-29}
$$

Because the add, subtract, input, *and output* commands are collectively represented by the logical symbol I_1, the logic for the output operation is included in equation 11-29. Examining this equation, we see that it might be possible to make the logical function carried out by A_w exactly the same for all four of these instructions, if certain steps were taken. First, it would be necessary to change the input logic to flip-flop C_3 so that it is set during the input and output commands. Second, the input logic to flip-flop Y must be changed so that it remains in the "zero" state during the input and output commands. And finally, the logic for flip-flop A_2 must be modified so that it remains in the "zero" state during the whole of the output command. With these modifications made, the logic of equation 11-29 could be written as shown in equation 11-30, and made a part of J_{Aw}.

$$I_1C_3A_1t_0 + I_1C_3\bar{t}_0(\bar{A}_1\bar{A}_2Y + \bar{A}_1A_2\bar{Y} + A_1\bar{A}_2\bar{Y} + A_1A_2Y) \tag{11-30}$$

Of course, it would be necessary to modify the operation of flip-flops C_3, Y, and A_2 in order to accomplish this simplification, and when the modification is complete it must be examined to see whether there is indeed a net savings over the original configuration, or whether the modifications required introduced more complexity than was eliminated from the inputs to A_w.

OTHER GENERAL-PURPOSE COMPUTERS

As was indicated at the beginning of this chapter, the difference between the simple computer we have designed here and the general-purpose computer used as an accounting machine and in scientific calculations is a difference in degree rather than in kind. The differences are, however, illuminating and it will be worthwhile to discuss them briefly.

1. *Differences in number representation.* In addition to binary machines there are, among others, decimal machines and machines in which numbers are represented in "floating point" form, i.e., by means of a coefficient and an exponent. In addition to serial computers, there are parallel ones.

2. *Differences in order structure.* For obvious reasons, the computer which we have designed has an order code which is designated single address. There are also machines with instructions having two, three, and four addresses. A typical order for a four-address machine might be "divide the operand from address a by the operand from address b, store the quotient in address c, leave the remainder in the accumulator and obey next the instruction in storage location e." Note that in designing a computer having such an order code, the designer would have to provide logic for as many as four accesses to the memory for a single order.

3. *Differences in operations performed.* In addition to add, subtract, jump, shift right, transfer, transfer clear, input, and output as provided in our simple computer, most practical computers also have the commands, multiply, divide, shift left, and sometimes the logical operations "and" and "or."

4. *Differences in size.* A practical computer, generally speaking, has many thousands of storage locations rather than hundreds. In addition, it usually has a large number of input and output channels, each consisting of a number of bits in parallel rather than a single bit. These various input, output, and memory devices are all accessible to the programmer through commands which enable him to select whichever device he needs to use.

THE DESIGN OF A LIMITED-PURPOSE COMPUTER

As has been mentioned before, the computer which will be described and designed in this part of the chapter is not intended to be a practical machine. It is intended to illustrate the advantages in simplicity and speed which can be obtained from a computer designed to do a specific job.

Functional Description

Suppose we are to provide a machine which is to compute continuously the square root of 100 different positive binary numbers. These numbers are to be supplied to the computer from an outside source, and they have the characteristic that they do not change very rapidly—

a full-scale change in 10 seconds is the fastest expected. Each of the numbers (a_i) which is to be operated on is expressed to a precision of twenty binary digits.

Of the many rules which may be used for taking the square root of a number, one very simple iterative procedure requires no multiplications or divisions other than by a power of two. Let us suppose we keep a record of two numbers, x_{ni}^2 and x_{ni}, for each of the 100 a_i's. Let us further suppose that it is our objective to make a_i and x_{ni}^2 as nearly equal as possible, so that x_{ni} will be as nearly equal to the square root of a_i as is possible. We then proceed by comparing a_i with x_{ni}^2.

If $a_i \geq x_{ni}^2$, take

$$x_{(n+1)i} = x_{ni} + 1$$

and

$$x_{(n+1)i}^2 = x_{ni}^2 + 1 + 2x_{ni} \qquad (11\text{-}31)$$

On the other hand, if $a_i < x_{ni}^2$, take

$$x_{(n+1)i} = x_{ni} - 1$$

and

$$x_{(n+1)i}^2 = x_{ni}^2 + 1 - 2x_{ni} \qquad (11\text{-}32)$$

We see, then, that if we have x_{ni}^2 and x_{ni} stored in a computer together with a_i, we can continuously modify x_{ni} by adding unity to it or subtracting unity from it until it becomes equal to the square root of a_i. In order to know whether we must add or subtract unity, we compare

Table 11-16

n	a_i	$x_{ni}^2 = x_{(n-1)i}^2 + 1 \pm 2x_{ni}$	x_{ni}	$2x_{ni}$
0	4 8 2 6 2 1	4 7 0 5 9 6	6 8 6	1 3 7 2
1		4 7 1 9 6 9	6 8 7	1 3 7 4
2		4 7 3 3 4 4	6 8 8	1 3 7 6
3		4 7 4 7 2 1	6 8 9	1 3 7 8
4		4 7 6 1 0 0	6 9 0	1 3 8 0
5		4 7 7 4 8 1	6 9 1	1 3 8 2
6		4 7 8 8 6 4	6 9 2	1 3 8 4
7		4 8 0 2 4 9	6 9 3	1 3 8 6
8		4 8 1 6 3 6	6 9 4	1 3 8 8
9		4 8 3 0 2 5	6 9 5	1 3 9 0
10		4 8 1 6 3 6	6 9 4	1 3 8 8
11		4 8 3 0 2 5	6 9 5	1 3 9 0

etc.

a_i with x_{ni}^2 according to the rules of equations 11-31 and 11-32; and we find the *new* value of x_{ni}^2 (corresponding to the new value for x_{ni}) by adding unity to the old value and then either adding or subtracting twice x_{ni}. The process is illustrated in Table 11-16, where we assume that originally some particular $a_i = 482{,}621$, and $x_{0i} = 686$. After nine applications of the rule described above, x_{ni}^2 becomes greater than a_i. From then on, x_{ni} oscillates about the square root of a_i.

Designing the Computer

Framework. In Figure 11-9 a schematic diagram of the computer is shown. It consists of three shifting or circulating registers (one for

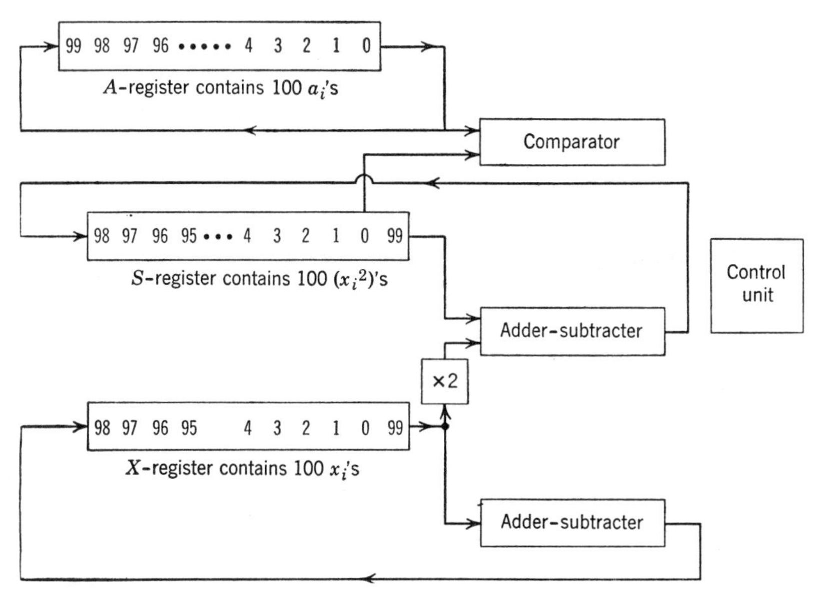

FIGURE 11-9

x_i, one for x_i^2, and one for a_i), each containing 100 twenty-bit numbers; a comparator to be used in comparing a_i with x_{ni}^2; an adder-subtracter to modify x_{ni}^2 by adding $(1 \pm 2x_{ni})$ to it; an adder-subtracter to modify x_{ni} by adding (± 1) to it; and a control unit which determines which action should be taken to modify x_{ni} and x_{ni}^2.

In operation, the computer continually carries out a comparison and modification of the kind described in equations 11-31 and 11-32. At any given time, one iteration is being carried out on one set of numbers a_i, x_{ni}, x_{ni}^2 as they are shifted out of their circulating registers. $x_{(n+1)i}$ and $x_{(n+1)i}^2$ are then stored back in their registers, and are not modified until an iteration has been performed on each of the ninety-nine *other* numbers. Thus one iteration is performed on each number once per circulating-register cycle. If 6000 such cycles are performed per minute (a reasonable number for a magnetic drum computer), 1000 will be performed in 10 seconds. A number a_i may thus change from zero to full-scale ($2^{20} \approx 10^6$) in 10 seconds, and the computer will be able to keep up with it, for its square root will change from zero to $2^{10} \approx 10^3$. The computer therefore satisfies adequately the maximum change-rate requirement specified in the functional description.

Word description. The framework of Figure 11-9 is expanded in Figure 11-10 to show additional detail. The A-, S-, and X-registers

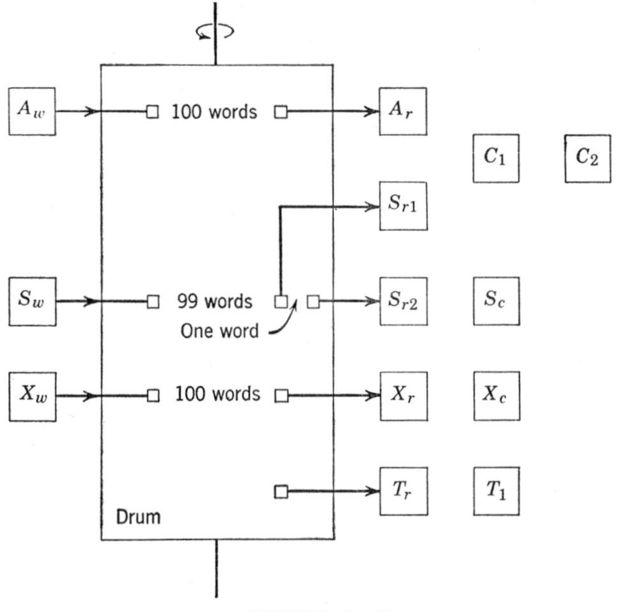

FIGURE 11-10

are three drum circulating registers having write amplifiers A_w, S_w, and X_w and read amplifiers A_r, S_{r1}, S_{r2}, and X_r. The relative timing of information from the various read amplifiers is indicated in Figure 11-11, where it will be seen that, at any time, two numbers a_i and ${x_{ni}}^2$ may be read, least significant bit first, from A_r and S_{r1} for comparison purposes. At the same time, the word $({x_{n(i-1)}}^2)$ which was compared with a_{i-1} during the previous word-time is read from S_{r2}, another read amplifier on the S channel of the drum one word-time away from amplifier S_{r1}. The number $x_{n(i-1)}$ is meanwhile read from amplifier X_r.

Space bits labeled "sp" separate words on all three channels. The T_r and T_1 signals at the bottom of Figure 11-11 will be discussed later.

A description of computer functions breaks down conveniently to a description of three operations.

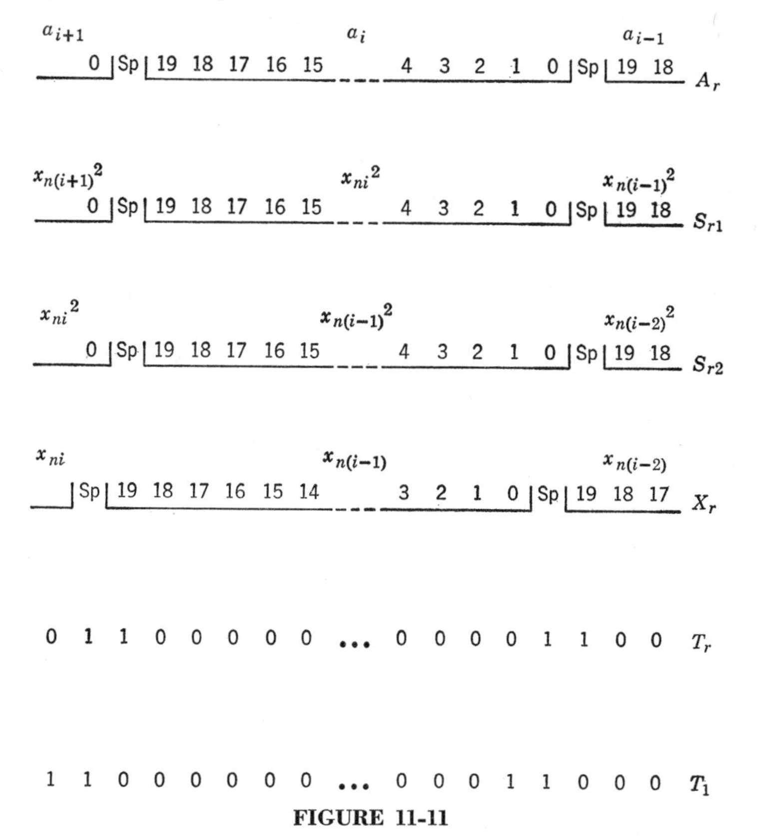

FIGURE 11-11

1. *Comparison of* a_i *and* x_{ni}^2. During every word-time, the number read from A_r is compared with that read from S_{r1}. Flip-flop C_1 is used to effect the comparison, and since we are dealing with serial binary numbers the comparator of equation 9-91 may be used. At the end of every word-time $C_1 = 0$ if $a_i \geqq x_{ni}^2$, and $C_1 = 1$ if $a_i < x_{ni}^2$. At the end of each word-time, when the space bits are in A_r and S_{r1}, C_1 must be reset to "zero" ready for the next comparison, and at that same time C_1 is shifted into C_2. Meanwhile, the number from A_r is copied without change into A_w.

2. *Modification of* $x_{n(i-1)}^2$. While a_i and x_{ni}^2 are being compared, flip-flop C_2 contains the result of the comparison of a_{i-1} and $x_{n(i-1)}^2$.

If $C_2 = 0$, we must add $(1 + 2x_{n(i-1)})$ to $x_{n(i-1)}^2$, thus forming $x_{(n+1)(i-1)}^2$ which is to be written in S_w. If $C_2 = 1$, we must add $(1 - 2x_{n(i-1)})$ to $x_{n(i-1)}^2$ and store the result in S_w. As will be seen from Figure 11-11, $x_{n(i-1)}$ appears from X_r shifted one bit to the left with respect to $x_{n(i-1)}^2$ read from S_{r2}. This shift corresponds to a multiplication by two, and thus permits us to add (or subtract) the number from X_r directly to (or from) the number from S_{r2}. Whether we add or subtract $2x_{n(i-1)}$, we must add unity at the same time. Flip-flop S_c is provided as a carry flip-flop for the addition (or subtraction).

3. *Modification of* $x_{n(i-1)}$. If $C_2 = 0$, we must add unity to $x_{n(i-1)}$; if $C_2 = 1$, we must subtract unity. In either case, flip-flop X_c is provided for the carry, and the result must be stored in X_w.

Truth tables

1. *Comparison of* a_i *and* x_{ni}^2. In Table 11-17, the comparison logic is shown. The logic follows exactly the word description given above, and no additional comment is required.

Table 11-17

Bit in A_r	C_1 J	C_1 K	C_2 J	C_2 K	A_w J	A_w K
sp	0	1	C_1	$\bar{C}_1$	A_r	$\bar{A}_r$
0	$S_{r1}\bar{A}_r$	$\bar{S}_{r1}A_r$	0	0	"	"
1	"	"	"	"	"	"
2	"	"	"	"	"	"
3	"	"	"	"	"	"
...	...		...		...	
18	"	"	"	"	"	"
19	"	"	"	"	"	"
sp	0	1	C_1	$\bar{C}_1$	"	"
0	$S_{r1}\bar{A}_r$	$\bar{S}_{r1}A_r$	0	0	"	"
1	"	"	"	"	"	"
2	"	"	"	"	"	"
3	"	"	"	"	"	"
...	...		...		...	
18	"	"	"	"	"	"
19	"	"	"	"	"	"
sp	0	1	C_1	$\bar{C}_1$	"	"
0	$S_{r1}\bar{A}_r$	$\bar{S}_{r1}A_r$	0	0	"	"

2. *Modification of* $x_{n(i-1)}^2$. During any word-time that $C_2 = 0$, we must add the number from X_r to the number from S_{r2} and at the same

time add unity to their sum. This is obviously easily done by providing an ordinary binary adder and setting its carry flip-flop equal to "one" before the addition begins. However, during any word-time that $C_2 = 1$, we must *subtract* the number in X_r from the number in S_{r2} and at the same time add unity. Although the sum and difference logic is the same for the two operations (and therefore the logic for write amplifier S_w will be the same), the carry is different. In particular, the logic for the operation $(X - Y + 1)$, where X and Y are random numbers, would be somewhat complicated. However, our problem is greatly simplified because the two numbers we are to operate on are related: one is the square of half the other. In Table 11-18a it is

Table 11-18

x	··· 0 0	··· 0 1	··· 1 0	··· 1 1
x	··· 0 0	··· 0 1	··· 1 0	··· 1 1
	··· 0 0	··· 0 1	··· 0 0	··· 1 1
	··· 0 0	··· 0 0	··· 1 0	··· 1 1
x^2	··· 0 0	··· 0 1	··· 0 0	··· 0 1

(*a*)

(S_{r2})	x^2	··· 0 0	··· 0 1
(X_r)	$2x$	··· 0	··· 1
	1	1	1
(S_w)	$x^2 - 2x + 1$	··· 0 1	··· 0 0
	borrow	0	0

(*b*) $C_2 = 1$

(S_{r2})	x^2	··· 0 0	··· 0 1
(X_r)	$2x$	··· 0	··· 1
	1	1	1
(S_w)	$x^2 + 2x + 1$	··· 0 1	··· 0 0
	carry	0	1

(*c*) $C_2 = 0$

shown that the square of any even binary number ends in 00, while the square of any odd one ends in 01. Therefore, during the two bit-times that the two least significant bits of $x_{n(i-1)}{}^2$ are in S_{r2}, there can be only two possible configurations read from S_{r2} and X_r, and these are shown in Table 11-18b and c.

Examining these two configurations in Table 11-18b, where $(x^2 - 2x + 1)$ is formed, we find the following rule is applicable: the least significant bit of ${x_{(n+1)(i-1)}}^2$ is the complement of the least significant bit of ${x_{n(i-1)}}^2$; the next least significant bit is zero; and the remaining bits may be formed by subtracting $2x_{n(i-1)}$ from ${x_{n(i-1)}}^2$, where the subtraction starts with a borrow of zero. The logic for this operation is shown in Table 11-19. Note that the borrow flip-flop S_c is reset at bit 1 time, and subsequently contains the carry for subtraction. The difference is written in S_w.

Table 11-19. $C_2 = 1$

Bit in A_r	S_w		S_c	
	J	K	J	K
sp				
0	$\bar{S}_{r2}$	S_{r2}		
1	0	1	0	1
2	$\bar{S}_{r2}\bar{X}_r S_c + \bar{S}_{r2} X_r \bar{S}_c + S_{r2}\bar{X}_r\bar{S}_c + S_{r2} X_r S_c$	$\bar{S}_{r2} X_r S_c + S_{r2}\bar{X}_r S_c + S_{r2} X_r \bar{S}_c + \bar{S}_{r2}\bar{X}_r\bar{S}_c$	$X_r\bar{S}_{r2}$	$\bar{X}_r S_{r2}$
3	"	"	"	"
4	"	"	"	"
5	"	"	"	"
...	...		...	
18	"	"	"	"
19	"	"	"	"
sp				
0	$\bar{S}_{r2}$	S_{r2}		
1	0	1	0	1
2	$\bar{S}_{r2}\bar{X}_r S_c + \bar{S}_{r2} X_r \bar{S}_c + S_{r2}\bar{X}_r\bar{S}_c + S_{r2} X_r S_c$	$\bar{S}_{r2} X_r S_c + S_{r2}\bar{X}_r S_c + S_{r2} X_r \bar{S}_c + \bar{S}_{r2}\bar{X}_r\bar{S}_c$	$X_r\bar{S}_{r2}$	$\bar{X}_r S_{r2}$
3	"	"	"	"
4	"	"	"	"
5	"	"	"	"
...	...		...	
18	"	"	"	"
19	"	"	"	"
sp				
0	$\bar{S}_{r2}$	S_{r2}		
1	0	1	0	1

Returning now to Table 11-18c, where $(x^2 + 2x + 1)$ is formed, we can state another rule similar to the one used in deriving Table 11-19: the least significant bit of ${x_{(n+1)(i-1)}}^2$ is again the complement of the least significant bit of ${x_{n(i-1)}}^2$; the next least significant bit is again zero; and the remaining bits may be formed by adding $2x_{n(i-1)}$ to $x_{n(i-1)}^2$, where the addition begins with a carry equal to the contents of X_r at bit 1 time. The logic for the sum is then the same as that for the difference, as given in Table 11-19. The carry logic is shown in Table 11-20.

Table 11-20. $C_2 = 0$

Bit in A_r	S_c	
	J	K
sp		
0		
1	X_r	$\bar{X}_r$
2	$X_r S_{r2}$	$\bar{X}_r \bar{S}_{r2}$
3	"	"
4	"	"
5	"	"
...	...	
18	"	"
19	"	"
sp		
0		
1	X_r	$\bar{X}_r$
2	$X_r S_{r2}$	$\bar{X}_r \bar{S}_{r2}$
3	"	"
4	"	"
5	"	"
...	...	
18	"	"
19	"	"
sp		
0		
1	X_r	$\bar{X}_r$

3. *Modification of* $x_{n(i-1)}$. The logic for the formation of $(x_{n(i-1)} \pm 1)$ is shown in Table 11-21. The logic for sum and difference is again identical, and the carry logic requires that $K_{Xc} = \bar{X}_r$ for addition ($C_2 = 0$), and $K_{Xc} = X_r$ for subtraction ($C_2 = 1$). Note that X_c is set at bit 0 time, so that the addition (or subtraction) does not begin until bit 1 time, when the least significant bit of $x_{n(i-1)}$ is in X_r (see Figure 11-11).

We must now review Tables 11-17, 11-19, 11-20, and 11-21 to determine what timing control signals are required. For the logic of Table 11-17 we see it is only necessary to distinguish the space bit from all others; for Tables 11-19 and 11-20 we must be able to distinguish the

Table 11-21

Bit in A_r	X_w		X_c	
	J	K	J	K
sp				
0			1	0
1	$X_r\overline{X}_c + \overline{X}_rX_c$	$X_rX_c + \overline{X}_r\overline{X}_c$	0	$\overline{C}_2\overline{X}_r + C_2X_r$
2	"	"	"	"
3	"	"	"	"
4	"	"	"	"
...	...		...	
18	"	"	"	"
19	"	"	"	"
sp				
0			1	0
1	$X_r\overline{X}_c + \overline{X}_rX_c$	$X_rX_c + \overline{X}_r\overline{X}_c$	0	$\overline{C}_2\overline{X}_r + C_2X_r$
2	"	"	"	"
3	"	"	"	"
4	"	"	"	"
...	...		...	
18	"	"	"	"
19	"	"	"	"
sp				
0			1	0

0 and 1 bits from all others; and for Table 11-21 bit 0 must be distinguished from all others. In sum, we must be able to distinguish four bit-times for each word: space bit, 0 bit, 1 bit, and all other bits. It would of course be possible to design a five flip-flop clock-pulse counter having twenty-one states, and to form appropriate logical signals for the required bit times. A much simpler solution requires only that a special channel be written on the drum containing two "ones" every word-time, as shown in Figure 11-11. Read amplifier T_r reads these bits and they are immediately shifted into flip-flop T_1. These two memory elements thus have four states which may be identified with the four necessary bit-times as follows:

T_r	T_1	
0	0	all bit-times except sp, 0, and 1
1	0	sp bit
1	1	0 bit
0	1	1 bit

Input equations. The input equations may now be written by collecting together and combining the timing and truth-table logic as follows:

$$J_{C1} = S_{r1}\bar{A}_r(\bar{T}_r + T_1)$$
$$K_{C1} = T_r\bar{T}_1 + \bar{S}_{r1}A_r \tag{11-33}$$

$$J_{C2} = C_1T_r\bar{T}_1$$
$$K_{C2} = \bar{C}_1T_r\bar{T}_1 \tag{11-34}$$

$$J_{Aw} = A_r$$
$$K_{Aw} = \bar{A}_r \tag{11-35}$$

$$J_{Sw} = \bar{S}_{r2}T_rT_1 + \bar{T}_1(\bar{S}_{r2}\bar{X}_rS_c + \bar{S}_{r2}X_r\bar{S}_c + S_{r2}\bar{X}_r\bar{S}_c + S_{r2}X_rS_c)$$
$$K_{Sw} = S_{r2}T_rT_1 + \bar{T}_rT_1 + \bar{T}_1(\bar{S}_{r2}X_rS_c + S_{r2}\bar{X}_rS_c + S_{r2}X_r\bar{S}_c + \bar{S}_{r2}\bar{X}_r\bar{S}_c) \tag{11-36}$$

$$J_{Sc} = C_2\bar{T}_1X_r\bar{S}_{r2} + \bar{C}_2T_1X_r + \bar{C}_2\bar{T}_1X_rS_{r2}$$
$$K_{Sc} = C_2T_1 + C_2\bar{T}_1\bar{X}_rS_{r2} + \bar{C}_2T_1\bar{X}_r + \bar{C}_2\bar{T}_1\bar{X}_r\bar{S}_{r2} \tag{11-37}$$

$$J_{Xw} = X_r\bar{X}_c + \bar{X}_rX_c$$
$$K_{Xw} = X_rX_c + \bar{X}_r\bar{X}_c \tag{11-38}$$

$$J_{Xc} = T_r$$
$$K_{Xc} = \bar{T}_r(\bar{C}_2\bar{X}_r + C_2X_r) \tag{11-39}$$

$$J_{T1} = T_r$$
$$K_{T1} = \bar{T}_r \tag{11-40}$$

Comments on the Limited-Purpose Computer

Before the design of this computer could be said to be complete, it would of course be necessary to provide logic for inserting new values for the various a_i, for extracting desired solutions x_i, and possibly for setting new initial values for x_i and x_i^2 (one initial value easily obtained is $x_i = x_i^2 = 0$). However, at this point the computer has served its purpose—that of illustrating how a limited-purpose computer can be designed—and therefore the input-output functions will not be introduced.

In appraising the value of this computer, and of limited-purpose computers in general, the following conjecture should be discussed: given any fixed set of operations which must be performed, it is possible to design a limited-purpose computer which can carry them out faster and cheaper than a general-purpose computer can. It is clear, for example, that (given a memory large enough to hold all necessary operands and instructions) the general-purpose computer designed in the first part of this chapter could carry out the iterations of equations 11-31 and 11-32. It should also be clear that the general-purpose computer is more complicated and expensive than the limited-purpose one, and that it would perform the iterations more slowly: several drum revolutions would be required for it to perform even one iteration, where the limited-purpose computer does 100 iterations in a single revolution. This speed and simplicity have been achieved at a complete sacrifice of flexibility. In general, the lack of flexibility means that the control unit for a limited-purpose computer can be exceedingly simple, for the sequence of operations it performs is fixed and the data to be operated on can be arranged so that it becomes available as required without the necessity for addresses and memory selection.

In conclusion, it seems appropriate to point out that the future applications of digital computers—in scientific, business, and engineering calculations, and in the control of such varied things as automobile traffic or chemical processes or machine tools—will depend to a large extent on the ingenuity and resourcefulness of engineers competent in logical design. Both limited- and general-purpose computers have a place in these applications, and the logical and system designer must be aware of the advantages and disadvantages of both.

BIBLIOGRAPHY

W. L. Van der Poel, "A Simple Electronic Digital Computer," *Applied Science Research,* **2,** Part B, 367–400 (1951).

C. F. West and J. E. Deturk, "A Digital Computer for Scientific Applications," *Proc. I.R.E.,* **36,** no. 12, 1452–1460 (Dec. 1948).

M. V. Wilkes, and J. B. Stringer, "Micro-Programming and the Design of Control Circuits in an Electronic Digital Computer," *Proc. Cambridge Philosophical Society,* 230–238 (Apr. 1953).

W. H. Dunn, C. Eldert, and P. V. Levonian, "A Digital Computer for Use in an Operational Flight Trainer," *I.R.E. Trans. on Electronic Computers,* **EC-4,** no. 2, 55–63 (1955).

R. R. Johnson, "An Electronic Digital Polynomial Root Extractor," *Proc. of the Western Joint Computer Conference,* 119–123, 1955.

E. E. Bolles and H. L. Engel, "Control Elements in the Computer," *Control Engineering,* **3,** no. 8, 93–98 (1956).

A. W. Burks and I. M. Copi, "The Logical Design of an Idealized General-Purpose Computer," *J. Franklin Institute*, **261**, no. 3, 299–314 (1956).
W. E. Smith, "A Digital System Simulator," *Proc. of the Western Joint Computer Conference*, 31–36, 1957.
J. M. Frankovich and H. P. Peterson, "A Functional Description of the Lincoln TX-2 Computer," *Proc. of the Western Joint Computer Conference*, 146–155, 1957.
S. P. Frankel, "The Logical Design of a Simple General Purpose Computer," *I.R.E. Trans. on Electronic Computers*, **EC-6**, no. 1, 5–14 (Mar. 1957).
R. J. Mercer, "Micro-Programming," *J. Association for Computing Machinery*, **4**, 157–171 (1957).

EXERCISES

1. Suppose the first sixty storage locations of the general-purpose computer described in this chapter contain sixty numbers labeled $a_0, a_1, a_2, \cdots, a_{58}, a_{59}$. Devise a program which sums these numbers in pairs, putting each sum in the location formerly occupied by the first of the two operands. That is, $(a_0 + a_1)$ is to be put in location zero, $(a_2 + a_3)$ in location two, $(a_4 + a_5)$ in location four, etc., until $(a_{58} + a_{59})$ is put in location fifty-eight. It is of course not possible to use more than 128 words of storage for operands plus instructions. The machine should stop when the computation is complete, and it may be assumed that the numbers a_i are such that $|(a_i + a_{i+1})| < 1024$. Write the detailed computer code corresponding to the program you have devised. (*Hint:* Each basic calculation requires three commands—two additions and a transfer—and thirty such sets would occupy ninety storage locations. These ninety, in addition to the sixty required for the basic data, would use more computer memory than is available. This difficulty may be overcome by starting with the three basic commands and planning for the program to modify them by operating on them as if they were numbers. For example, if we start with the three commands "Add zero; add one; transfer-clear zero," we can change them to "Add two; add three; transfer-clear two" by using the computer to add two to each command in succession, putting the modified commands back in their former locations.)

2. Two numbers are stored in locations zero and one of the general-purpose computer described in this chapter. Devise a program which forms the product of these two numbers. The routine should put the most significant ten bits of the product in location two, and the least significant ten in location three. If the product is negative, it should be stored in two's complement form and its sign bit will be the sign bit of location two. The sign bit of location three should always be zero.

3. Write the simplest minterm-type expression for equations
a. 11-2 *c.* 11-8 *e.* 11-36
b. 11-3 *d.* 11-16 *f.* 11-37

4. Write the simplest maxterm-type expression for equations
a. 11-8
b. 11-37

5. Add one flip-flop to the bit-time counter for the general-purpose computer. Design the counter so that the various bit-time combinations indicated in Figure 11-7 are distinguished as efficiently as possible. Write the simplest input equations for the counter flip-flops. Estimate the net savings in diodes obtained by using this new counter in place of the one described by equations 11-24 through 11-27. Assume that equations 11-1 through 11-27 must all be minterm-type expressions (second order).

6. Make the modifications necessary in equations 11-2, 11-6, and 11-22 so that equation 11-30 may be incorporated into equations 11-3. Can K_{Aw} be simplified, as well as J_{Aw}?

7. Design a serial, decimal, general-purpose computer having the following order-code:

1. *Subtract the contents of storage location n from the number in the accumulator.*
2. *Transfer the number in the accumulator to storage location n.*
3. *Shift the number in the accumulator one decimal digit to the right.*
4. *If the number in the accumulator is positive or zero, execute next the instruction in storage location n. Otherwise proceed as usual.*

The computer has a 100-word magnetic-drum memory, arranged in five channels of twenty words each. A word consists of three decimal digits (expressed in the 8-4-2-1 code) and a sign bit. A single space bit separates words. Negative numbers are represented by their ten's complements. The accumulator is a drum circulating register, and any other drum registers may be supplied as required. Normally (except for the Jump Command) instructions are carried out in the sequence in which they appear in the memory.

8. Suppose a transient failure causes the limited-purpose square-root computer of this chapter to make a single error. Before this failure, $a_{40} = 15{,}200$; $x_{40}^2 = 15{,}129$; and $x_{40} = 123$. After the error x_{40}^2 is accidentally changed to 21,000 and x_{40} becomes 124. Assuming that a_{40} remains unchanged, what will happen to x_{40}? If a new value of $a_{40} = 1$ is inserted, what will happen to x_{40}?

12 / *Miscellaneous practical matters*

When the equations have been written and checked, the logical design may be said to be complete, but there remains a great deal of work which must be done before the logical designer's job is truly over. This chapter is intended to round out the picture of the responsibilities of the logical designer. The fact that the chapter is short does not mean that the subjects covered are not important, but only that they are not wholly within the province of this book.

The extra functions of the logical designer fall naturally into two parts: getting the computer built, and getting it to work after it is built.

FROM EQUATIONS TO CONSTRUCTION

The complete equations derived by the logical designer assume that all memory elements are capable of supplying whatever power is required to the decision elements to which they are connected. The first step which must be taken in construction, therefore, is that of computing, from the equations, the current which must be supplied (or absorbed) by each memory element to (or from) its associated gating circuits. For a computer composed of several hundred flip-flops, where a given flip-flop may appear in a dozen or more scattered equations and perhaps in different levels in the different equations, this may be quite a computation. When the calculation is complete, there will in general be many flip-flops which have bigger loads than the circuits can supply, and the logical designer will either have to rewrite the equations, dividing the load among some of the less-used memory elements and adding more elements if necessary, or he must arrange to insert power amplifying devices at appropriate places in the computer.

The next step is to lay out the circuits in the arrangement they will have in the computer frame or cabinet. The layout is made with the object of keeping major components (registers, the adder, the control unit, memory selection equipment, etc.) in neat and orderly arrays so that the maintenance man is able to understand how the circuits correspond to the logic. Another important characteristic of the layout is that connecting wires should be short, so that stray capacitance is minimized. When the layout is complete, stray capacitance should be measured and its effect calculated (if this has not already been done), for the extra current necessary to charge and discharge these capacitances may overload some flip-flop already supplying near-maximum power.

It is of course difficult for the wireman to connect circuit components together just by looking at the logical equations. The final step necessary in preparation for construction is therefore the drawing of wiring diagrams (or their equivalent), showing the exact position of every interconnecting wire on the computer. It is the logical designer's job to oversee the construction of these diagrams, making sure that no errors slip in, and helping to devise a scheme of representing connections which makes it easy for maintenance personnel to find a wire which corresponds to some term in an equation, or to discover what term a given wire is intended to represent. The establishment of this correspondence between logic and equipment is very important, and will be mentioned again in the next section when the maintenance manual is discussed.

FROM CONSTRUCTION TO OPERATION

When construction is complete on a machine as complicated as a computer, a check-out period of days or weeks will be necessary, during which faulty circuits will be discovered and replaced, and wiring mistakes located and corrected. Such work is ordinarily the job of a maintenance crew, not of the logical designer. However, when the *first* model of a computer has been constructed, the process of getting it to work is even more difficult, being complicated by errors in the original logic, in loading calculations, and in the construction of wiring diagrams from equations. Almost all of these errors will be eliminated by the time the second model of the computer is constructed, but they present a formidable problem in the first model, and one which the logical designer must help to solve.

It is obviously important that steps be taken throughout design to prevent mistakes in logic, loading, and wiring. It is certainly necessary that the work of each individual be checked by at least one other, and it may be worthwhile for some details to be worked out independently by two different people, and the results checked one against the other. Unfortunately, no amount of checking seems to uncover all errors, and many go undiscovered until construction is complete. The logical designer must therefore work closely with the maintenance crew in diagnosing and rectifying the peculiarities which turn up.

In preparation for the check-out process, the logical designer has two further responsibilities to discharge between the time the wiring diagrams are complete and the time the computer is ready to be tested. The first is an obligation to write a description of the operation of the whole computer in a form in which it can be readily understood and learned. This description will probably contain all the equations, together with term-by-term explanations for each one. It will also include information enabling the maintenance man to find the correspondence between equations and equipment, as was mentioned above. It will be used by the first maintenance crew as an aid in understanding and operating the system, and will be the basis of future maintenance manuals. If many copies of the computer are to be built and sold, the problem of providing maintenance is a difficult one, and is out of the hands of the logical designer. At any rate, the maintenance technique most often used is that of replacing large units of equipment and repairing them away from the machine, so that a detailed logical description of the operation of the computer need be used only for the most obscure failures.

The other responsibility borne by the logical designer is that of providing testing programs or problems for the computer, and of helping to work out testing procedures. The purpose of a test program is first to detect and second to isolate and locate a computer fault. Of all the individuals concerned with the design of the computer, the logical designer is the person best able to point out weak design points (e.g., heavily loaded flip-flops) and best qualified to plan programs which pinpoint the location of faults. Needless to say, the logical designer should work with experienced programmers when writing these routines. Similarly, he should work with expert maintenance men in helping to devise testing procedures for checking out the first model of a computer (which will involve the detection, location, and correction of errors in logic); checking out subsequent models of a logically correct computer, just after construction is complete; and performing routine maintenance on a computer after it is in the hands of the customer.

APPENDIX

I / *The solution of simultaneous Boolean equations*

Suppose that $x_1, x_2, x_3, \cdots, x_n, a_1, a_2, a_3, \cdots, a_q$ are Boolean variables, and we are given a set of equations

$$f_i(x_1, x_2, \cdots, x_n, a_1, a_2, \cdots, a_q) = g_i(x_1, x_2, \cdots, x_n, a_1, a_2, \cdots, a_q) \qquad i = 1, 2, 3, \cdots, p \quad (1)$$

How can we solve these equations explicitly for the x_i in terms of the a_i only?

First, note that each of the p equations above may be rewritten in the form

$$f_i\bar{g}_i + \bar{f}_i g_i = 0 \qquad i = 1, 2, 3, \cdots, p \qquad (2)$$

Furthermore, these p equations may be added together, with the result

$$\sum_{i=1}^{p} (f_i\bar{g}_i + \bar{f}_i g_i) = 0 \qquad (3)$$

Since equation 3 is in general a function of all x_i and of all a_i, it may be written in the form

$$\sum_{i=0}^{2^n-1} A_i m_i = 0 \qquad (4)$$

where each A_i is some function of $a_1, a_2, \cdots, a_q$, and m_i is a minterm in the variables $x_1, x_2, \cdots, x_n$. That is,

$$m_0 = \bar{x}_1\bar{x}_2\bar{x}_3 \cdots \bar{x}_{n-1}\bar{x}_n$$

$$m_1 = \bar{x}_1\bar{x}_2\bar{x}_3 \cdots \bar{x}_{n-1}x_n$$

$$m_2 = \bar{x}_1\bar{x}_2\bar{x}_3 \cdots x_{n-1}\bar{x}_n$$

$$m_3 = \bar{x}_1\bar{x}_2\bar{x}_3 \cdots x_{n-1}x_n$$

$$\cdot \quad \cdot \quad \cdot \quad \cdot \quad \cdot \quad \cdot \quad \cdot \quad \cdot \quad \cdot \quad \cdot \quad \cdot \quad \cdot \quad \cdot$$

$$m_{2^n-1} = x_1x_2x_3 \cdots x_{n-1}x_n$$

It is thus always possible to write any collection of Boolean simultaneous equations in the form of a single equation similar to equation 4.

Example 1. For the T flip-flop of Chapter 5,

$$g_1Q + g_2\bar{Q} = \bar{T}Q + T\bar{Q}$$

Applying equation 2, we find

$$(g_1Q + g_2\bar{Q})\overline{(\bar{T}Q + T\bar{Q})} + \overline{(g_1Q + g_2\bar{Q})}(\bar{T}Q + T\bar{Q}) = 0$$

$$(g_1Q + g_2\bar{Q})(TQ + \bar{T}\bar{Q}) + (\bar{g}_1Q + \bar{g}_2\bar{Q})(\bar{T}Q + T\bar{Q}) = 0$$

$$g_1QT + g_2\bar{Q}\bar{T} + \bar{g}_1Q\bar{T} + \bar{g}_2\bar{Q}T = 0$$

$$(\bar{g}_1Q + g_2\bar{Q})\bar{T} + (g_1Q + \bar{g}_2\bar{Q})T = 0 \tag{5}$$

Equation 5 is in the form of equation 4, where

$$\begin{aligned} A_0 &= \bar{g}_1Q + g_2\bar{Q} \\ A_1 &= g_1Q + \bar{g}_2\bar{Q} \end{aligned} \tag{6}$$

Example 2. For the R-S flip-flop of Chapter 5,

$$\begin{aligned} g_1Q + g_2\bar{Q} &= S + \bar{R}Q \\ RS &= 0 \end{aligned} \tag{7}$$

Applying equation 2, we find

$$(g_1Q + g_2\bar{Q})\overline{(S + \bar{R}Q)} + \overline{(g_1Q + g_2\bar{Q})}(S + \bar{R}Q) = 0$$

$$(g_1Q + g_2\bar{Q})\bar{S}(R + \bar{Q}) + (\bar{g}_1Q + \bar{g}_2\bar{Q})(S + \bar{R}Q) = 0$$

$$R\bar{S}(g_1Q + g_2\bar{Q}) + \bar{S}g_2\bar{Q} + S(\bar{g}_1Q + \bar{g}_2\bar{Q}) + \bar{R}\bar{g}_1Q = 0$$

The two parts of equation 7 may now be combined as indicated by equation 3, and put in the form of equation 4:

$$\bar{R}\bar{S}(\bar{g}_1Q + g_2\bar{Q}) + \bar{R}S(\bar{g}_1Q + \bar{g}_2\bar{Q}) + R\bar{S}(g_1Q + g_2\bar{Q}) + RS = 0 \tag{8}$$

Equation 8 is now in the form of equation 4, where

$$\begin{aligned} A_0 &= \bar{g}_1Q + g_2\bar{Q} \\ A_1 &= \bar{g}_1Q + \bar{g}_2\bar{Q} \\ A_2 &= g_1Q + g_2\bar{Q} \\ A_3 &= 1 \end{aligned} \tag{9}$$

Let us now attack the basic problem by solving a Boolean equation in one unknown, x. Equation 4 is then

$$A_0\bar{x} + A_1x = 0 \tag{10}$$

and we want to find an equation for x in terms of A_0 and A_1. Since x can be a function of only these two variables, we must have

$$x = c_0\bar{A}_1\bar{A}_0 + c_1\bar{A}_1A_0 + c_2A_1\bar{A}_0 + c_3A_1A_0 \tag{11}$$

where c_0, c_1, c_2, and c_3 are arbitrary Boolean quantities. The complement of x is then

$$\bar{x} = \bar{c}_0\bar{A}_1\bar{A}_0 + \bar{c}_1\bar{A}_1A_0 + \bar{c}_2A_1\bar{A}_0 + \bar{c}_3A_1A_0 \tag{12}$$

Substituting equations 11 and 12 into 10, we find

$$\bar{c}_1\bar{A}_1A_0 + \bar{c}_3A_1A_0 + c_2A_1\bar{A}_0 + c_3A_1A_0 = 0$$

Therefore

$$\bar{c}_1\bar{A}_1A_0 + c_2A_1\bar{A}_0 + A_1A_0 = 0 \tag{13}$$

Since the Boolean sum of the three terms in equation 13 is zero, *each term* must equal zero. That is,

$$\left.\begin{aligned}\bar{c}_1\bar{A}_1A_0 &= 0\\ c_2A_1\bar{A}_0 &= 0\end{aligned}\right\} \tag{14}$$

$$A_1A_0 = 0 \tag{15}$$

Equation 15 states that there is no solution to equation 10 unless $A_1A_0 = 0$. From equations 14 we deduce that the most general solution to equation 10 will be found when $\bar{c}_1 = 0$ or $c_1 = 1$, and $c_2 = 0$.

Substituting these values into equation 11, and remembering that $A_1A_0 = 0$, we find

$$x = \bar{A}_1A_0 + c_0\bar{A}_1\bar{A}_0$$

which (by virtue of equation 15) reduces to

$$x = A_0 + c_0\bar{A}_1 \tag{16}$$

Example 3. Applying this result to the T flip-flop of equations 5 and 6 above, we first test to see whether $A_1A_0 = 0$.

$$A_1A_0 = (g_1Q + \bar{g}_2\bar{Q})(\bar{g}_1Q + g_2\bar{Q}) = 0$$

We may therefore substitute equations 6 into 16, and find

$$\begin{aligned}T &= \bar{g}_1Q + g_2\bar{Q} + c_0\overline{(g_1Q + \bar{g}_2\bar{Q})}\\ &= \bar{g}_1Q + g_2\bar{Q} + c_0(\bar{g}_1Q + g_2\bar{Q})\\ &= \bar{g}_1Q + g_2\bar{Q}\end{aligned} \tag{17}$$

which checks with the result of Chapter 5.

We may apply the same technique to the solution of an equation in two unknowns, x_1 and x_2. Equation 4 then becomes

$$A_0\bar{x}_1\bar{x}_2 + A_1\bar{x}_1x_2 + A_2x_1\bar{x}_2 + A_3x_1x_2 = 0 \qquad (18)$$

We begin by writing x_1 and x_2 as general functions of A_3, A_2, A_1, and A_0, as follows:

$$x_1 = \sum_{i=0}^{15} c_{1,i}(A_3A_2A_1A_0)_i$$
$$x_2 = \sum_{i=0}^{15} c_{2,i}(A_3A_2A_1A_0)_i \qquad (19)$$

where $c_{1,i}$ and $c_{2,i}$ are again arbitrary constants.

Substituting equations 19 into 18, and setting all terms equal to zero, we arrive at the following results:

$$A_3A_2A_1A_0 = 0 \qquad (20)$$

(This is a necessary and sufficient condition for the existence of a solution to equation 18.)

$$c_{1,14} + c_{2,14} = 0 \qquad \text{Therefore } c_{1,14} = c_{2,14} = 0$$
$$c_{1,13} + \bar{c}_{2,13} = 0 \qquad \text{Therefore } c_{1,13} = 0 \quad c_{2,13} = 1$$
$$c_{1,12} = 0$$
$$\bar{c}_{1,11} + c_{2,11} = 0 \qquad \text{Therefore } c_{1,11} = 1 \quad c_{2,11} = 0$$
$$c_{2,10} = 0$$
$$\bar{c}_{1,9}\bar{c}_{2,9} + c_{1,9}c_{2,9} = 0$$
$$c_{1,8}c_{2,8} = 0$$
$$\bar{c}_{1,7} + \bar{c}_{2,7} = 0 \qquad \text{Therefore } c_{1,7} = c_{2,7} = 1$$
$$c_{1,6}\bar{c}_{2,6} + \bar{c}_{1,6}c_{2,6} = 0$$
$$\bar{c}_{2,5} = 0$$
$$c_{1,4}\bar{c}_{2,4} = 0$$
$$\bar{c}_{1,3} = 0$$
$$\bar{c}_{1,2}c_{2,2} = 0$$
$$\bar{c}_{1,1}\bar{c}_{2,1} = 0$$

This set of equations shows what conditions must hold on all "arbitrary" constants, if equations 19 are to be a solution to 18. Some of the equations clearly state that certain constants must be "zero" or "one," as shown. We can find explicit expressions for the remaining constants by applying the result of equation 16. For example, since $c_{1,4}\bar{c}_{2,4} = 0$, we can consider this to be an equation in a single unknown $c_{2,4}$, and can solve it as follows:

$$c_{1,4}\bar{c}_{2,4} + 0 \cdot c_{2,4} = 0$$

$$A_0 A_1 = c_{1,4} \cdot 0 = 0$$

$$c_{2,4} = c_{1,4} + K_4 \bar{0} = c_{1,4} + K_4$$

In a similar way, we find

$$c_{2,9} = \bar{c}_{1,9}$$

$$c_{2,8} = K_8 \bar{c}_{1,8}$$

$$c_{2,6} = c_{1,6}$$

$$c_{2,2} = K_2 c_{1,2}$$

$$c_{2,1} = \bar{c}_{1,1} + K_1$$

Substituting all these constants into equations 19, and assuming that $A_3 A_2 A_1 A_0 = 0$, we find

$$\begin{aligned} x_1 = A_1 A_0 &+ c_{1,0}\bar{A}_3\bar{A}_2\bar{A}_1\bar{A}_0 + c_{1,1}\bar{A}_3\bar{A}_2\bar{A}_1 A_0 \\ &+ c_{1,2}\bar{A}_3\bar{A}_2 A_1\bar{A}_0 + c_{1,4}\bar{A}_3 A_2\bar{A}_1\bar{A}_0 \\ &+ c_{1,5}\bar{A}_3 A_2\bar{A}_1 A_0 + c_{1,6}\bar{A}_3 A_2 A_1\bar{A}_0 \\ &+ c_{1,8} A_3\bar{A}_2\bar{A}_1\bar{A}_0 + c_{1,9} A_3\bar{A}_2\bar{A}_1 A_0 \\ &+ c_{1,10} A_3\bar{A}_2 A_1\bar{A}_0 \end{aligned} \tag{21}$$

$$\begin{aligned} x_2 = A_2 A_0 &+ c_{2,0}\bar{A}_3\bar{A}_2\bar{A}_1\bar{A}_0 + (\bar{c}_{1,1} + K_1)\bar{A}_3\bar{A}_2\bar{A}_1 A_0 \\ &+ K_2 c_{1,2}\bar{A}_3\bar{A}_2 A_1\bar{A}_0 + c_{2,3}\bar{A}_3\bar{A}_2 A_1 A_0 \\ &+ (c_{1,4} + K_4)\bar{A}_3 A_2\bar{A}_1\bar{A}_0 + c_{1,6}\bar{A}_3 A_2 A_1\bar{A}_0 \\ &+ K_8 c_{1,8} A_3\bar{A}_2\bar{A}_1\bar{A}_0 + c_{1,9} A_3\bar{A}_2\bar{A}_1 A_0 \\ &+ c_{2,12} A_3 A_2\bar{A}_1\bar{A}_0 \end{aligned} \tag{22}$$

Equations 21 and 22 are thus the most general solution to equation 18.

Example 4. Applying this result to the R-S flip-flop of equations 8 and 9 above, we first test to see whether $A_3 A_2 A_1 A_0 = 0$.

$$A_3 A_2 A_1 A_0 = 1 \cdot (g_1 Q + g_2 \bar{Q})(\bar{g}_1 Q + \bar{g}_2 \bar{Q})(\bar{g}_1 Q + g_2 \bar{Q}) = 0$$

We may therefore substitute equations 9 into 21 and 22. Noting first that $A_1A_2 = \bar{A}_1\bar{A}_2 = \bar{A}_3 = 0$, we find:

$$\begin{aligned} R &= A_1A_0 + c_{1,10}\bar{A}_2A_1\bar{A}_0 \\ &= A_1A_0 + c_{1,10}\bar{A}_2A_1 \end{aligned} \tag{23}$$

$$\begin{aligned} S &= A_2A_0 + c_{2,12}A_2\bar{A}_1\bar{A}_0 \\ &= A_2A_0 + c_{2,12}A_2\bar{A}_1 \end{aligned} \tag{24}$$

Next, we see from equations 9 that

$$\begin{aligned} \bar{A}_2A_1 &= \bar{g}_1Q + \bar{g}_2\bar{Q} \\ A_2\bar{A}_1 &= g_1Q + g_2\bar{Q} \\ A_1A_0 &= \bar{g}_1Q \\ A_2A_0 &= g_2\bar{Q} \end{aligned}$$

Substituting these into equations 23 and 24, we find

$$\begin{aligned} R &= \bar{g}_1Q + c_1\bar{g}_2\bar{Q} \\ S &= g_2\bar{Q} + c_2g_1Q \end{aligned}$$

APPENDIX

II / *The relationship between binary and reflected codes*[*]

Suppose that two n-bit binary numbers are given by

$$B = \sum_{i=0}^{n-1} b_i 2^i \qquad B + 1 = \sum_{i=0}^{n-1} c_i 2^i$$

Let us convert each of these numbers to another code, using the formulas

$$\begin{aligned} g_{n-1} &= b_{n-1} & h_{n-1} &= c_{n-1} \\ g_j &= b_j \bar{b}_{j+1} + \bar{b}_j b_{j+1} & h_j &= c_j \bar{c}_{j+1} + \bar{c}_j c_{j+1} \\ & & j \neq n - 1 & \end{aligned} \tag{1}$$

Thus the two numbers are represented by a sequence of digits in each code as follows:

Number	B	$B + 1$
Binary code	$b_{n-1}b_{n-2} \cdots b_3b_2b_1b_0$	$c_{n-1}c_{n-2} \cdots c_3c_2c_1c_0$
G-code	$g_{n-1}g_{n-2} \cdots g_3g_2g_1g_0$	$h_{n-1}h_{n-2} \cdots h_3h_2h_1h_0$

We shall show that the G-code for $(B + 1)$ differs from the G-code for B in only one bit-position.

[*] This derivation is based on one given in H. J. Gray, P. V. Levonian, and M. Rubinoff, "An Analog-to-Digital Converter for Serial Computing Machines," *Proc. I.R.E.*, **41,** no. 10, 1462–1465 (Oct. 1953).

First suppose that $b_0 = 0$. Then $c_0 = 1$ and $b_j = c_j$ for $j \neq 0$, since the two binary numbers differ by unity. Therefore

$$g_0 = b_1 \qquad h_0 = \bar{c}_1 = \bar{b}_1$$

$$g_j = b_j\bar{b}_{j+1} + \bar{b}_j b_{j+1} \qquad h_j = c_j\bar{c}_{j+1} + \bar{c}_j c_{j+1} = b_j\bar{b}_{j+1} + \bar{b}_j b_{j+1}$$

$$j = 1, 2, 3, \cdots, (n-2)$$

$$g_{n-1} = b_{n-1} \qquad h_{n-1} = c_{n-1} = b_{n-1}$$

Thus $g_j = h_j$ for all j except $j = 0$.

Next, suppose that $b_0 = 1$. Then $c_0 = 0$ and $b_j = \bar{c}_j$ up to and including some particular digit-pair b_p, c_p. Thereafter, $b_j = c_j$. That is,

$$b_j = \bar{c}_j \qquad j = 0, 1, 2, \cdots, p$$

$$b_j = c_j \qquad j = (p+1), (p+2), \cdots, (n-1)$$

Therefore

$$g_0 = \bar{b}_1 \qquad h_0 = c_1 = \bar{b}_1$$

$$g_j = b_j\bar{b}_{j+1} + \bar{b}_j b_{j+1} \qquad h_j = c_j\bar{c}_{j+1} + \bar{c}_j c_{j+1} = \bar{b}_j b_{j+1} + b_j\bar{b}_{j+1}$$

$$j = 1, 2, 3, \cdots, (p-1), (p+1), (p+2), \cdots, (n-2)$$

$$g_p = b_p\bar{b}_{p+1} + \bar{b}_p b_{p+1} \qquad h_p = c_p\bar{c}_{p+1} + \bar{c}_p c_{p+1} = \bar{b}_p\bar{b}_{p+1} + b_p b_{p+1}$$

$$g_{n-1} = b_{n-1} \qquad h_{n-1} = c_{n-1} = b_{n-1}$$

$$n - 1 \neq p$$

Thus again, $g_j = h_j$ for all j except $j = p$.

Summarizing, we have shown that a code derived from a binary code by use of the rule of equations 1 changes in only one bit position from one number to the next higher or lower one.

Equations 1 show how to derive a G-code from a binary code. We may rewrite it to show how to derive the binary code from the G-code by solving equations 1 for b_j, using the method developed in Appendix I. Since

$$g_j = b_j\bar{b}_{j+1} + \bar{b}_j b_{j+1}$$

$$\bar{g}_j(b_j\bar{b}_{j+1} + \bar{b}_j b_{j+1}) + g_j(b_j b_{j+1} + \bar{b}_j\bar{b}_{j+1}) = 0$$

Factoring b_j, we find

$$b_j(\bar{b}_{j+1}\bar{g}_j + b_{j+1}g_j) + \bar{b}_j(b_{j+1}\bar{g}_j + \bar{b}_{j+1}g_j) = 0$$

Applying formulas 15 and 16 of Appendix I, we see

$$b_j = b_{j+1}\bar{g}_j + \bar{b}_{j+1}g_j$$

which is equation 8-18.

Index